THE MAYFLIES OF FLORIDA

By

LEWIS BERNER

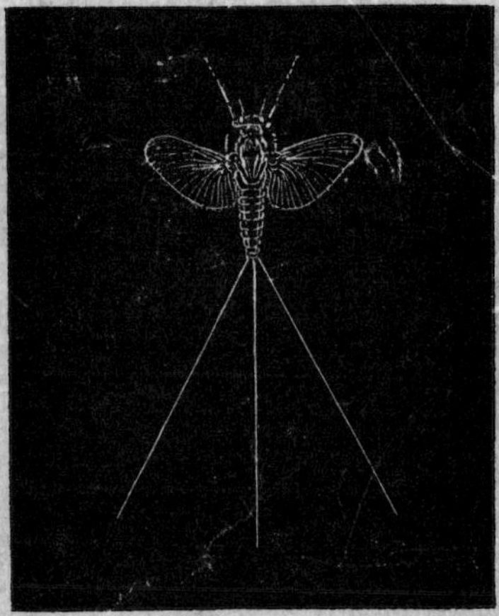

UNIVERSITY OF FLORIDA PRESS

GAINESVILLE, 1950

UNIVERSITY OF FLORIDA STUDIES

BIOLOGICAL SCIENCE SERIES

VOLUME IV, NUMBER 4

THE MAYFLIES OF FLORIDA

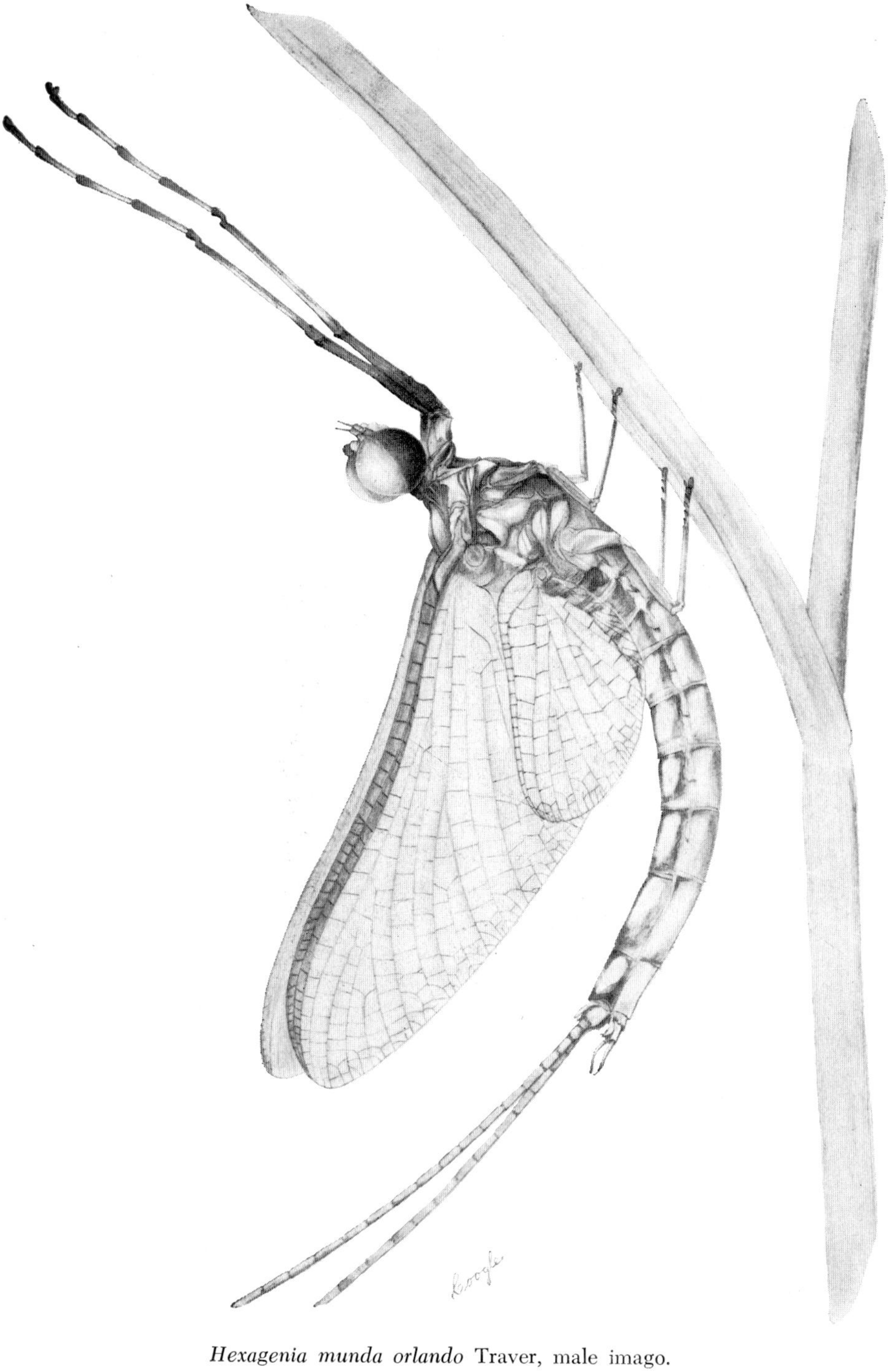

Hexagenia munda orlando Traver, male imago.

THE MAYFLIES OF FLORIDA

By

LEWIS BERNER

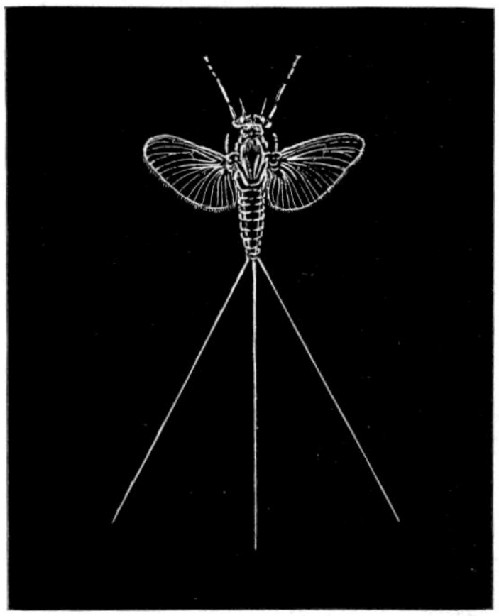

UNIVERSITY OF FLORIDA PRESS

Gainesville, 1950

Printed in the United States of America by
Pepper Printing Company
Gainesville, Florida

PREFACE

The present investigation was begun in 1937 after it was learned that almost nothing was known of mayflies in Florida. It was soon apparent that the study of these insects was far more complex than first appeared and that relatively few of the species inhabiting the state were recorded in the literature. As the insects became better known, wider vistas were opened and facets of the problem, which previously were only faint shadows, now began to manifest themselves. The real task soon became clear—it would consist of several phases, chiefly identifying those species of Ephemeroptera occurring in Florida, mapping their distribution, studying their ecology and habits, and learning as much as possible of their life histories. In all, seven years were devoted to this endeavor, in the course of which nearly all of Florida was covered, about fifty thousand specimens were collected and examined, numerous species of mayflies were reared, and much ecological information was obtained.

This work is an attempt to bring together all available data on Florida mayflies so that future students of the Ephemeroptera will have available a ready source of information on the group as it occurs in this southernmost extension of the Coastal Plain of the United States. Perhaps the material included here may serve as a starting point for some of the many problems still confronting the fresh-water biologist in Florida.

I wish to express my sincere appreciation to the many people who have helped in one way or another in this study. In particular, I should like to thank Professor T. H. Hubbell, of the University of Michigan, for his continual encouragement in this work, for many helpful criticisms, and for examination of the manuscript; Professor J. S. Rogers, also of the University of Michigan, whose aid, encouragement, and advice have been invaluable; and Professor H. H. Hobbs, of the University of Virginia, for numerous suggestions and discussions of the problem, aid in collecting, and help in preparing the illustrations. Various members of the faculty of the Department of Biology of the University of Florida have offered many suggestions and participated in discussions which have been very elucidating.

To the many friends who have collected mayflies for me, I wish to express my gratitude. In particular, Dr. H. H. Hobbs, Dr. F. N. Young, Dr. L. J. Marchand, and Mr. W. A. McLane have assiduously collected all ephemerids with which by chance or otherwise they have happened to come in contact.

All half-tone illustrations were prepared by Miss Esther Coogle, Staff Artist, Department of Biology, University of Florida, who spent countless hours perfecting techniques and laboring over the drawings. She was also most helpful in preparing the line drawings for publication by eradicating the discoloration caused by storage during my period of service in the army. Mr. William

Brudon, Staff Artist, Museum of Zoology, University of Michigan, illustrated the distribution of mayfly nymphs on a rock.

The staff of the University of Florida Press has worked diligently in helping me get the manuscript in final form for publication. Dr. Lewis F. Haines, Director of the Press, has done a most careful job of editing the paper. Miss Penelope Griffin has read and reread the manuscript many times in an attempt to eradicate any possible discrepancies that may have crept into the writing. To these two people in particular, and to the other members of the Press staff, I should like to take this means of expressing my gratitude.

And finally, I wish to thank Dr. F. P. Ide, of the University of Toronto, Dr. H. T. Spieth, of the College of the City of New York, and Dr. J. R. Traver, of the Massachusetts State College, for examining and verifying identifications of many of the Florida species of mayflies.

LEWIS BERNER

University of Florida
October 1, 1949

CONTENTS

ILLUSTRATIONS

Text Figures

Plates

MAPS

INTRODUCTION

Few orders of insects have been so little studied in Florida as the mayflies, or Ephemeroptera. The fragile, delicate-winged adults, requiring special methods of collection and preservation, have not been favorite objects of entomological study; and the same is true of the small and often inconspicuous aquatic nymphs of these insects.

Although the Ephemeroptera include no enemies of agriculture, and at first glance would appear to have little or no economic importance, such an impression is altogether erroneous. The entire economy of aquatic life is intimately bound up with the presence and abundance of mayflies, and it has been demonstrated time and again that these insects, in both immature and adult stages, constitute an important item in the diet of many species of fish, or in that of other organisms that in turn are eaten by fish. The words of the late Dr. Frison (1935: 284-85) apply with particular force to the study of the ephemerids. He says: "The insect life of our inland streams (and lakes) affords a most interesting and profitable field for investigation; and the sooner we learn to place and evaluate this fauna correctly, the sooner we will be in a position to formulate conclusions and generalizations of importance in our efforts to understand our aquatic resources and to forward their intelligent use."

One of the most beautifully written descriptions of the life and mating of these insects is given in poetic prose by William Bartram (1794) in his *Travels,* in which he refers to the mayflies seen along the St. Johns River. Nearly sixty years passed before Ephemeroptera were again noted in the state, when Francis Walker, in 1853, described the tiny *Caenis diminuta* from the St. Johns Bluff on the St. Johns River. Not until 1900 was the group once more reported from Florida; in that year Dr. Nathan Banks described a new species, *Callibaetis floridanus,* from south Florida. In 1931 Dr. J. R. Traver described *Hexagenia weewa* and *Hexagenia orlando* from the state. *Oreianthus* sp. No. 1 was recorded from Florida by Dr. Traver in 1935, and in 1937 she gave a full description of the nymph. I recorded nine additional species of mayflies in 1940 and nine others in 1946, while Spieth (1941) listed three species of *Hexagenia* from the state, bringing the total number of published records to twenty-four species. In the present paper, twenty-four additional species are discussed, several of which are new. However, descriptions and names of the new species are being withheld because the species are not represented in my collection by those stages which can be described advantageously.

Only a single paper dealing with the biology of Florida mayflies in any other than a taxonomic sense has appeared. This paper (Berner, 1941) describes and discusses ovoviviparity in the mayflies of the genus *Callibaetis,* with particular reference to the Florida species.

The Biology of Mayflies by J. G. Needham, J. R. Traver, and Yin-Chi Hsu has summarized our knowledge of the mayflies of North America north of

Mexico, and has brought it up to date (1935). Naturally, there must be limitations to a book of this sort, and a certain number of adverse criticisms have been leveled at it, some quite unjustified. The book is extremely useful and has stimulated interest in this formerly much neglected order. The greatest gap is found in the taxonomic section, which deals almost entirely with the adult insects. The majority of species are described only as imagoes, and many of these only in the male sex, although sexual dimorphism is frequently exhibited among the Ephemeroptera. Nevertheless, the compilation of all available information on North American mayflies has greatly lightened the task of persons beginning a study of this fascinating group of insects.

HISTORICAL REVIEW

The history of work on the Ephemeroptera was very thoroughly reviewed by Morgan (1913), who allotted a short paragraph to each worker that had contributed to a knowledge of the mayflies up to 1911. Needham in *The Biology of Mayflies* briefly discussed the earlier Ephemeropterists, but did not attempt any detailed analysis of the present trends in the study of Ephemeroptera.

A large part of the work on North American mayflies has been of a taxonomic nature, but several important studies on morphology, ecology, embryology, and life histories have been made. Taking up the account with the year 1911, to which date Morgan's summary extends, the following important contributions deserve mention.

J. G. Needham published a study of the genus *Hexagenia* in 1920, dealing with taxonomic, as well as ecological problems. Taxonomic studies by this author appeared occasionally during the years 1924-1932, until his interest in the Odonata came to occupy his time completely. However, the data on mayflies that he had accumulated over a period of thirty-five years were not to be set aside lightly, and these were brought together in 1935 in *The Biology of Mayflies*, of which Needham is senior author.

Nathan Banks, formerly Curator of Insects at the Museum of Comparative Zoology, has retired. He published taxonomic papers in 1914, 1918, and 1924.

W. A. Clemens, no longer working with the mayflies, has published papers dealing with a fairly wide range of subjects. In 1913, 1915, 1922, and 1924 he dealt with taxonomy and life histories. His most important paper, "An Ecological Study of the Mayfly *Chirotenetes*," appeared in 1917, and is a thorough and enlightening piece of work. Three other papers by this author describe mayflies in relation to their availability as fish food.

Ann H. Morgan in her more recent studies has treated physiological problems, mainly oxygen relationships. Her first paper in 1911 was taxonomic with some ecological discussions, and in 1913 she published "A Contribution to the Biology of Mayflies." Her two other papers dealing with mayflies include a study of venation (1911) and a description of the mating flight of a South American species of *Campsurus* (1929).

Helen E. Murphy has contributed a single paper dealing with North American mayflies, in which she describes the complete life history of *Baetis vagans* (1922).

G. S. Dodds and F. L. Hisaw in 1924 published "Adaptations of Mayfly Nymphs to Swift Streams," a paper which has proved to be extremely important. The former author also published a paper on the mayflies of Colorado (1923).

R. J. Tillyard (deceased), although publishing almost exclusively on exotic Ephemeroptera, has contributed to the knowledge of North American mayflies chiefly through his elucidation of wing venation and paleoentomological work.

F. M. Carpenter has specialized in the study of paleoentomology. His first paper relating to fossil mayflies appeared in 1933.

J. A. Lestage, of Belgium, has worked principally with European mayflies in a taxonomic sense; however, he has made studies of certain families which also occur in North America, and has included our forms in his discussions.

Georg Ulmer, nearly all of whose publications have been in German, has worked very little with the North American fauna. His studies have been entirely taxonomic, and his paper listing the genera of the mayflies of the world in the form of a key is most useful. He has described two genera of mayflies from North America, *Habrophlebiodes* and *Tricorythodes*.

Ferris Neave published an extremely interesting study on the migratory habits of *Blasturus cupidus* in 1930. His other works (1929, 1932, and 1934) have been of an ecological nature and have proved to be very enlightening. He is no longer working with the Ephemeroptera.

J. McDunnough has described more species of North American Ephermeroptera than any other taxonomist. His first paper dealing with the group appeared in 1921, and since that date numerous descriptions of new species have regularly been published. Through Dr. McDunnough's work, the mayfly fauna of the northern part of North America is now rather well known, and its taxonomy appears to be established on a firm basis.

F. P. Ide first published a taxonomic and ecological treatment of certain Canadian mayflies in 1930, and since has written several other papers which are principally of a taxonomic nature. His most important contributions, however, are his excellent studies on the postembryonic development of mayfly nymphs and the effect of temperature on the distribution of mayfly nymphs in a stream. Ide's recent work summarizes the results of quantitative investigations on the insect fauna of streams with particular emphasis on the ephemerids.

H. T. Spieth has published papers of a taxonomic nature as well as studies on the coloration of mayflies and the rearing of nymphs. His paper on "The Phylogeny of Some Mayfly Genera" (1933) has been exceedingly useful to students of evolution and has aroused much interest in phylogenetic work among the Ephemeropterists. His most recent work is largely taxonomic.

Jay R. Traver is second only to Dr. McDunnough in the number of species she has described. Her first taxonomic paper on the genus *Hexagenia* was published in 1931, and since then numerous descriptions of new species have appeared. Dr. Traver's work in North Carolina was particularly interesting, for it clearly showed how poorly the fauna of the Southeast is known. One of her most important works is the taxonomic portion of *The Biology of Mayflies*. Her more recent papers have dealt with exotic mayflies.

Earle F. Lyman has been studying the Ephemeroptera from an ecological viewpoint and has published his observations on habits.

R. H. Daggy has carried on taxonomic work with the midwestern species of mayflies.

The present tendencies among the Ephemeropterists are still taxonomic to a great extent, but phylogeny, distribution, and ecological studies are beginning to move into the foreground. Because of an awakening realization of the fundamental importance of our natural resources, ecological studies of the Ephemeroptera, as well as of other aquatic insects, will probably soon overshadow the present taxonomic research, but before such studies can be made, a taxonomic system with a firm and logical basis must be erected. Such a system is within sight, but until many more regions are thoroughly investigated, a sound and comprehensive classification cannot be attained. Particular emphasis needs to be paid the immature stages, for it is in the nymphal form that the mayflies spend almost their entire life, and it is this stage that is important in the economy of waters.

THE ORDER EPHEMEROPTERA

The order Ephemeroptera is a natural grouping of primitive and pterygote insects. In general, the forewings are delicate, membranous, and triangular in outline, and usually have intercalary veins and many cross veins; the hind wings are much smaller than the forewings and may sometimes be wanting. The wings are normally held vertically over the back when the insects are at rest. The mouth parts of the adult are vestigial and the alimentary tract has undergone modifications fitting it as an aerostatic organ. There are ten abdominal segments, and from the posterior end of the abdomen either two or three caudal filaments arise. The immatures are aquatic and undergo a hemimetabolous development. The mouth parts of the nymphs are well developed for chewing. They possess two or three caudal filaments, and gills are present on certain of the abdominal segments. Mayflies are the only insects which molt after they have attained functional wings; prior to this final molt they are spoken of as subimagoes.

The imago, or final adult form, in most instances can be easily differentiated from the subimago by several characteristics. The wings of the subimago are dull and grayish, whereas those of the adult are normally glistening and iridescent; the hind margin of the wings of the subimago are ciliated but those of the adult are usually bare; the body of the subimago is dull and that of the adult is glossy; the forelegs of the subimago are relatively short, but they become much elongated in the imago, particularly the male; the caudal filaments are somewhat hairy and relatively short in the subimago and usually lose this hairy condition when they elongate in the imaginal stage. Many of the males have divided compound eyes, which are less prominent and rather dull in the subimago and very prominent and glossy in the imago.

The eggs* of nearly all mayflies are laid in fresh water, where varying periods are required for development to take place. The freshly hatched nymphs possess no gills, but within one to three molts these structures make their appearance; the mouth parts, however, are all present, though generalized and simple in form. Soon the nymphs assume the body shape and structures which they retain over most of their nymphal life. In the few species of which the life history has been completely worked out, there are at least twenty-five preadult molts, and in most cases more than this number are thought to occur, some species requiring as many as forty-five.

The nymphs of mayflies are adapted to their environments in many and various ways, though they all have certain features in common, such as gills, caudal filaments, and chewing mouth parts (a species recently described by Spieth lacks the molar area on the mandibles). The burrowing species exhibit some of the most remarkable adaptive characteristics; in these nymphs the legs have become flattened, digging structures—the forelegs are used for pushing the silt aside, the hind legs for passing it back out of the burrow. The frontal process of the head and the greatly elongated tusks, or rami, of the mandibles first encounter the silt in the digging process, loosening it so that the legs may complete the task of excavation and removing the material. The gills of these burrowers are feathery and thoroughly penetrated by tracheae; hence there is a particularly efficient mechanism for obtaining oxygen. Even the feeding habits have become modified to such an extent that these nymphs, much like earthworms, eat the organic materials encountered in their burrowing; however, they seem to feed somewhat selectively, as surprisingly little sand is found in their digestive tracts.

Numerous other distinctive modifications of form and structure are encountered among those nymphs which do not burrow, and they are directly related to the habits and habitats in which the nymphs normally live. Dodds and Hisaw (1924) have shown that nymphs which live on the undersides of rocks in midstream, where there is almost no current, have flattened bodies, whereas the bodies of those inhabiting the swiftest waters are torpedo-shaped. Still others with rounded, streamlined bodies dwell in ponds, but here it is the ability to swim and dart about rapidly that is correlated with the streamlined form.

I have found that there is a definite correlation between claw size and habitat. Those species which dwell in ponds and in very slow-flowing water have thin, attenuated claws which lightly touch the object to which the insect is clinging, but they do not form a strong clamp. A nymph with such long claws would probably have difficulty in maintaining itself in a strong current, for the claws

*Ovoviviparity has been noted in *Callibaetis* and *Cloeon.*

could scarcely have sufficient leverage to support the insect in this situation. The mayflies living in swifter currents have short, thick claws which clamp firmly into the object on which the insect is living. The stronger the current, in general, the shorter and thicker the claws. Mayflies from rapids have very short thick claws, often with strong teeth on the underside, and in one species that I have examined this pectination has developed so markedly that the entire tarsal claw has become a comblike structure and the major claw is no longer distinct.* For the most part, the size of the tarsal claws is fairly uniform within a genus, but this is not always so. In one genus, *Centroptilum,* there are two Florida species: *viridocularis* which occupies slow-flowing water; and *hobbsi* which inhabits swift-flowing water. In *viridocularis* the claws are three-fourths as long as their respective tarsi; in *hobbsi* the claws are only one-half as long as their respective tarsi.

The nymphal leg structure, even in the nonburrowers, is correlated with the habitat. Those species inhabiting ponds and relatively quiet waters have rather elongated limbs which stand out from the body, holding it above the object to which the nymph is attached. The sprawlers have shorter, thicker legs, the femora being somewhat broadened; the legs are held out horizontally from the body and do not serve to lift the insect from the surface on which it is crawling. The clingers of swift water have thickened, shortened legs with a correspondingly greater development of the muscles. Their legs are held horizontally with the femora close to the body; the latter are thin along the anterior edge but relatively thicker posteriorly, thus presenting lessened resistance to the current.

The caudal filaments show special features in the freely swimming and darting mayfly nymphs. In these species, there is a great development of hairs medially from the lateral filaments and laterally from the median filament so that these hairs overlap to form a paddle. By rapidly flicking their abdomens and straightening their tails, the insects develop very efficient swimming organs which enable them to move rapidly through the water. Those species which swim little or not at all have a correspondingly small development of hairs on the caudal filaments, and in some of these nymphs the tails are almost bare.

PRIMITIVE AND SPECIALIZED CHARACTERS

It is frequently said that mayflies are primitive insects, and it is certainly true that many primitive characteristics are exhibited by this order. Snodgrass (1935: 12) writes: "During the Carboniferous period, or probably earlier, a group of winged insects evolved a mechanism in the wing base for flexing the wings horizontally over the back when not in use. The descendants of this group (Neopterygota) include the majority of modern winged insects, while the more primitive nonwing-flexing insects are represented today by only two

*An undescribed nymph from Mexico.

orders (Odonata and Ephemerida), both of which have descended from Carboniferous times but are not closely related to each other." The earliest known fossil insects identifiable as mayflies were taken from Permian shales. These early mayflies were long-winged, long-bodied insects with slender legs and three long tails. The fore and hind wings were of nearly equal size and the wing veins were very well developed. Fossil mayfly nymphs taken from Jurassic beds in Siberia seem to have gills on nine abdominal segments.

The forewings of the more generalized mayflies possess nearly all the veins of the archetype venation. The costal vein runs to the humeral brace, and from thence outward to the edge of the wing; the foremargin consists of a slightly thickened, folded wing membrane beyond the costa. The subcosta extends from the wing base to the apex of the wing, and the radius branches a short distance from the wing base to give rise to R_1, which runs parallel to the subcosta, and to the radial sector, which in turn divides. The radial sector forms R_2, R_3, and R_4 plus R_5; R_4 and R_5 make up a single vein and the terminology is used only to prevent confusion. The media is represented by both its primitive branches, the anterior media (lost in the Neopterygota) and the posterior media; and the cubitus is likewise represented by the anterior and posterior cubitus. The Ephemeroptera are the only modern insects in which both branches of the media persist. The number of anal veins varies from one to three according to the genus.

In the hind wings of the more primitive genera the venation is but little modified from that of the forewings; however, with the shortening of the wing the area occupied by Rs has become smaller and R_3 has disappeared.

The genitalia of mayflies are likewise primitive. Imms (1931: 41) states that "a very primitive condition is found among Ephemeroptera, where the penis lobes, instead of being fused to form an aedeagus, are relatively large and entirely free, with the parameres closely applied to them." A styliger plate is borne on the terminal end of the ninth sternite, and forceps originate at the posterior edge of the styliger plate. The penes are believed to arise from the ninth segment and protrude between the ninth and tenth segments. According to Spieth (1933: 73), "the styliger plate of the mayflies is interpreted by Crampton and Walker as representing coxites, which have fused together to form a single structure. This structure, however, is separate from the ninth sternite. A parallel condition is to be found, according to Walker, in the Phasmoidea and Acridoidea, except that in these groups the styli are lacking." The forceps of the male Ephemeroptera are homologues of styli, and in all mayfly genera, except *Caenis, Brachycercus,* and *Campsurus,* are segmented. In the females, the oviducts open to the outside or into a common vestibule formed as a fold of the thin membrane joining the seventh and eighth segments. The sternum of segment seven is produced backward to form a structure known as an egg valve, but no true ovipositor is formed; however, in some of the Leptophlebine mayflies there is the beginning of such a structure (fig. 86).

The abdomen of mayflies, although possessing only ten distinct segments, is nevertheless considered to be quite primitive. The usual number of segments in the abdomen of modern insects is ten or eleven, but embryological evidence indicates that the number of segments of the early insects was twelve. The generalized form of an abdominal segment is approximately retained in those larval forms which preserve rudiments of the abdominal appendages. In mayflies each gill-bearing segment is distinctly divided into a dorsum and a venter by large lateroventral lobes supporting the gill. These lobes represent the bases of abdominal appendages. Snodgrass (1935: 273) concludes that there is little doubt that the gills of mayflies are appendicular parts of abdominal limbs, of which the supporting lobes are the bases. He states that "the gill stalk or gill plate, by its position on the basis and its basal musculature, suggests that it is the homologue of the stylus of the thysanuran abdominal appendages. The gill basis is very evidently the equivalent of the stylus-bearing plates of the Machilidae, though since it is immovable, there are no body muscles inserted upon it."

In order to arrive at any conclusions as to the primitiveness of the mouth parts of the Ephemeroptera, it is necessary to turn to the nymphal stages. The mayfly mouth parts are of the mandibulate type throughout the order, consisting of a labrum, hypopharynx, a pair of mandibles, a pair of maxillae, and a labium. The mandibles of most mayflies are very generalized in possessing a ventral incisor area and an inner or dorsal molar surface. Between the incisor and molar areas there is a small structure called the lacinia mobilis, the function of which is unknown. This structure has been interpreted by some as a mandibular palp and a prostheca, but according to Crampton it cannot represent a true palp. This author has suggested that the mandibles of mayflies in some respects approach those found in the Crustacea. A pair of superlinguae is attached as lateral lobes of the hypopharynx. Superlinguae are best developed in apterygote insects and in some of the more primitive members of the Pterygota. The galea, lacinia, stipes, and palpifer are fused in the maxillae of all modern mayflies and form a structure called the galea-lacinia. In some of the more primitive genera it is thought that a line of separation can be seen between the galea and lacinia. On the whole, the maxillae of mayflies must be considered as rather specialized.

The primitive condition of the tarsus of adult pterygote insects is believed to be five-segmented; tarsi having fewer than five segments have probably been specialized by the loss or fusion of some of the segments. In the adults of some of the more primitive mayfly genera, the tarsi are composed of five freely movable segments, but in most mayflies there is a certain degree of fusion. The legs, therefore, cannot be used as reliable indicators of the phylogenetic position of the Ephemeroptera with relation to other insects.

The gills of mayfly nymphs are considered to be homologous with abdominal legs—appendages which have been modified for respiratory purposes. Snodgrass states that there is no question but that the gills arise from limblike rudiments of the embryo, and that their structure and musculature suggest that they are parts at least of true abdominal appendages. The most primitive of the modern mayfly genera bear gills on the first seven abdominal segments, and the earliest fossils known had these structures on the first nine segments. The retention of these abdominal structures appears to indicate primitiveness, even though they are present in highly modified and variable forms.

A consideration of the characters discussed above will also show that mayflies, while generalized in many respects, are at the same time highly specialized in certain features. As Needham (1935: 207-8) has said: "The group as a whole, while retaining a good many primitive characters, has gone a long way in specializing on lines of its own. There are no very primitive mayflies. There is no one genus or species that can be set down as nearest to the ancestral form in all particulars. The lines run criss-cross here as elsewhere; and do not lend themselves to a serial arrangement."

One of the principal modifications and specializations undergone by the wings has been the cephalization of the flight mechanism, with relative enlargement of the forewings and corresponding reduction of the hind wings until, in some genera, the latter have entirely disappeared. In the specialized genera, venation tends to become reduced, with certain of the longitudinal veins becoming detached, some even being shifted at their bases and reattached. Paralleling this modification of the longitudinal veins, there is a tendency toward reduction in the number of cross veins in the forewings, and in the highly specialized genera they may be almost completely absent. At the same time that the cross veins are becoming reduced in number, a series of marginal veinlets develops, occupying the interspaces between the longitudinal veins. The presence of these veinlets is a mark of greater specialization than the irregular marginal network (the remains of the archedictyon) found in the most primitive mayflies; in the higher genera even the veinlets may be lost.

The hind wings indicate degree of specialization very clearly, for in the Ephemeroptera there exists a whole series of species ranging from those with well-developed metathoracic wings to those showing complete absence of such structures. In certain of the highly specialized groups the hind wings are reduced to tiny threadlike rudiments, which are sometimes so small as to be difficult to find, and which could hardly be of any use as organs of flight. The venation of these threadlike wings consists only of one to three longitudinal veins.

The form of the forewings in the primitive genera of Ephemeroptera is somewhat triangular. With increasing specialization, they have become more elongated and narrower, but in those genera with greatly reduced hind wings there is a tendency for the secondary development of a triangular wing with a well-developed anal angle, perhaps in compensation for the partial or complete loss of the hind wing. The return to a triangular wing shape reaches its culmination in the highly modified Caeninae, which completely lack metathoracic wings.

Not only is there a correlation between specialization and reduction in size of the metathoracic wings, but closely linked with this reduction is a decrease in body size. It is very likely that as wing size decreased a diminution in body size was a natural consequence. The amount of wing surface for support of the insect being lessened, the necessity for less body weight or an increase in wing surface of the other pair of wings was encountered. Both conditions have resulted. A diminution of body size parallels the diminution in size of metathoracic wings, with concomitant changes in shape of the mesothoracic wings but with no relative increase in resulting wing surface. In *Caenis*, which has entirely lost the hind wings and in which the body is thick and heavy by comparison with that of the Baetinae, the forewings have become enlarged and broadened considerably at the anal angle. Although certain other genera of mayflies have lost their metathoracic wings (*Pseudocloeon* and *Cloeon*), there has been no such relative increase in size of the forewings. In these dipterous mayflies, the body is lightened to accommodate the lessened wing surface.

There are few specializations in the genitalia of mayflies; however, in the more advanced genera there appears to be a fusion of segments of the forceps, and in *Caenis* and *Brachycercus* this fusion has reached its culmination, the claspers consisting of but a single segment. The males of the more specialized species also show a fusion of the paired penes, although the vasa deferentia still open to the outside separately, and in the Baetinae the penes have become internal.

Although the Leptophlebiinae are not the most highly specialized mayflies, certain of the females have the sternites of the seventh and eighth segments modified to form a simple ovipositor. The ovipositor is really nothing more than a prolongation of the egg valve of other mayflies, and its value in oviposition is entirely unknown.

All mayflies supposedly have ten abdominal segments; however, in *Baetisca* and the European *Prosopistoma* there seems to be some amalgamation of the segments, although it is claimed that ten segments can be distinguished. The first abdominal segment of *Baetisca* is intimately related to the metathorax, and it is only with difficulty that this segment can be differentiated. Spieth (1933) considers that gills in the nymphs of *Baetisca* are found on segments 1 through 5, but Traver (1935) claims that these structures occur on segments 2 through 6. The latter states that segment 1 is almost completely fused with the thorax, a statement with which I am in agreement. Thus this condition of lessening the number of abdominal segments, although indicating specialization, is found in a genus which has many primitive characters.

The mouth parts of immature mayflies are relatively unspecialized, the greatest modifications occurring in the burrowing species, where each mandible gives rise to a long tusk used in digging (fig. 40). The fusion of the parts of the maxillae indicates a specialization in these structures. Spieth states (1933: 81) that "even the most primitive maxilla of the mayflies (as is patent from the nature of the lacinia-galea) is specialized as compared with more generalized insect types." The degeneration evinced by the mouth parts of adult mayflies is a highly modified condition paralleled in relatively few other insects. The mouth parts of the adults are reduced in size, asymmetrical in form, and lack sclerotization. Internally, the musculature degenerates. As a whole the mouth parts are shrunken very closely together into a single whitish mass beneath the clypeus.

One of the most interesting specializations shown by mayflies is the enormous development of the eyes in males. Baetid males have the compound eyes completely divided into two parts, a lower rounded structure and a much larger "turbinate" portion capping the smaller part. The ommatidia of the turbinate eyes have become greatly elongated, and the shape of these eyes is extremely varied, ranging from a caplike, flattened structure closely hugging the lower eye to one which extends vertically from the head for some distance. In many species the turbinate eyes are brightly colored in shades varying from yellow to orange to brown. It is thought that this exceptional development of the male eye is correlated with the habit of aerial mating, for in the mating flight the male approaches the female from below. It has been pointed out by Cooke (1940) that the specializations have reached such a state of development that if a male (of *Stenonema vicarium*) be approached by a female from below, she will be completely ignored, probably because she is invisible to him; however, if the female is above the male, she is immediately seized and copulation ensues.

The legs of mayflies, although primitive in some respects, are really highly modified organs. Adult mayflies do very little walking and no running whatsoever. In such groups as the genus *Campsurus*, for example, the legs (except the forelegs of the male) have become vestigial, and the adult can no longer alight but must remain on the wing during its entire imaginal life. The forelegs of male mayflies are elongated and have acquired a reversible joint at the base of the tarsus as an adaptation for seizing the female during the mating flight. The more specialized mayflies show a tendency toward a reduction in number of tarsal segments in the two posterior pairs of legs by fusion of the basal segments with the tibia. The alteration of the alimentary tract to suit it for its aerostatic functions is a specialized character common to all mayflies.

The modification of the mouth parts, the overdevelopment of the ovaries, the enlargement and emptying of the alimentary canal, the enormous development of the eyes of the males, the elongation of the forelegs of the males, and the development of elaborate copulatory organs all fit the adult mayfly for efficient mating and the certitude of continuing the species.

FAMILIES OF MAYFLIES

There have been many arrangements of mayflies in various families and much shifting back and forth of the genera. As yet, there is no general agreement as to what constitutes a family and what a subfamily in the Ephemeroptera. For the sake of uniformity, I shall follow the taxonomic system set forth by Needham, Traver, and Hsu (1935), in which three families of mayflies are recognized to occur in North America, north of Mexico. The familial groups are arranged systematically in Table 1. An asterisk after the subfamily name indicates that representatives are found in Florida.

TABLE 1

FAMILIES AND SUBFAMILIES OF THE EPHEMEROPTERA

FAMILY	SUBFAMILY
Ephemeridae	Ephoroninae Ephemerinae* Potamanthinae Campsurinae* Neoephemerinae*
Heptageniidae	Heptageninae*
Baetidae	Ametropinae Metretopinae* Siphlonurinae* Leptophlebiinae* Baetiscinae* Ephemerellinae* Caeninae* Baetinae*

Ulmer's key (1933) to the Ephemeroptera of the world lists three suborders—Ephemeroidea, Baetoidea, and Heptagenioidea—which in general correspond to the families listed above. These in turn are divided into fourteen families. His families, in some cases, do not coincide with the subfamilies of Needham and Traver. Spieth (1933) used superfamilies in his discussion of the phylogeny of North American mayflies, as follows: superfamily Siphlonuroidea, including the families Siphlonuridae, Heptageniidae, Baetidae; superfamily Ephemeroidea, including the families Leptophlebidae, Ephemeridae, Ephemerellidae; superfamily Caenoidea, including the family Caenidae; and superfamily Baetiscoidea, including the family Baetiscidae.

The usage of the terms suborder, superfamily, family, and subfamily as higher categories thus appears to be merely a matter of convenience, subject to the whims of each particular worker. However, consistency in usage would seem to be desirable, and whether the term family or subfamily is used to designate these categories is a matter of choice. The higher categories of the taxonomic system set up by Needham and Traver appear to be adequate in the present status of our knowledge.

Much more phylogenetic study is necessary before the positions of several of the mayfly genera in the families and subfamilies can be satisfactorily determined. Certain genera exhibit characteristics which would place them in either of two subfamilies, and this should be indicated in discussions of these genera. Fortunately, there are no genera of North American mayflies which cannot be placed rather definitely in the higher categories, except perhaps *Isonychia,* which has been included both in the Heptageniidae and Baetidae. In this paper, *Isonychia* is considered as belonging to the Baetidae, but the placement is debatable.

Ulmer's 1933 key to the Ephemeroptera of the world included 115 genera, and since the publication of this work, several changes and additions have been made. At present, 51 genera are recognized as occurring in North America, north of Mexico. Since the publication of *The Biology of Mayflies, Acentrella* has been re-erected, *Oligoneuria* discovered in the drainage of the Mississippi, *Lachlania* recorded from Canada, and a new genus, *Traverella,* described from western North America.

When Traver wrote her taxonomic review of the mayflies of North America in 1935, she included 507 species, and since the publication of this work, many additional ones have been described. There are still many parts of North America which have hardly been touched, and when these are carefully studied, the number of species known from this continent will undoubtedly be greatly increased.

WING VENATION OF MAYFLIES

The wing venation of mayflies* has been reviewed by Spieth (1933) and by Needham (1935). These two authors disagree in their interpretation of one important set of veins. Needham and his followers consider that the radius is divided into R_1, Rs, R_2, R_3, R_4, and R_5, and that the anterior media is missing from the wings of modern ephemerids. Tillyard, Carpenter, Spieth, and others adopt the view that the radius is divided into R_1, Rs, R_2, R_3, and R_4 plus R_5, and that the anterior media is present. The essential difference between these two views rests upon the interpretation of the homologies of one vein in the mid-wing, which divides to produce two branches. Needham has named this branched vein the outer fork of the posterior branch of the Rs (OF); Tillyard and his supporters consider it to be the anterior media. In the present paper the term anterior media is used to designate the vein in question. In all other respects, there is full agreement between the two groups of workers concerning the generalized nature of the venation of mayflies.

Mayflies are unique among pterygote insects in completeness of fluting of the wings. There is a regular alternation of high and low (convex and concave) veins. A very important feature of the mayfly wing, as shown by Spieth, is that all convex veins belong to the dorsal surface of the wing and all concave veins to the ventral surface. His findings (1933: 60) indicate that "the cross veins belong principally to the dorsal surface, i.e., at the base where they join the concave veins a stump of the cross vein is attached to the main vein, while on the dorsal surface the cross veins are always complete and vigor-

*The names of the wing veins of mayflies have been discussed on page 7.

ously developed. . . . It is important to note that veinlets at the edge of the wings are always restricted to the dorsal surface."

The triadic type of branching is also a characteristic feature of mayfly venation. When a longitudinal vein forks, there is interpolated between the two branches a third vein of opposite position, which does not reach the base of the fork. For example, the anterior media, a convex vein, forms a fork just beyond the middle of the wing (called the outer fork [OF] of the radius by Needham). The two branches of this fork are convex; however, the intercalary vein lying between the two branches is concave. This system of fluting is primarily an adaptation for radial strengthening of the wing, and the cross veins serve chiefly to hold the ridges and furrows in place. In the more specialized mayflies, which have small hind wings or none at all, there is a decrease in the number of cross veins, while in those species which have large wings with prominent fluting, the cross-venation is well developed, reaching a maximum in *Ephoron* in which there are several hundred cross veins in the forewings.

The origin of the various wing veins has been studied principally by proponents of the Comstock-Needham system. Ann Morgan (1912) investigated the origin of the definitive adult venation, but later studies by Tillyard and others show that some of her results were erroneous. She adopted for mayflies the same interpretation of the relationship of radial sector and media as that proposed by Needham for the Odonata wing, with Rs crossing M. More recent work, however, lends no support to the idea that this occurs in either order. Wing venation and tracheation in Ephemeroptera are completely reviewed in *The Biology of Mayflies.*

TAXONOMIC CHARACTERS

Perhaps taxonomists have not been sufficiently thorough in their treatment of mayflies. Only external structures or features have thus far been used taxonomically, and until students of mayflies overcome their inordinate desire to preserve perfect specimens, the approach by study of internal anatomy will not be used. Of course, the difficulties involved in an investigation of the internal anatomy of small organisms must also be considered, and when large numbers of individuals are to be examined such studies would obviously be out of the question. It might be thought, at least, that in a group so long studied as the Ephemeroptera there would be little left to learn about the external morphology of these insects; some of their structures are, however, still imperfectly known, even though used to a great extent in taxonomy.

Of primary importance in the separation of families, genera, and even species, the wing veins of mayflies are still not satisfactorily homologized. The problems of venation have been discussed above in the considerations of phylogeny and venation. The wing veins of each genus of mayflies (particularly the longitudinal veins) are very constant in structure and position—so much so that only in few groups can specific differentiation be found, and then only by reference to minute details. The highly modified hind wings are very useful in differentiating genera, and in many cases the shape of the costal angulation, the number of longitudinal veins, and the general shape and relative proportions of these wings are helpful even in distinguishing species.

The structure of male genitalia has been widely employed by taxonomists throughout the field of entomology and even in the study of other animals. These reproductive structures are particularly useful in the differentiation of ephemerid species and have been widely applied. Many of our more recently described species have been established on genitalic differences alone, but there are usually other characters which parallel such genitalic divergences. Each genus has its own peculiar penial shape, and within the genus there may be much variation. One of the best examples of this is seen in *Paraleptophlebia*, where the penes of each species are distinctive. The shape, number of segments, and relative proportions of the forceps or claspers are also very useful in differentiating genera and, in some cases, species. The reproductive apparatus of the female mayfly is so poorly developed externally that it is of no use taxonomically, except in those few genera which possess a rudimentary ovipositor.

Osgood Smith studied the eggs of a number of genera of mayflies and found that within a genus the sculpturing and accessory structures are very uniform. He was able to construct a key to the known eggs of North American mayflies, which has been incorporated in *The Biology of Mayflies.*

The number of tarsal segments in the posterior pair of legs of adult mayflies has proved useful as a familial character. Within some families, the length of tarsal segments of the forelegs of the males is used to distinguish genera, and in some cases this character has been relied on to such an extent that it is impossible to identify a female to genus, unless males can be definitely associated with her. Much further work is needed in these groups in order to clarify the situation. Within the genera, leg structure is little used taxonomically, although I believe that relative proportions of the various parts will prove useful when sufficient measurements are made.

Another character frequently used, and rightfully so, is eye structure, but this character again is applicable only to males. Sexual dimorphism in mayflies is so pronounced that unless one is familiar with the groups, males and females of the same species may easily be placed in different genera. Among the male ephemerids, there appear all gradations in structure of the compound eyes from a simple type to those which are completely divided and enormously enlarged into bizarre shapes. The division, or lack of division, of the male eye is used as a criterion for separating the Baetidae from the Heptageniidae, and in some cases this character is also used as a generic differentiator. There appears to be a definite correlation between eye size and shape and the night- or day-flying habits of the species. Those mayflies mating at dusk or after dark have small eyes which are approximately the same size in both sexes, or only slightly larger in the males. In *Palingenia* (European) the female is approached by the male as she floats on the surface of the water, and in this genus, the eyes of the male are a little smaller than those of the female. The males of the day-flying species, on the other hand, have huge eyes which are far larger than those of the females.

The colors found in mayflies vary from white through yellow, orange, red, and brown, to black. All their colors, which are generally drab, aid in making the mayflies as inconspicuous as possible. Spieth has pointed out that the more primitive species have particularly dull and subdued colors. He found that in most mayflies the entire exoskeleton is transparent, and that if color is present in the chitin, it is always some shade of olive brown. The tissues immediately underlying the exoskeleton are often impregnated with pigments which are the principal cause of the distinct adult color pattern. Spieth found that the white colors in the Ephemeroptera are due to two distinct factors: (1) in all adults, a chalky white substance is present underneath the exoskeleton and epidermis but external to the musculature; and (2) certain species have a milky appearance which occurs not only in the body but also on the wings. This is a physical color, which disappears when the insect is immersed in liquids of the same refractive index as that of the white structure. The two types of white combine to form the color pattern of those adults showing this second type of coloration. He also pointed out that some of the oranges, bright yellows, and greenish yellows present in certain species are probably chlorophyll derivatives, since they deteriorate in dried specimens and are completely destroyed by preservatives.

The disposition of pigments in sexually dimorphic species is particularly interesting. The wings and body of both sexes of these forms are brilliant. In the males, however, the pigments are no longer evenly distributed in each abdominal segment but are concentrated in the first and the last four, with the intervening ones hyaline white. There may be a definite color pattern overlying the white of segments 2 through 6, but this is never conspicuous. In nearly all cases, this strongly localized coloration is limited to specialized genera of mayflies in which the individuals are less than 10 millimeters in length. It is probable that the glistening wings and the hyaline segments help to render the insects inconspicuous, because there is such a great reduction in the amount of dark color exposed to the view of predators. The females of these species have uniformly colored abdomens, but it has been suggested that if the entire exoskeleton of these insects were transparent, the light-colored muscles as well as the egg masses would show through and the females would be easily seen.

Differences in maculation are reliable, in general, for distinguishing species, but the great degree of variability in some species causes misgivings whenever color pattern alone is used, unless it is absolutely distinctive. The genus *Stenonema* is one of the chief groups in which color has been used for the establishment of new species; however, it should be employed with caution. In Florida, there occurs a species of *Stenonema* which by color pattern alone might fit into any one of three species of the *interpunctatum* group. Spieth (1947), knowing of the high degree of seasonal variability in color patterns in this genus, has partially revised *Stenonema,* placed several species in synonymy, and erected subspecies to indicate the close relationships of many of these forms.

To distinguish between species, colorational differences have probably been employed more than any other character or group of characters. Spieth (1938) made a study of coloration and its relation to seasonal emergence in the Ephemeroptera, and from his study it appears that some of the species, which have been described solely on the basis of maculational differences, are really only seasonal forms of the same species. As an example of this, I have been told by Drs. Spieth and Lyman and Mr. Jenkins, who have worked in the Great Lakes region, that *Stenonema tripunctatum* shows seasonal variation in coloration exceptionally well. Traver has described several species differentiated from *S. tripunctatum* by colorational variations, but the other workers mentioned above, through observations during the entire emergence period of the mayflies of a particular region, agree that all of these are merely seasonal forms of the one species.*

Size of adults, including both wing length and body length, has been used to some extent for the determination of species; however, so many factors modify size that it is not considered a safe criterion by itself for the erection of species. Wing length is much less variable than is body length, for the latter is subject to shrinkage, elongation, swelling, and other distortions when the insect is killed. Although there has been very little taxonomic use of relative proportions of wings, this may be worthy of future study.

The families of mayflies as here recognized appear to be natural groupings, for the evidence from the adults is substantiated by the structure of the immatures. Needham (1935: 208) states that

> . . . in this order the struggle for existence has fallen largely upon the nymphs, which are better equipped to meet competitors. These show greater divergences in adaptation to their several types of habitat. These have differentiated on lines of their own, independently of the adults, and tell their own story. It goes without saying that our interpretations of nymphal and adult evolution will, when correct, be in agreement. One must corroborate the other; for nature preserves or eliminates species as wholes.

Mouth parts and head shape are the familial characters used to separate the families of mayflies in the nymphal stages. The subfamilies, on the other hand, are to a large extent separated by the structure of the mouth parts and the structure of the gills, of which the latter are probably the more important. Gill shape and structure are extremely varied. The gills may be foliaceous, single or recurved; they may be highly tracheate or completely lack air tubes; they may be present on seven abdominal segments or absent from some of them; they may lie flat on the abdomen or be so shifted that they serve as suckers on the venter of the insect; they may be modified so that some form protective

*Spieth in 1947 revised the *tripunctatum* complex in an attempt to clarify the problem of seasonal coloration changes.

covers for others or all may be completely exposed; they may be strongly muscled so that they can be vibrated very rapidly or so flabby that such reactions are impossible; and they may be entirely confined to the abdomen or, in a few cases, may also occur on the thorax or even on the maxillae. In short, the gills are probably the most variable structures to be found in nymphal Ephemeroptera.

Since hind wings are absent in the adults of some species of mayflies, the immature stage of these naturally lacks hind wing pads. The presence or absence of these metathoracic hind wing pads is used as a taxonomic character.

The caudal filaments (two cerci and one median tail) are very useful in studies of the nymphal forms. In some species, they are very flexible organs; in others they are sturdy swimming structures. The cerci of many of the genera are densely clothed with hairs medially, and the median tail of these genera has heavy growths laterally; in many others the three tails are almost bare, having only a light covering of short hairs.

Color pattern, while frequently used for species determination, is much less frequently employed in nymphs than in the adults. Often when such characters are found to distinguish species, structural differences parallel them.

A COMPARISON OF FLORIDA MAYFLIES WITH THE NORTHERN FAUNA

TAXONOMY.—The genera of mayflies occuring in Florida, except *Oreianthus*, are found over the entire eastern part of North America, and even *Oreianthus* is known as far north as North Carolina. It is of interest to note that many of the genera, such as *Ephemera, Choroterpes, Habrophlebia, Habrophlebiodes, Pseudocloeon*, and *Tricorythodes*, have never before been recorded from the Coastal Plain, but all of these are now known to occur in Florida. *Tricorythodes* was previously recorded no closer to Florida than Texas and West Virginia; *Choroterpes* only as close as Texas and northern Ohio.

Thirteen of the Florida species are found also in southern Canada, Ohio, New York, and elsewhere in the North. There are also several other Florida species, very similar to species described from the Northern region, which differ from them only in minor colorational or genitalial characters. All but *Hexagenia bilineata* and *Pentagenia vittigera* of these wide-ranging species are small forms. This conforms with the generalization (for which no explanation is at present forthcoming) that the smallest mayflies are, on the whole, the most widely disseminated—a generalization that seems to be world-wide in application.

Six species hitherto unknown are here recorded from the state, but five of them are known only from the nymphal stage. In addition, thirteen new Florida species have been described, making a total of nineteen species discovered during the course of this investigation. The total list of forty-eight species from the state seems large for such a small and (from the standpoint of mayflies) ecologically limited region.

Actually, Florida is poor in number of species as compared to those areas that can boast of mountainous, hilly, and coastal regions which are all within relatively small boundaries. Traver (1932, 1933, 1937) found more than one hundred different species in North Carolina, but her collections included specimens from three physiographic provinces. Ide (1935) found fifty-five species in one stream in Ontario, and I have been informed that subsequently he has taken more than one hundred species from a single stream. No such concentrations of species are found in any Florida stream; in these waters conditions for mayflies do not compare in favorability with the rapid, rocky, and well-aerated streams of Ontario. In our most populous streams the maximum number of species found is only nineteen, and even this is exceptional (Sweetwater Creek, Liberty County).

In summary, it appears that the Floridian Ephemeropteran fauna is mostly of Southern origin; however, there are certain Northern elements which have entered the region and which have succeeded there remarkably well.

COMPARATIVE ABUNDANCE.—In Florida the actual number of individual mayflies present in a given situation is smaller than in Northern streams. This is partly explained by the relatively small number of suitable mayfly habitats in any particular stream; however, there is a difference in abundance even in corresponding situations. In mountain creeks in southern North Carolina where I have collected, the number of mayflies found on the underside of rocks greatly exceeded the number which might be found in analogous situations* in Florida.

In the smaller rivers and streams descending from the mountains along the east coast of Mexico, the number of individual insects on a single rock is amazing. Nearly two hundred mayflies were taken from one rock, approximately 15 x 15 x 8 inches, and on the same rock the caddisfly cases and blackfly larvae literally covered the surface. When a Florida stream is compared with such a river it can be seen that, even though mayflies are the predominant insects, they are so much less abundant that they can in no wise be as important in food chains as those of the rocky streams of the North and of Mexico.

Adults of the great majority of Florida species emerge throughout the year, and for this reason there are no great swarms (except of some of the burrowing forms). The flights, which are small, are composed of inconspicuous insects that gather in groups to mate. The literature indicates that many of the small Northern species collect in very large swarms to carry on their mating flight; but in the areas where such flights occur the emergence of those species forming the large swarms is limited to a short period. It has been stated that those species which emerge over a long time—the entire summer—do not form large

*It is necessary to employ the term "analogous situations" because in Florida the rock habitats are rarely present, and the insects use as substitutes submerged logs, boards, or any other available objects to which they can hold.

flights, and from these accounts it would seem that the flights of such species resemble in size those of the Florida ephemerids.

The burrowing mayflies, being more or less seasonal, emerge in great numbers in Florida, but do not form such tremendous swarms as occur in the more northerly parts of the country. The lake species, however, during their emergence period constitute a conspicuous part of the insect fauna in the Central Highlands of Florida, and at this time are extremely important in the food chains of lakes. The number of individuals in Florida lakes more closely approaches the abundance characteristic of Northern regions than do the numbers occurring in any other type of situation in this state. Only one stream among those examined—the Santa Fe River—may perhaps compare favorably in abundance of mayflies with some Northern streams.

COMPARATIVE ECOLOGY.—Naturally the conditions most affecting the distribution of mayfly species are the ecological factors of the aquatic environments where the nymphs occur. The great majority of mayflies are rheocolous, and the rate of flow of water is probably the most important single factor determining the presence or absence of a given species. The very swift, rocky streams which are preferred by many Northern species are absent in Florida, but there are numerous streams in the northwestern part of the state that have a constant and fairly rapid movement. Nevertheless, many species which occur in streams with a similar rate of flow in north Georgia and North Carolina are not found in them. The *interpunctatum* complex of *Stenonema* is represented in Florida by a single species, and this complex (perhaps even the same species as that found in Florida) also occurs in the mountain streams of North Carolina. Yet *Iron,* which in North Carolina occurs in association with the *Stenonema* nymphs, for some unknown reason does not seem to extend south of the region of Atlanta, Georgia. Perhaps *Stenonema* may possess a greater degree of tolerance, or there may be differences in the feeding habits of the two which prevent the spread of *Iron* into the Coastal Plain.

Ide has shown that the entire stream fauna is very much affected by thermal conditions, and water temperatures are very closely tied up with rate of flow of streams. In Ontario, according to this author, the number of mayfly species increases downstream owing to the higher temperatures that prevail in the lower stretches, the addition of species adapted to the higher temperatures being more rapid than the elimination of headwater species which require lower temperatures. His analysis of an Ontario stream in terms of temperature is very interesting; and although no similar studies have been made of Florida streams, Ide's results are not applicable to them since they have a nearly uniform temperature throughout. His conclusions would also suggest that the greatest number of species should be found in the southern part of North America, where higher temperatures prevail throughout the year, but the reverse is true.

Habitats within Florida streams are apparently more limited in extent and variety than in Northern ones. A great many of the Northern streams are filled with rocks and pebbles, and this condition evidently affords the optimum circumstances for great numbers of mayfly species. Combined with rocks are, of course, pools, vegetation, debris, and many other sites which offer refuge to the mayfly nymphs. Florida streams, devoid of rocks and with much barren, sandy bottom, can offer little variety in comparison, even though in these streams all available habitats are utilized.

Perhaps the greatest barriers to the wide dispersal of Northern mayflies in Florida are the large areas in which there are no constantly flowing streams. Thousands of square miles of the state, particularly in the coastal regions, are low and flat, and the grade of the lowlands is not sufficient to maintain permanently flowing creeks and rivers. Typical rheotropic species cannot exist for long in standing water, and even if a species were accidentally introduced while there was some flow, as soon as the flow ceased, the nymphs would probably die. The dry, sandy scrublands of southwest Georgia and of Florida also offer a serious barrier to less vagile species, and this has probably helped to keep the number of mayflies in Florida less than that in neighboring states and the northern part of the continent. Rogers (1933) found that the same conditions affected the distribution of craneflies. He states that "one of the most important barriers to the northern groups, the ranges of which extend into the Piedmont Province of Georgia and the Carolinas, is the monotonous, low pine-lands of the southern coastal plain with their dearth of clear, fairly rapid, pebble-bottom streams. . . ."

COMPARISON OF LIFE HISTORIES.—Many differences in behavior would be expected between the mayflies of the North and those of the South, but the most striking are found in life histories. Those species of Ephemeroptera which are known to occur both in Canada and in Florida show wide variations in emergence period, and might in consequence be considered different physiological subspecies. In Canada and the Northern states, all these species have a rather limited period of emergence during late spring and summer, but in Florida this is not the case. Here, every one of the species common to both North and South emerges throughout the year except during the short cold spells, and mating takes place at any time of the year unless cold weather is encountered, when the insects become lethargic.

Although Spieth had little definite information on the point, he stated rather accurately (1938c: 214) that "in the southern part of the United States, the length of the emergence period of the group is much longer. There seems to be no reason why in the tropics and subtropics there should not be some species emerging during each month of the year." The greater part of Florida cannot be considered subtropical, yet Spieth's conjecture holds true for the whole area.

Spieth (1938c: 214) makes one statement with which I cannot agree: "Regardless of the time of emergence during the year, each species has a definite limited period of emergence. In those species which have more than one generation each year, naturally there is more than one emergence period. Usually the period of emergence is relatively short." In Florida, the period of emergence of the great majority of species is not at all limited, although there may be many broods emerging during the year. I do not believe that results obtained in the North, upon which Spieth based his conclusions, can be accurately applied to the species inhabiting Florida.

Not only are many of the mayflies nonseasonal in Florida, but this has been found true of other groups of animals as well. Professor J. S. Rogers has told me that although the life histories of many craneflies are not seasonally limited

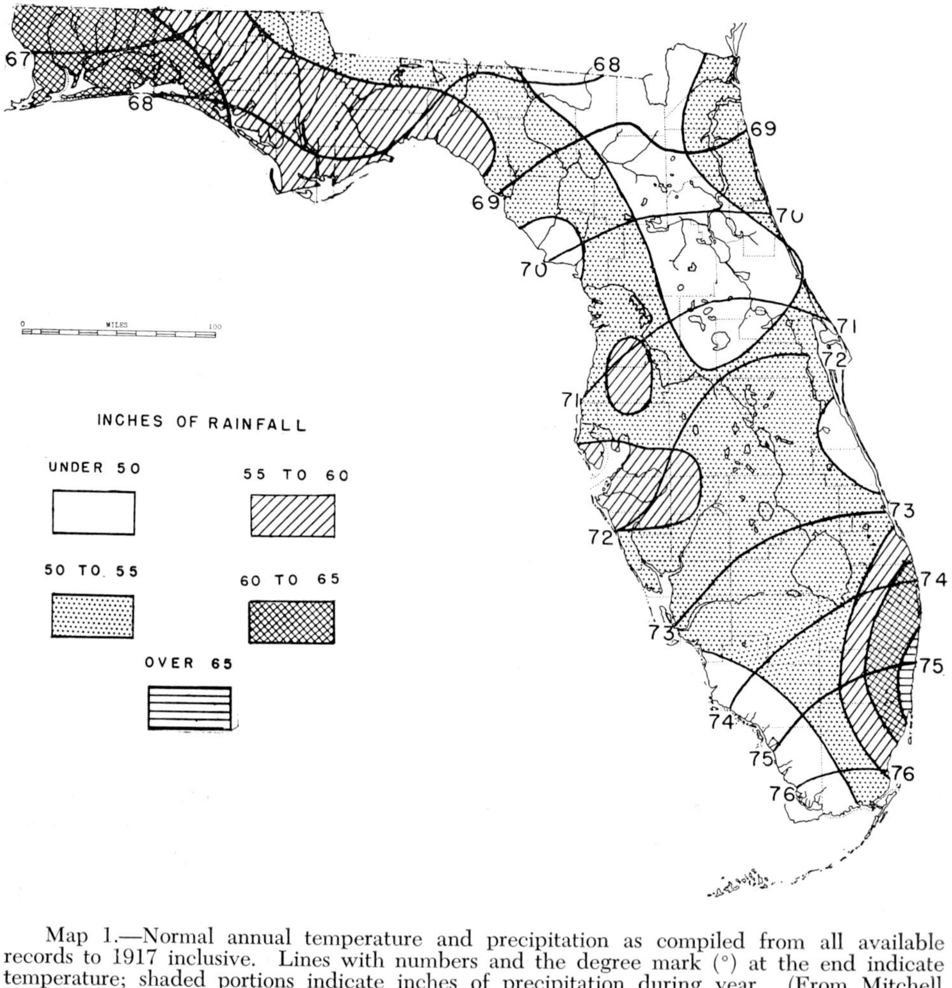

Map 1.—Normal annual temperature and precipitation as compiled from all available records to 1917 inclusive. Lines with numbers and the degree mark (°) at the end indicate temperature; shaded portions indicate inches of precipitation during year. (From Mitchell and Ensign, "The Climate of Florida.")

in Florida, they are definitely so limited in the North. This characteristic of nonseasonal emergence has been found to be particularly true in the southern part of the peninsula.

Reference to map 1 will show that average temperatures in Florida are suitable, even in winter, for the adult stage of insects which are primarily of Northern origin.

There are very few Florida mayflies which are definitely seasonal and these few are probably Southern in origin, whereas those species emerging throughout the year are Northern forms which have immigrated to Florida. This habit has been noted by specialists working with the Lycosid spiders, dragonflies, craneflies, and Orthoptera. The temperature factor appears to be linked with this unique behavior. It has been shown with respect to the mayflies that the colder the water inhabited by the immatures, the shorter the period of emergence. In Florida the lowest winter temperature of the water is probably equal to, or only slightly below, that of the Northern waters during the summer period of emergence. Consequently, mayflies in Florida are seldom, if ever, confronted with conditions which are unfavorable for transformation. For this reason a species (*Paraleptophlebia volitans*, for example) which in Canada emerges for about two weeks during the summer, can emerge throughout the year in Florida. The generalization that species of Northern origin are nonseasonal in Florida, and that seasonal forms in this state are all of Southern origin, holds good; but this is not a complete statement of the facts. Many of the species believed to be Southern in origin are nonseasonal like the Northern forms in this area. But this too may be explained on the basis of temperature if these Southern forms arose in the cold waters of the southern Appalachians or the hilly regions of the Piedmont. The only truly seasonal species are *Oreianthus* sp. No. 1, *Baetisca rogersi, Campsurus incertus, Hexagenia munda marilandica*, probably the other species of *Hexagenia, Ephemera simulans, Ephemerella trilineata*, and perhaps a few others of which I do not have adult specimens.

ZOOGEOGRAPHY

The topography of Florida has been described by Cooke (1939) in his *Scenery of Florida*. He divides the state into five natural areas (map 2) and these have been found to be closely correlated with differences in the mayfly fauna. The divisions of Florida are discussed in the section of the present paper treating the ecology of these insects (p. 28).

Since mayflies are so limited ecologically, their introduction into a new area is subject to many obstacles. The various factors influencing drainage are the dominant elements directing the movement of primarily aquatic insects. Hubbell and Stubbs have suggested (*in* Carr, 1940) that the following events and conditions have occurred in the more recent geological history of Florida:

1. Persistence of a land area in central Florida, in the form of large islands or a group of keys, at least since the beginning of the Pleistocene, and probably since Pliocene times.
2. The bridging of the gap to the mainland on one or more occasions (perhaps first in the Pliocene), followed by renewed insular isolation.
3. Final establishment of peninsular conditions during the Pleistocene.

4. More or less extensive marginal submergences in late Pleistocene, reducing much of the eastern margin of the peninsula to a coastal archipelago.
5. Persistence of a salt-water barrier between Florida and the West Indies at least since pre-Cenozoic times, and certainly throughout the period of derivation of the modern biota of the state.

It is unnecessary to hypothesize that such islands have existed in order to derive the ephemerid fauna. The winged stage of mayflies and the methods of dissemination of the insects would allow for rapid penetration of an area should ecological conditions become satisfactory for the maintenance of that species. On many of the existing Bahaman Islands, conditions today would not allow a very extensive development of mayflies. It is not at all unlikely that when Florida was rising from the sea as isolated islands, these islands were very similar to the present-day Bahamas. If such conditions persisted until land connections were formed, the introduction of mayflies into the region has been very recent, and, as is suggested below, the fauna was derived mainly through the larger river drainages. Quite likely, there were some mayflies inhabiting Florida islands, but they must have been pond forms and species which could withstand wide variations in ecological conditions. Those species which may be endemic to Florida could easily have arisen since the establishment of the peninsula in the Pleistocene.

As Florida rose, the streams draining the southeastern portion of the continent were extended onto this new area carrying with them their Northern faunas. In the western part of Florida, the Chattahoochee, which drains the foothills of the Appalachians, and the Flint, which drains the Piedmont of Georgia, came together to form the new Apalachicola River. As the Florida tributaries of this great stream began to erode back into the uplands, the conditions in their valleys tended to approach those of the more northern reaches of the river, and, as conditions in these tributaries became suitable, the mountain and Piedmont species came to occupy all available habitats within the small, rather rapidly flowing, sand-bottomed streams.

The Apalachicola River drainage has without doubt been the main highway of ingress to Florida for the great majority of animals which require flowing water or hardwood forests. Rogers (1933) found that the ecological conditions existing in the Apalachicola drainage would admirably explain the distribution of many Northern craneflies in Florida. Carr (1940) reports that the most extensive invasion of Florida by the Northern element is encountered in that portion of the panhandle which is drained by the Apalachicola River. Hubbell (1936: 354) states that "in this peculiar environment [the deep ravines of the Apalachicola region] a great many northern plants occur, evidently the remnants of a northern flora left as relicts from Pleistocene times in these deep, moist, cool ravines. . . ." It has also been noted that many Northern plants reach their southernmost limits in these ravines and that many typically Southern plants intermingle here with the Northern species. Not only is this true of plants, but such has proved to be the case in the craneflies, the Odonata, the Opilionids, the Orthoptera, the crayfish, the amphibians, and the reptiles.

Many of the species of mayflies which have entered the state have spread out from the Apalachicola drainage and now occupy rather wide ranges in Florida, but there are certain forms which seem to be more or less confined to this region.

The Apalachicola River was doubtless not the only route of entry from the North. It seems probable that the Suwannee and its tributaries may also have brought in certain Northern elements, such as *Brachycercus*. Very likely there has been some migration along the more easterly Coastal Plain region where the streams descend from the higher Piedmont. However, the forms which may have entered Florida over such routes are few in comparison with those that have come in by way of the western path. Perhaps some species have entered by more than one course, moving in from the west and meeting to form a unified population in northern Florida.

Traver's paper on the Ephemeropteran fauna of Puerto Rico (1938) and infrequent records of other species from the West Indies show that those species of mayflies found in Florida have but few affinities with the insular fauna. No species are shared in common, and the genera which occur in both regions are very widely distributed. Although other species of insects have become established both in Florida and the West Indies, the mayflies have been unable to do this principally because of the ecological conditions of south Florida. The great majority of forms described from the islands are inhabitants of streams, particularly of mountain streams, and even if these mayflies were accidentally introduced into Florida, establishment would be impossible in the South. If, by some rare chance, a female which had been fertilized were carried to one of the permanently flowing sand-bottomed streams, it is barely possible that the species might gain a foothold, but this has apparently not occurred.

Hobbs (1942) has hypothesized that the crayfish now occurring in Florida have been derived from migrants from the North or from the West; his evidence indicates that the Western element is much the larger. A few of the Northern species of mayflies do appear to have swung westward around the southern Appalachians and thence into the Gulf Coastal Plain and Florida. In this sense only, can any part of the Floridian fauna be considered of Western origin.

Among the factors that may be used in explaining the present distribution of rheotropic mayflies in Florida, temperature may have exercised a considerable influence. In west-central peninsular Florida—that is, in the Tampa region—the mean annual temperature of the air is 71.5° F., whereas in the northwestern part of the state, where the principal part of the continental fauna is concentrated, the average temperature is several degrees lower. Since temperatures in streams are less subject to variations than those in air, it is likely that average annual temperatures are more important to aquatic organisms than to terrestrial forms, for which the extremes of temperature are more likely to be the critical factors. It seems reasonable to suppose that the score or so of mayfly species common to peninsular Florida and the western part of the state must have relatively wide limits of temperature toleration. Those species having more restricted temperature toleration have remained confined to the northern and northwestern parts of the state.

DISPERSAL OF MAYFLIES.—The two most important factors involved in the dissemination of mayflies are wind and water. Of these, the latter, as far as the actual establishment of species is concerned, is by far the more important, for the immature stages of all Ephemeroptera (except a very few brackish water species) require fresh water.

The ability to fly allows for the greatest amount of movement of ephemerid species, although the adults are more or less confined to moist situations. The relatively short life span of the imagoes must necessarily limit the flight range of the insects; however, the area occupied by a species can be extended somewhat through voluntary flight of the adults.

Such small, feeble insects as mayfly imagoes are easily carried by the wind. This can be seen in the present distribution of the more diminutive genera, *Caenis, Baetis,* and others, which are found throughout the world. Dr. P. A. Glick (1939), of the U. S. Bureau of Entomology, has made a thorough and very interesting study of the distribution of insects, spiders, and mites in the air. By means of traps suspended from the wings of an airplane, he collected many thousands of insects from various strata of the air. His results for the Ephemeroptera are listed below in Table 2:

TABLE 2

THE DISTRIBUTION OF MAYFLIES AT VARIOUS ALTITUDES *

Height of Flight	Time of Collection	*Caenis hilaris*	*Caenis* sp.	*Hexagenia* sp.	*Ephemera* sp.	Undetermined sp.
3000 ft.	Day					
	Night		1			
2000 ft.	Day	1			1	
	Night					
1000 ft.	Day		1			
	Night					2
500 ft.	Day					
	Night		2			
400 ft.	Day					
200 ft.	Day		1	1		
50 ft.	Day					

*Glick, 1939.

The above figures illustrate the fact that mayflies can be carried to extreme heights, and, by means of horizontal air currents, doubtless to considerable distances. Perhaps the greatest obstacles to be surmounted in aerial distribution at great heights are the short adult life and the necessity of the fertilized female reaching fresh water. Even when these two obstacles are overcome, the chance of the ecological factors being satisfactory are but slight; current-loving species could hardly develop in a pond, and vice versa. The possibility of distribution through wind currents at great heights, therefore, seems to be somewhat remote, but the chance nevertheless exists and must be considered.

From his study, Glick concluded that size, weight, and buoyancy of an insect bear directly upon the height to which it may be carried by air currents. He found that many species represented at high altitudes were small insects. Temperature was undoubtedly the most important factor regulating the numbers of insects to be found in the air at any given time, and he found that the optimum range was from 75° to 79° F., surface temperature.

Dr. Glick points out that the intensity of air currents is a great factor in the distribution and dispersal of insects. Most insects were taken at the lower altitudes when the surface wind velocity was from five to six miles per hour, and fewest when it was calm.

Winds at low altitudes are of importance in transporting adult insects from one region to another but their effects are probably local; however, such local spreading continued over a long period would eventually greatly increase the range of species. A combination of strong winds at low altitudes, flight, and proper ecological situations would allow rapid distribution of mayflies.

Within the continental area as a whole, stream piracy probably has acted, and is acting, as an important agent for the dispersal of mayflies. By this means, species may spread from one drainage system to another and from one region to another, gradually coming to occupy very wide ranges. This would be of particular importance in the case of those species which, as adults, have but limited powers of flight and those which are unable to withstand desiccation.

Flood conditions also operate as an influence in the spread of a species during the immature stages. At the time of floods, stream velocity is greatly increased, and owing to this increase, rocks, logs, pebbles, and other objects to which the immatures cling are moved violently downstream. Doubtless, the greater portion of the animals on these objects are destroyed, but a few may survive to carry the species far from its original home. From the lower reaches of the stream to which the nymphs have been carried, the species may extend its range into new drainages by the flight of the winged stages.

It is barely possible that wading birds might play a part in transporting mayfly nymphs from one body of water to another. Many water-dwelling organisms have been transported successfully on the legs of such birds, and if the flights of the birds were short, necessitating only little time out of the water, it is possible that certain mayfly nymphs might survive a journey of this nature.

When the various possible distributional agencies are subjected to analysis, it can be seen that mayflies are principally disseminated by flight, by winds near the earth's surface, by floods, and more rarely by high air currents and stream piracy.

CLIMATIC DIVISIONS OF FLORIDA

C. H. Merriam divided Florida into two regions which appear to be useful in a general way for differentiating ecologically the flora and fauna of the Nearctic and Neotropical regions. He considered that part of Florida north of a line from St. Lucie Inlet on the east coast to Fort Myers on the west coast to be continuous with the Austral zone of the continental (Nearctic) portion of North America, the remainder of the state south of the line being Neotropical.

P. P. Calvert, likewise, has recognized two zones in Florida, separated by differences in the mean annual temperature. Most of peninsular Florida falls into his Zone III, with a mean annual temperature ranging from 68° to 77° F., whereas the western part of the state lies in his Zone II, with a temperature ranging from 59° to 68° F.

Rainfall in the Nearctic region of Florida is to some extent confined to a midsummer rainy season, but in the subtropical southern tip of the state such a season is more sharply defined. In the northwestern area, although there is some demarcation of a rainy season, the rainfall is more evenly distributed throughout the year. Byers (1930) has given in tabular form the average rainfall and temperature in various regions of the state in July and January, and this shows fairly well the tendencies described above.

The distribution of mayflies in Florida is chiefly dependent on the presence of running water, although temperature probably also exercises no little influence. Since great areas of the state are characterized by swamplike and sandy conditions and pine barrens with no development of streams, their Ephemeropteran fauna is extremely limited. This is particularly true of that portion which lies at the southern tip of the state and is generally designated the Neotropical region. The mayflies found here are not Neotropical in origin, but are true Nearctic species distributed throughout Florida; in fact, *Caenis diminuta* is just as common throughout the eastern part of much of the Nearctic as in this unique southern biological area.

HABITATS OF FLORIDA MAYFLIES

It is generally recognized that the mayfly population of a region is intimately related to the aquatic conditions of that region. From the standpoint both of the ecological distribution of mayflies and of physiography, Florida can be divided into five natural areas (map 2). The boundaries of these areas shown on the map are rather arbitrary. The aquatic conditions in the Coastal Lowlands overlap to a great extent those conditions found in the other areas, but these boundaries are useful in delimiting in a broad way the fresh-water situations as they occur in Florida.

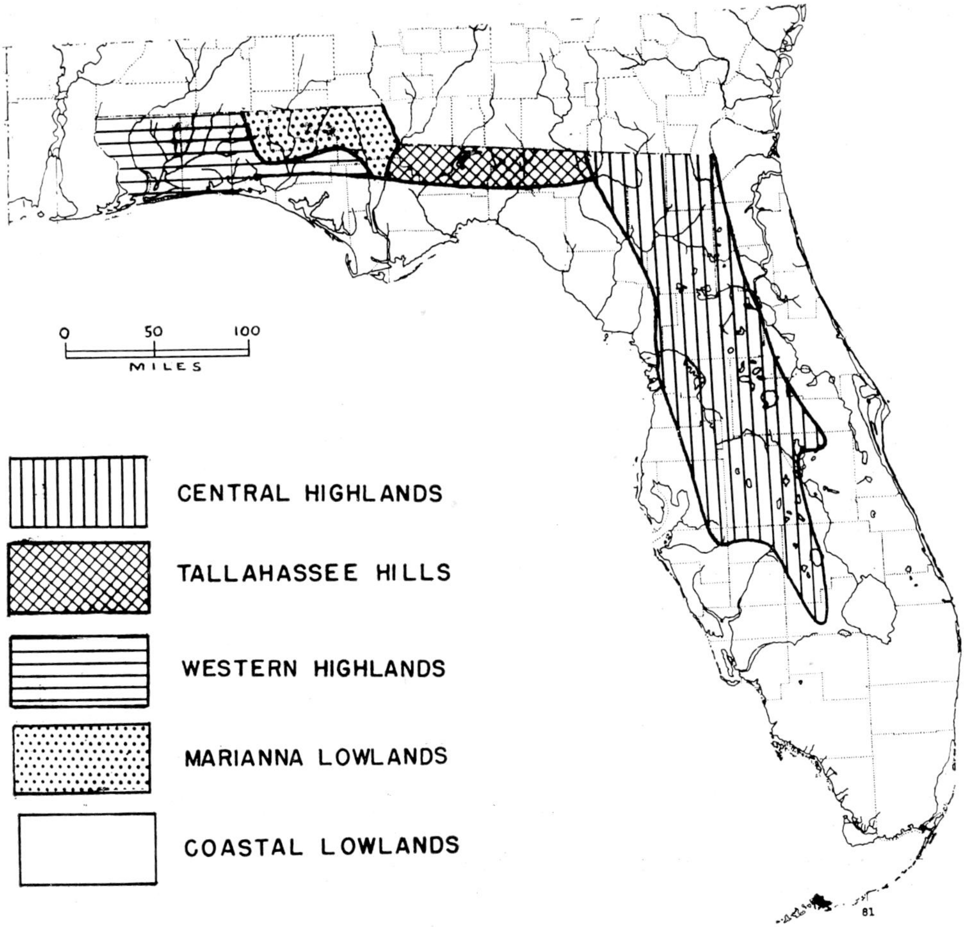

Map 2.—The physiographic divisions of Florida. (From Cooke, *The Scenery of Florida.*)

The Coastal Lowlands almost everywhere lie less than one hundred feet above sea level; the height of the other regions above sea level varies from one hundred to about three hundred feet. The drainage of each of these subdivisions of the state is closely linked with its topography.

The Coastal Lowlands are swampy and marshy, including, in their southern reaches, the great expanse of the Florida Everglades, and continuing northward into the mucklands of the Lake Okeechobee region, which in turn is continuous northward on the east and west with the lowlands of the coast.

The Central Highlands may be divided into two areas. The southern half is entirely a lake region where almost every depression has its own pond or lake; the northern part of this division not only includes lakes and ponds, but also numerous surface, as well as subterranean, streams of all sizes.

The Tallahassee Hills and Western Highlands are rolling areas composed of red-clay hills, well drained by many small streams. The Marianna Lowlands, on the other hand, are chiefly flatwoods, swampy areas, and rolling sandy hills covered by pine forests with few permanently flowing streams.

Several large rivers rising in Alabama and Georgia flow through northwestern Florida and empty into the Gulf of Mexico. In the east, the large St. Johns River flows north along the border between the Central Highlands and the Coastal Lowlands.

DESCRIPTIONS OF HABITATS

INTERMITTENT CREEKS

In Florida, there are relatively few intermittent streams. These few are, for the most part, merely connecting canals between swamps, ponds, or other bodies of water, the levels of which fluctuate continually according to the amount of precipitation. As far as mayflies are concerned, the intermittent streams are rather poor habitats, and only the very tolerant *Callibaetis* and *Caenis* nymphs can withstand the alternating conditions in the creeks produced by the continual changes in the levels of the bodies of water connected by the intermittent streams. Vegetation is usually abundant, but differs from the type found in constantly flowing waters. Also, the drying of parts of the canal bed produces conditions which the stream-inhabiting mayfly nymphs cannot for long endure, so that even if, by chance, a species should become established during the period of flow, the cessation of this flow would bring about the death of the immatures.

PERMANENT CREEKS

Permanently flowing streams are the most important of the mayfly habitats in Florida. The creeks of this area may arbitrarily be divided into the following categories:

Sand-bottomed creeks with little vegetation.
Sand-bottomed creeks choked with vegetation.
Silt-bottomed creeks with little vegetation.
Silt-bottomed creeks choked with vegetation.

Sand-bottomed creeks with little vegetation.—These are small, shallow, gently flowing streams with sandy beds. The creeks vary from as little as one foot to forty feet in width, and in depth from a couple of inches to as much as five feet. The bottom is composed of loose rolling sand which builds up in midstream into small ridges behind which small masses of debris accumulate. Large rocks almost never occur in Florida streams, but pebbles may be found imbedded in gravelly riffles where the water becomes quite shallow. There are occasional pools, but they are not a conspicuous element of the streams. The pools are usually small, quiet areas near the banks or at curves. Debris accumulates to a rather marked degree in some of the streams, almost any obstacle forming a nucleus for the accumulation of much leaf drift, sticks, and other objects. Tree trunks frequently form dams and give rise to riffles, while the tangle of branches and twigs provides a network in which much detritus becomes entangled. Silt accumulates near the shore and in places may produce rather thick deposits, in some streams even forming a layer more than two feet in thickness; however, the silt deposits in the sand-bottomed streams are usually sparse and of little consequence. Near shore, leaf drift becomes a fairly important habitat, for many insects are harbored in this material, and in the almost stagnant shore pools the leaf debris, interspersed with silt, may be several layers thick. The flow of water in the sand-bottomed streams is never rapid in the sense in which a Northern stream is said to be rapid; rather, the flow is gentle with the surface seldom breaking. Most of these streams are circumneutral to slightly acid, but some may be pronouncedly acidic. Nearly all of them have tinted waters which vary in shades from almost colorless to a strong tea color, according to the area drained and to the amount of rainfall. Most of the streams drain flatwoods, hammock lands, or swampy areas, and are fed by springs or diffuse seepage areas. Vegetation is almost completely absent from the streams except for a few scattered *Orontium* plants and an occasional clump of *Persicaria* near the quiet shore zone; however, these plants are not particularly important from the standpoint of furnishing habitats for mayflies.

Sand-bottomed creeks choked with vegetation.—This sort of stream occurs mostly in the northwestern part of Florida beyond the Apalachicola River. The beds of the creeks are composed of fairly loose sand, but instead of being almost bare, as in the type of stream mentioned above, they are covered with dense growths of *Vallisneria, Sagittaria,* and *Potamogeton,* which in turn are thickly covered with algae. There may be some debris collected in the slower areas near shore, but in midstream the vegetation is swept clean, although an occasional partially submerged log may lie among the plants. These streams are usually not much more than twenty feet across and may be as much as five feet in depth at the center. In their deeper parts, the vegetation tends to disappear, and may be entirely absent from exceptionally deep stretches. The rate of flow is moderate, but seldom strong enough to cause any marked disturbance of the water surface. Mainly, the creeks drain scrublands and high pine and hammock country, and the water is much lighter in color than that of the sand-bottomed creeks with little vegetation; however, the water is usually definitely acidic, with a pH approximating 6.0. Silt deposits are not so pronounced in these streams as in those of the first class, and debris along the shore is likewise less in quantity.

Silt-bottomed creeks with little vegetation.—Silt-bottomed streams are rather common in the northern part of the Central Highlands of Florida. The stream bottom is covered with a layer of silt overlying the sand and varying from a few inches to several feet in depth. The rate of flow is comparatively slower than that of the sand-bottomed streams, but it is steady and quite perceptible. The water is definitely acidic and usually a rather strong tea color. The silt bottom is frequently overlain by layers of leaves and strewn with much other debris, but there are few or no plants in the stream proper. Near shore, *Persicaria* and various sedges and grasses may be present, but they are not especially abundant. This type of creek averages about twenty feet in width and from a few inches to three feet in depth.

Silt-bottomed creeks choked with vegetation.—These streams are not particularly common in Florida, and are confined mostly to the northwestern part of the state. The streams are shallow, one to three feet in depth, frequently wide, very meandering in their courses, and sometimes braided. The vegetation is dense, and may include plants which are characteristic of more slowly flowing or even stagnant water, such as *Isnardia, Persicaria,* and *Pontederia.* Other plants also found in the course of the stream include *Vallisneria, Potamogeton,* and *Sagittaria,* as well as many algae. The silt in the streams is very loose and may be as much as three to four feet in thickness. This material is soft, fluffy, and somewhat sticky, and at the slightest disturbance stirs up and clouds the water. There is usually a broad flood plain, and during high water the streams spread out widely over it until all vegetation is completely submerged (except larger bushes and trees). There is a constant flux of channels due to this flooding. In the shallower zones, the rate of flow of the water is negligible, and during a rather severe cold spell, I noticed that the surface of one of the streams was frozen over in the shallow zones.

RIVERS

Stagnant rivers.—Stagnant rivers are confined to the southern portion of the Florida peninsula, and many of them have been dredged within the last few years to serve as drainage canals for the Everglades. Strictly speaking, these rivers are not stagnant, but the flow is so slight that only rheotropic organisms which are very tolerant to stagnation can exist in them. The Miami River furnishes an ideal example of such a waterway. Until the drainage plans were carried out the Miami River was, in places, a shallow, moderate- to swift-flowing stream cutting through the limestone which forms the bedrock for south Florida. Charles B. Cory in his book *Hunting and Fishing in Florida* describes and illustrates the Miami River as it was in 1895, and paints a picture which certainly differs enormously from the present state: "The Miami river, which runs from the Everglades into Biscayne Bay, is probably the only river in Florida which has a fall or rapid worthy of the name. For about a half a mile at the head of this stream there is considerable fall. At this point the river is shallow and not navigable for boats, and has a very rapid current, in which 'Cavalia' (*Caroux hippus*) are numerous and may be taken with an artificial fly." Dredging has completely eliminated all rapids from the river and flow is not perceptible, even during the rainy season. The salt water of Biscayne Bay backs up into the river and produces a brackish condition some distance from the mouth of the canal.

These stagnant rivers have fairly heavy growths of aquatic vegetation near their shores, but they become deep rapidly, and in the deep regions the plants are quite limited. The rivers vary from one hundred to two hundred feet in width in the widest places, and in depth usually range from fifteen to twenty feet. Some of the rivers show a pronounced flow during the rainy season and may even be subject to floods. Since the digging of the drainage canals, however, this is the exception rather than the rule. There is very little deposit of silt in these streams and the bottom is mostly bare limestone.

Slow-flowing, deep rivers.—This category includes nearly all of the larger rivers of Florida, such as the Suwannee and the Apalachicola. These streams are large and deep, and drain very extensive areas; the Suwannee drains the Okefenokee Swamp and the Apalachicola rises in the foothills of the Appalachians and the Piedmont of Georgia and Alabama. The larger rivers have continuous flow, and during excessive rains they spread out over their flood plains. In the large rivers, vegetation is limited and occurs principally near the shores in protected places where the current is slow. This vegetation is more characteristic of standing water than of streams. There may be some rocks near shore in the shallow water, and silt may be deposited in protected areas, but for the most part the bottoms are hard clay or limerock. The Apalachicola River is frequently very muddy, owing to the heavy burden of silt it brings down from the Georgian highlands. The Suwannee is much clearer, although the water may be strongly colored by the various organic acids coming from the swamps drained by the river. The St. Johns River has a much slower current than the other two streams, and north of Lake George the flow is slight and is affected by the tides. This river has masses of water hyacinths growing at the shore, and conditions are not very different from those found in the lakes of the Central Highlands which support growths of this plant. Nearer its headwaters, there is *Vallisneria* and *Sagittaria* in the stream and the flow is more noticeable. Much of the water of the St. Johns is of swamp origin, and this is reflected in the brownish coloration of the stream.

Larger calcareous streams.—This category includes streams which are, in general, smaller than those included above and which are definitely basic in their reactions. Most of these streams rise from springs and form such small rivers as Silver River, Wakulla River, and Santa Fe River. They are clear, cool, and moderately flowing, with rather dense growths of vegetation in the stream proper. Vegetation is composed chiefly of *Vallisneria, Sagittaria, Nais, Isnardia,* mosses, and many algae, which in the shallower zones form dense mats completely covering the floor of the stream. The water may be colorless if the source is confined to springs, but if swamp waters also contribute to the stream the water may be tinged with brown. Many of these rivers have sand bottoms, and there may be thick deposits of calcareous silt in the quieter shallow zones. Leaf drift and other debris become entangled in fallen trees and other catch-alls, and such debris forms an important habitat for many mayflies and other organisms. In some of the rivers there are outcrops of limestone, and in the Santa Fe River, specifically, there are many loose rocks which are of great importance as habitats for aquatic organisms. The depth of these streams varies from three feet to twenty feet or more, the width from seventy-five to three hundred feet. Faunistically, this type of stream is the richest found within the boundaries of Florida, both in numbers of individuals and in species. Carr (1940: 25) states that "optimum conditions [for fluvial organisms] apparently exist in those rivers which run over ledges of exposed limestone, or which receive most of their water from calcareous springs."

DITCHES AND PUDDLES

Roadside ditches.—These are extremely rich situations for aquatic organisms if there is permanently standing water in them. The ditches are extremely varied in appearance but the fauna is fairly constant. In some of them, there are dense growths of *Pontederia,* which is the predominant plant. The water is shallow in the zone of pickerelweed, but may become much deeper beyond this region; in the deeper parts other vegetation which can stand greater degrees of submergence is present. The bottom of the ditches is usually covered with grasses and may have heavy growths of *Globifera umbrosa, Isnardia,* and *Persicaria.* Algae form dense mats in some of the ditches along with *Utricularia, Ceratophyllum, Potamogeton,* and *Myriophyllum.* At times during the summer the water may become very warm; in winter it is often quite cold, with ice occasionally forming over the surface. The water in the ditches is usually acid but may be basic; the lowest pH recorded was slightly above 4.0. The depth of the water varies from a few inches to as much as four or five feet, and the width of the ditch (that part containing the water) may be from one to fifteen or more feet.

Puddles.—In this category are placed those small and transitory bodies of water left by the retreat of a stream or formed by heavy rains. There is no, or very little, aquatic vegetation in such puddles, and it is only rarely that immature insects are found in them. Such animals are usually stream relicts which soon perish, for the puddles disappear rapidly during dry weather; however, I have found mayflies in such places, and for that reason I am including this type of situation. In some of the puddles there may be several layers of leaf debris. Silt accumulates between the leaves and algae begin to grow, continuing as long as the water remains.

PONDS

The ponds of Florida may be divided into several types, but nearly all of them share one characteristic in common—they seldom have surface streams draining into or from them.

Sinkhole ponds.—These ponds are formed by the dissolution of underlying limestone. Rainwater, percolating downward to the water table, dissolves vertical chimneys in the limerock into which the surface cover may collapse gradually or suddenly, producing a steep-walled, open sink. Where the cover is thicker or less compact a saucer- or funnel-shaped depression may result. There are many of these ponds of sinkhole origin in Florida, and particularly around the Gainesville area. Though some of them are dry, the great majority have standing water which maintains a fairly constant level because the water table is high enough to supply the ponds continually. The sides of the ponds both above and below the water level are steep and the zone of rooted aquatic vegetation is very limited. The sides of the sinkholes above the water are usually covered with vegetation which extends from the edge of the water up to the rim of the depression. In many of the sinkholes a narrow sand beach may be formed where the slope of the sides levels off, but the shore zone extending into the water from this beach is very narrow, and the drop to deep water rapid.

There are two chief types of sinkhole ponds: (1) ponds with the surface free of vegetation; and (2) ponds with their surfaces covered by vegetation. The first of these is one of the common types of sinkholes encountered in peninsular Florida. The margin of the pond has a fairly rich growth of both submergent and emergent vegetation, which extends outward to the region where the drop off to deep water occurs. The plants are principally *Saccrolepis striata, Persicaria, Mayaca fluvitalis, Juncus,* and some *Typha,* as well as numerous species of algae. This sudden drop to deeper water begins at a depth of about four feet and extends to a depth of about ten feet; thence the drop continues more gradually to about twenty feet. In many of the ponds there may be deeper holes such as that found in Lake Mize, discussed by Harkness (1941), which over a very small area reaches a depth of nearly eighty feet. In the shallower portions the bottom is sandy and covered with patchy growths of *Utricularia.* There are only small deposits of silt in the shallower zones, but in the very deepest parts the bottom may be thickly covered with black dense silt, in which practically anaerobic conditions exist and from which larger organisms are almost entirely absent. The water varies from colorless to a strong tea color, depending on the type of region drained by the pond. Many of the ponds are rather turbid, and this turbidity, combined with strongly tinged water, allows little penetration of light to the deeper parts. The water varies in reaction from pond to pond, being acidic, circumneutral, or slightly basic.

The second type of sinkhole pond includes two principal varieties: (1) those covered with water hyacinths, *Piaropus crassipes;* and (2) those covered with duckweed, *Lemna minor,* mud-marys, *Bruneria punctata, Azolla caroliniana,* and *Salvinia auriculata.* The first variety of pond may have a marginal ring of open water with the water hyacinths concentrated over most of the remaining area, or the entire surface may be covered with these plants. If there is some open water, submergent vegetation may take root and become fairly well established, but usually the water hyacinths are the only plants of any importance in the ponds.

The duckweed-covered ponds seldom have any submergent vegetation growing in them, for the entire surface is covered with a single layer of the plants, which cuts off light from any vegetation that might attempt growth on the pond bottom. When there is any wind the duckweed piles up and leaves clear areas on the water surface, but as soon as the wind dies down, the plants again spread evenly over the surface. Amphipods *(Hyallela azteca)* living among the duckweed are exceedingly numerous in the ponds, but bottom organisms are scarce, probably because of the small amount of available oxygen. The bed of the duckweed-covered pond has rather thick accumulations of black silt overlaying the sandy bottom. Almost anaerobic conditions exist in the silt which is heavily loaded with hydrogen sulphide. The sinkhole ponds are usually about one hundred to three hundred feet in diameter, and are mostly round in shape. Some of the larger ponds are formed by the coalition of two or more sinkholes.

Fluctuating ponds.—These constitute one of the more frequently encountered types of Florida ponds. Water fills shallow basins—some mere depressions in the original sea-laid sands, the majority produced by solution of underlying limestone or by wind action. The level of the water in these hollows varies with the amount of rain and surface runoff. Though there is a great fluctuation in the area occupied by the water because a slight rise or fall causes marked spread or retreat of the margins, the depth usually does not change greatly because of the shallowness of the pond. True aquatic vegetation follows the rise and fall of the water to some extent, but it is chiefly confined to the part of the pond below the more permanent water level. Rogers (1933) states that

> . . . these periodic and sporadic fluctuations [due to rainfall] prevent the development of any permanent vegetation within the broad zone of fluctuation. As the water recedes in the dry season, it leaves a belt of saturated earth in which some grasses and sedges spring up among mats of stranded algae and other aquatic vegetation. . . . The permanent water is usually filled with submerged, floating, and emergent aquatic plants and may support extensive floating mats of green algae.

The vegetation includes many true aquatics such as *Pontederia, Utricularia, Persicaria, Globifera,* and *Typha,* as well as *Hydrocotyl,* maiden cane (*Juncus*), and other semiaquatics. Many of the ponds have coverings of *Castalia,* and some support *Nymphaea,* but very few are covered with water hyacinths. Plants are not confined to a shore zone as in the sinkhole ponds, but may extend much farther out and, in some instances, may completely penetrate the ponds. The bottom of the fluctuating pond is mucky, but the layer of silt is not deep. The water is subject to rapid changes in temperature and is much influenced by external weather conditions. There is normally a slight tint to the water and it is usually acidic.

Temporary woods ponds.—These ponds are formed in depressions in hammocks and are of a more or less temporary nature; however, the vegetation present is such as to indicate its habitual occupation of the depression. In the ponds and around their margins are frequently found black gum (*Nyssa biflora*), button bush *(Cephalanthus occidentilas)*, willow *(Salix niger)*, and bladderworts (*Utricularia* sp.), and if the water remains for a sufficient length of time, other more succulent vegetation will become established. After the water has been standing in the depression long enough to allow growth of algae, many of the usual pond animals begin to migrate to, and become established in, the pond.

Sporadic ponds.—The sporadic ponds, which are of a rather temporary nature, may disappear for a brief period or even permanently during the dry season. The ponds are shallow and may be large or small, depending on the depression and the drainage of the region.

Jerome Sink.—Because of the unusualness of this situation, it deserves special classification. A description of the sink is included in the discussion of the ecology of *Choroterpes hubbelli.*

LAKES

According to the findings of Mr. S. Stubbs, formerly of the Florida State Geological Survey, nearly all lakes in Florida are the result of solution of underlying limestone; however, such lakes as Lake Okeechobee seem rather to occupy basins which are natural depressions in the surface formed as the land rose during the last geologic period. Cooke (1939: 100-101) states:

> Many Florida lake basins are simple sinks that have always been tributary to the groundwater supply and never have had a surface outlet. Others at one time or another formed part of the surface drainage and therefore were connected with the river system. Their history is complicated, for the fluctuations of sea level and the consequent fall or rise of the water table gave opportunity alternately to deepen the basin or to flood them. Some of them appear to have been estuaries during higher stages of the sea, for the sand-covered terraces around them stand within the limits of altitude of the coastal terraces.

Cooke concludes that Lake Okeechobee, Lake Istokpoga, Lake Kissimmee, Crescent Lake, Lake George, and others originated as hollows in the sea bottom.

Sand-bottomed lakes.—The sand-bottomed lakes, by far the commonest type of lake found in Florida, form the most conspicuous bodies of water in the Central Highlands. There are literally thousands of these lakes in this region, and they are particularly numerous in Lake County. The lakes vary in size from a quarter of a mile to nine or ten miles in width, but the depth is not correspondingly great and most of them are shallow, not more than forty or forty-five feet at their deepest holes. There is little vegetation except at the margin, and this is chiefly *Saccrolepis striata, Utricularia,* and algae. Wave action is slight, but it is probably sufficient to prevent the growth of other aquatic vegetation along the sandy shore. The *Saccrolepis* extends out into the lakes to a depth of five or six feet, and beyond this zone vegetation is limited to submergent plants such as *Utricularia,* which lies on the sand bottom as far out as there is sufficient light penetration for normal photosynthetic processes to take place. In this region, some silt is intermingled with the sand, but much of the sand is bare. Beyond a depth of about twenty feet, silt accumulates and covers the bottom sand with a layer several inches deep; with increase in the depth of the lake there is a direct increase in the amount of silt. In the very deep parts of the lakes, the silt may reach a depth of a foot or more, and in this region it is a very thick, black, fine, and oozy mud in which few organisms live. Along the shore some debris collects, but is not abundant; this debris forms a habitat for many insects which in Florida are normally found in streams. Probably the wave action in this region is sufficient to oxygenate the water and produce conditions which simulate those found in the moderately flowing streams of the state. The water of the lakes varies from very clear to strongly tea colored and the turbidity is also variable according to the lake. Most of the lakes are circumneutral, ranging not much more than 0.5 on either side of a pH of 7.0.

Silt-bottomed lakes.—This type is well exemplified by Newnan's Lake near Gainesville. The lake is large, being about ten miles in length and about three miles wide. It is bordered by a ring of cypress trees which extend from dry land into the water. Also margining the lake are water hyacinths which form dense growths between the bases of the cypress trees and extend out into the lake as far as there is a protected zone. With every shift in the direction of the wind, water-hyacinth plants break free and float across the lake, pushed onward by the wind. Frequently, during a period of sustained moderate winds, the lake is dotted with numerous floating "rafts" of water hyacinths; eventually the plants pile up along one shore until the wind again changes and sends them sailing back to the opposite side. The continual rain of dead water hyacinths onto the bottom of the lake has given rise to a thick layer of loose fluffy silt which completely covers the bottom to a depth of several feet. Almost no living organisms can be found in these bottom deposits because conditions are not conducive to life. The lake is shallow; in the middle it is not much more than ten feet in depth, and over most of its area it is shallower than this. The water has a definite brownish tinge and is rather turbid.

Numerous other lakes in the Central Highlands also belong in this same category. Orange Lake, one of the larger lakes of the region, is similar to Newnan's Lake in its type of bottom, but it is deeper, has much rooted vegetation in the form of water bonnets, and is somewhat exceptional in having numerous floating islands of vegetation. Some of the floating islands are large enough to support trees, but the great majority are composed of small clumps of vegetation which float about, changing location with each change in direction of the wind.

According to Rogers (1933), "these lakes show great variations in their aquatic vegetation; in many, the succession is directly toward swamp conditions with extensive development of cypress along the muddy shores; in others there is a distinct development and zonation of marsh vegetation before the shallow water is invaded by cypress or hardwood swamps."

Disappearing lakes.—Certain large lakes in the northern part of Florida go dry during periods of drought. Lake Iamonia, which went completely dry in 1938, is an ideal example of a disappearing lake. There is one near Lake City which is said to go dry "once every seven years." These lakes are shallow, and in most respects are similar to the silt-bottomed lakes in supporting an abundant growth of water hyacinths and in being ringed by cypress. Although at their maximum extent the disappearing lakes are fairly large, they vary greatly in size with seasonal fluctuation in depth.

MARSHES

Marshes are very common in peninsular Florida, particularly in the lower regions. They may be very limited in extent or quite large, according to the size of the original basin. Many ponds and lakes have become converted into marshes, and many others are in the process of transition. The water is shallow and vegetation extends throughout, growing very profusely. The most predominant plants are emergent and include cattail, pickerelweed, maiden cane, saw grass, water lilies, smartweed, and various grasses. Submergent plants are prominent including *Isnardia, Globifera, Myriophyllum,* and many algae. The water is rather warm during the summer and may freeze over during prolonged periods of cold. The filling of lake and pond basins is rapid and the great amount of decaying vegetation quickly builds up deposits of peat, which finally replace the water of the marshes. The level of the water is subject to great fluctuations according to the amount of rainfall, and at times the marshes go completely dry; however, during the greater part of the year, water remains in them. The marshes are definitely acidic, some having a pH ranging below 3.6; but this condition is local, and at different points in the same marsh the pH may range from 3.6 to 6.0 or higher. The organisms found in the marshes are not essentially different from those occupying similar habitats in ponds and along lake margins, for conditions in the marshes are very much like those of the pond margins. The principal differences lie in the fact that emergent vegetation occurs throughout the water rather than being confined to a shore zone, and in shallowness of the marshes which are not much more than three feet deep at their deepest point.

Another type of marsh found in the central part of Florida is very similar to the above, except that vegetation is principally *Persicaria* and saw grass (*Mariscus jamaicensis*), with *Pontederia* much more limited in extent and with no *Nymphaea* present; however, the organisms inhabiting these marshes are identical with those found in the other types.

The Everglades are so distinctive that they require discussion under a special category. Cooke (1939: 55-56) states:

> . . . The Everglades occupy a nearly level plain, which slopes from 15 feet above sea level at the south shore of Lake Okeechobee to sea level at the tip of the Peninsula, a distance of more than one hundred miles. On the west the Everglades merge into the Big Cypress Swamp, which presumably is a few

feet higher. On the south and southwest the Everglades are bounded by mangrove swamps, which separate them from the open waters of Florida Bay and the Gulf of Mexico. . . .

The Everglades, as the name implies, are open grassy meadows. Here and there clumps of trees lend variety to the landscape. . . .

The Everglades differ from most swamps and boggy places in the scarcity of trees and in their lack of ordinary mud and clay. The entire Everglades are underlain by hard limestone, which is cushioned in the lower, generally flooded parts by deposits of peat. . . . Where peat is absent, bare limestone shows at the surface.

Since drainage ditches have been put through the Everglades, the land, during the dry periods, becomes very susceptible to burning and much of the peat has been destroyed by fire in the past few years. During the rainy season much of the Everglades is covered to a depth of two or three feet by the water which overflows from the drainage canals. When this flooding occurs, the normal biocoenoses of the canals are no longer confined to these channels but spread uniformly over the inundated areas; however, the density is thin compared with the concentration of organisms within the canals themselves. Only pond, swamp, and stagnant water animals can maintain themselves in this environment, and the mayflies associated with this region, *Callibaetis floridanus* and *Caenis diminuta,* are typical pond and swamp inhabitants which are found throughout Florida.

Saw grass is the predominant vegetation and grows profusely throughout the marsh. The lack of shade trees allows the water to become very warm outside the canals, but in the canals themselves it is several degrees cooler. There is little submergent vegetation where the overflowing water has spread over the glades, but after it has stood for some time algae become noticeable.

SWAMPS

Cypress swamps.—Cypress swamps are very numerous in Florida, and some of these, such as the Big Cypress Swamp, occupy huge areas. The cypress swamps are formed in shallow depressions in flatwoods regions, and over most of the year standing water is present. Not only does cypress occur in them but black gum is also very common. Herbaceous vegetation is, however, limited, though there may be clumps of *Persicaria, Utricularia,* and some sphagnum along with masses of algae. The water is usually tinted by humic acids, and its reaction is definitely acidic. Depth of water varies considerably, but is seldom over three or four feet.

Bayheads.—Bayheads are very similar to cypress swamps, but the vegetation is much denser, and the bayheads usually form the headwaters of small creeks. The plants consist mostly of small trees and shrubs and also include some cypress. The other trees are principally sweet bay, black gum, wax myrtle, red maple, loblolly bay, titi, Virginia willow, and dahoon holly. The growth of the shrubs is so dense as to be almost impenetrable in places. Water stands in the depressions for the major part of the year, and it is only during the very driest seasons that it entirely disappears from the bayheads.

SPRINGS

According to Cooke (1939: 88-89):

> Most of the large springs of Florida are artesian. The water flowing from them rises through deep, generally vertical holes in limestone, some of which holes open into caverns, presumably nearly horizontal, through which flow underground rivers. . . .
>
> The cavity through which water ascends to an artesian spring is generally a former sink in which the direction of motion of the water has been reversed by the rise of the water table. If the water table were to fall below the mouth of the cavity, the spring would cease to flow and would revert to the form of a sink, provided the tubular cavity leading to the spring does not penetrate an impervious stratum, which might confine the water below it under pressure.

The runs of some of the larger springs have been discussed under larger calcareous streams, but the majority of them are much smaller. Vegetation is very dense just below the exit of the spring, but around the spring itself there is nothing but bare sand. Immediately beyond the periphery of the "boil," *Chara, Myriophyllum, Ceratophyllum, Vallisneria, Sagittaria,* and *Isnardia* become very abundant. The surface of the vegetation close to the spring is usually covered with a coating of calcium carbonate deposited from the water as the bicarbonate exposed to the air changes to the carbonate. The low oxygen content of the water is reflected in the small populations of truly aquatic insects in this region; however, snails of the genus *Goniobasis* are exceedingly numerous on the vegetation, and *Ampullaria* occurs frequently on the bottom sands. Approximately a quarter of a mile below the head of the springs a more abundant insect fauna becomes noticeable, and in this region the plants are quite free of the calcium carbonate. The water of the springs is crystal clear, cool, and definitely alkaline.

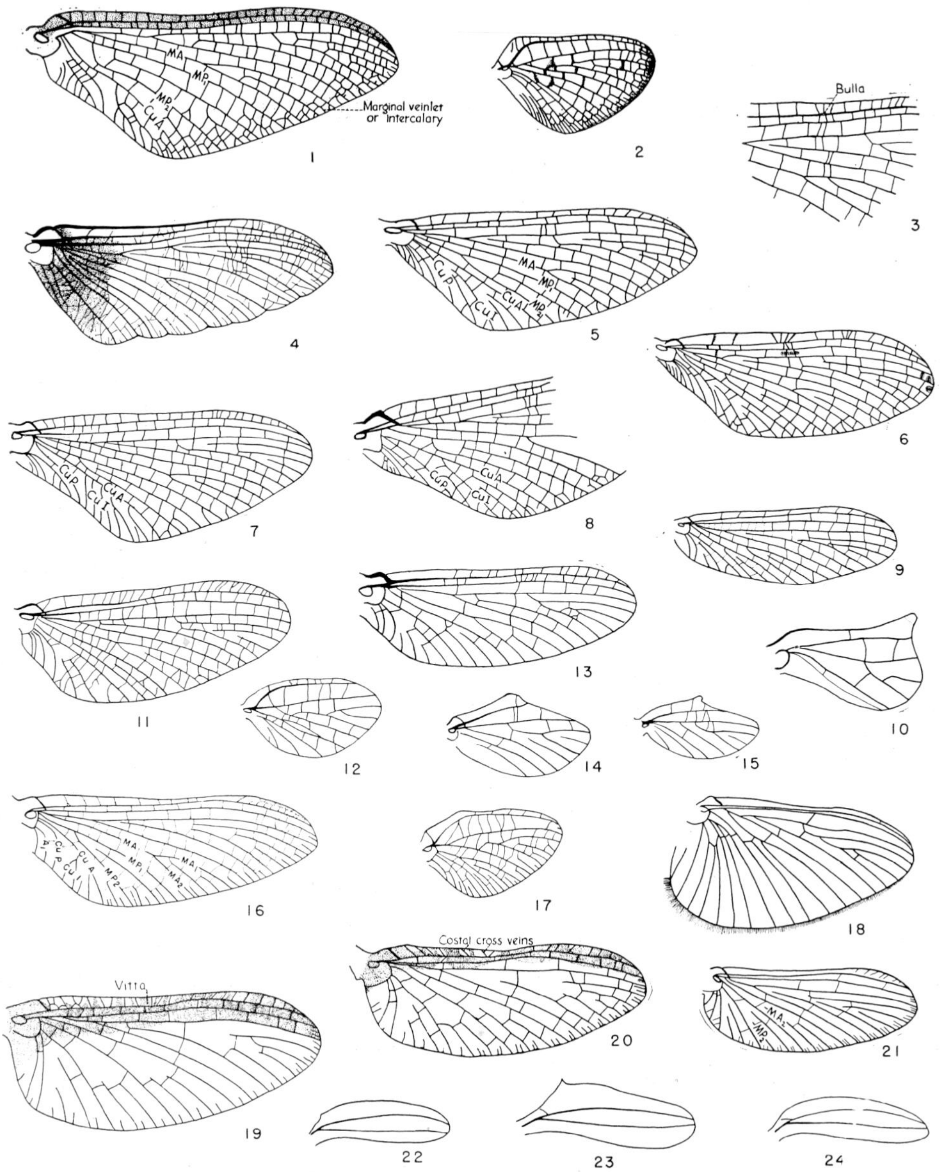

Fig. 1.—*Hexagenia munda orlando,* forewing of male.
Fig. 2.—*H. m. orlando,* hind wing of male.
Fig. 3.—*Ephemera simulans,* midsection of forewing.
Fig. 4.—*Baetisca rogersi,* forewing of male.
Fig. 5.—*Stenonema smithae,* forewing of male.
Fig. 6.—*S. proximum,* forewing of male.
Fig. 7.—*Isonychia pictipes,* forewing of male.
Fig. 8.—*Siphloplecton basale,* forewing of male.
Fig. 9.—*Habrophlebiodes brunneipennis,* forewing of male.
Fig. 10.—*H. brunneipennis,* hind wing of male.
Fig. 11.—*Paraleptophlebia volitans,* forewing of male.
Fig. 12.—*P. volitans,* hind wing of male.
Fig. 13.—*Choroterpes hubbelli,* forewing of male.
Fig. 14.—*C. hubbelli,* hind wing of male.
Fig. 15.—*Habrophlebia vibrans,* hind wing of male.
Fig. 16.—*Ephemerella trilineata,* forewing of male.
Fig. 17.—*E. trilineata,* hind wing of male.
Fig. 18.—*Caenis diminuta,* forewing of male.
Fig. 19.—*Callibaetis floridanus,* forewing of female.
Fig. 20.—*C. pretiosus,* forewing of female.
Fig. 21.—*Baetis spinosus,* forewing of male.
Fig. 22.—*B. spinosus,* hind wing of male.
Fig. 23.—*B. spiethi,* hind wing of male.
Fig. 24.—*Acentrella ephippiatus,* hind wing of male.

*KEY TO FLORIDA MAYFLIES**

ADULTS†

1 Veins CuA and MP strongly divergent at base; MP_2 strongly bent toward CuA basally (fig. 1). Hind tarsi with 4 segments........ Ephemeridae......(2)

1′ Veins CuA and MP only slightly divergent at base; MP_2 more nearly symmetrical (fig. 5). Hind tarsi with 4 or 5 segments........(9)

2 (1) Basal costal cross veins weak. Costal angulation of metathoracic wings approximately 90 degrees. Median carina on prosternum........(Neoephemerinae)......*Oreianthus* sp. No. 1 Traver

2′ Basal costal cross veins normally developed (fig. 1). Costal angulation of metathoracic wings greater than 90 degrees (fig. 2). No median carina on prosternum........(3)

3 (2′) Marginal veinlets present (fig. 1)........(4)

3′ Marginal veinlets absent........(Campsurinae)......*Campsurus incertus*

4 (3) Cross veins somewhat crowded at and below bulla (fig. 3); wings with pattern of dark spots. Tails 3, of equal length....*Ephemera simulans*

4′ Cross veins not crowded at bulla; wings with no such distinct pattern of dark spots (fig. 1). Tails 2 (plus short or vestigial middle filament)........(5)

5 (4′) Penes tubelike, tips not incurved........*Pentagenia vittigera*

5′ Penes broader at base than at apex, tips more or less incurved........ *Hexagenia*......(6)

6 (5′) Penes slender, somewhat beak-shaped (fig. 26); abdomen banded longitudinally with dark and light areas........*Hexagenia bilineata*

6′ Penes more or less hooklike (fig. 25); abdominal markings not as above........(7)

7 (6′) Body of female canary yellow; dark abdominal markings limited; forewing 24-28 mm. Male prominently marked; hind wing with wide purplish-brown border, large dark spots near middle; many cross veins of forewings widely margined with purplish brown; distal segment of middle and hind legs dark on under side. Genitalia as shown in fig. 27........*Hexagenia munda marilandica*

7′ Body of female and male heavily marked with purplish brown. Hind wing with or without wide purplish-brown border and brownish spots in middle; if brown border and blackish spots present, outer segment only of middle and hind legs dark on underside........(8)

8 (7′) Hind wing without prominent purplish-brown border and no large blackish spots (fig. 2). Outer segment of middle and hind legs with dark markings on underside. Female tawny. Inhabits lakes. See plate III........*Hexagenia munda orlando*

*Locality records for each species will be found under the discussion of the individual species.

† See p. 4 for characters differentiating imago from subimago.

8′ Hind wing of male with prominent purplish-brown border and large blackish spots; many cross veins of forewings widely margined with purplish brown. Outer segment only of middle and hind legs dark. Abdomen of female yellowish white; prominently marked with purplish brown. Inhabits streams .... *Hexagenia munda elegans*

9 (1′) Hind tarsi with 5 freely movable segments. Venation not greatly reduced; cubital intercalaries in 2 parallel pairs, long and short alternately (fig. 5). Eyes of male simple .... Heptageniidae (Heptageninae) *Stenonema* .... (10)

9′ Hind tarsi with 3 or 4 freely movable segments. Cubital intercalaries not as above. Venation sometimes greatly reduced (fig. 18). Eyes of male often divided .... Baetidae .... (12)

10 (9) Brown bar in forewing uniting 2-4 cross veins in space between R_1 and R_2. Brown spot at tip of forewing just below R_2 (fig. 6). Penes with spines along lateral border (fig. 30). Body of male yellow; of female yellow, but if eggs present, abdomen is brilliant orange .... *Stenonema proximum*

10′ Wing without brown bar. No brown spot at tip of forewing (fig. 5). Penes without spines on lateral borders (fig. 28). Both sexes white .... (11)

11 (10′) Dark spiracular spots present on abdomen. Tails narrowly annulate with purplish brown at joints. Posteromedial portion of penes angulate (fig. 28). See plate I .... *Stenonema smithae*

11′ Dark spiracular spots absent. Tails not banded. Posteromedial portion of penes rounded (fig. 29) .... *Stenonema exiguum*

12 (9′) Hind wings absent. Fork of MA very deep; posterior margins of wings ciliate; no marginal intercalaries (fig. 18). Eyes of male simple, neither divided nor grooved; widely separated .... Caeninae .... (13)

12′ Hind wings present or absent. Fork of MA in forewing normal or detached basally; margin of wings ciliate or not; marginal intercalaries present or absent in forewing (fig. 16 and fig. 21). Eyes of male divided or simple in genera lacking hind wings .... (15)

13 (12) Prosternum twice as wide as its length; forecoxae widely separated; second antennal segment three times the length of the basal segment .... *Brachycercus maculatus*

13′ Prosternum two to three times longer than broad; coxae much closer together; second segment of antennae not much more than twice as long as basal segment. Wing as shown in fig. 18 .... *Caenis* .... (14)

14 (13′) Stigmatic dark streaks on posterior tergites only. Head pale. Thorax light brown. Short fine black streak on dorsal edge of each femur .... *Caenis hilaris*

14′ Stigmatic streaks present on anterior tergites. Head dark shaded. Thorax light brown. Hind femora with a dark apical band. See plate XVI .... *Caenis diminuta*

15 (12′) A_1 of forewings ends in outer margin; no cubital intercalaries; basal third of forewing and basal 4/5 of hind wing suffused with orange-brown (fig. 4). Eyes of male not divided .... (Baetiscinae) .... *Baetisca rogersi*

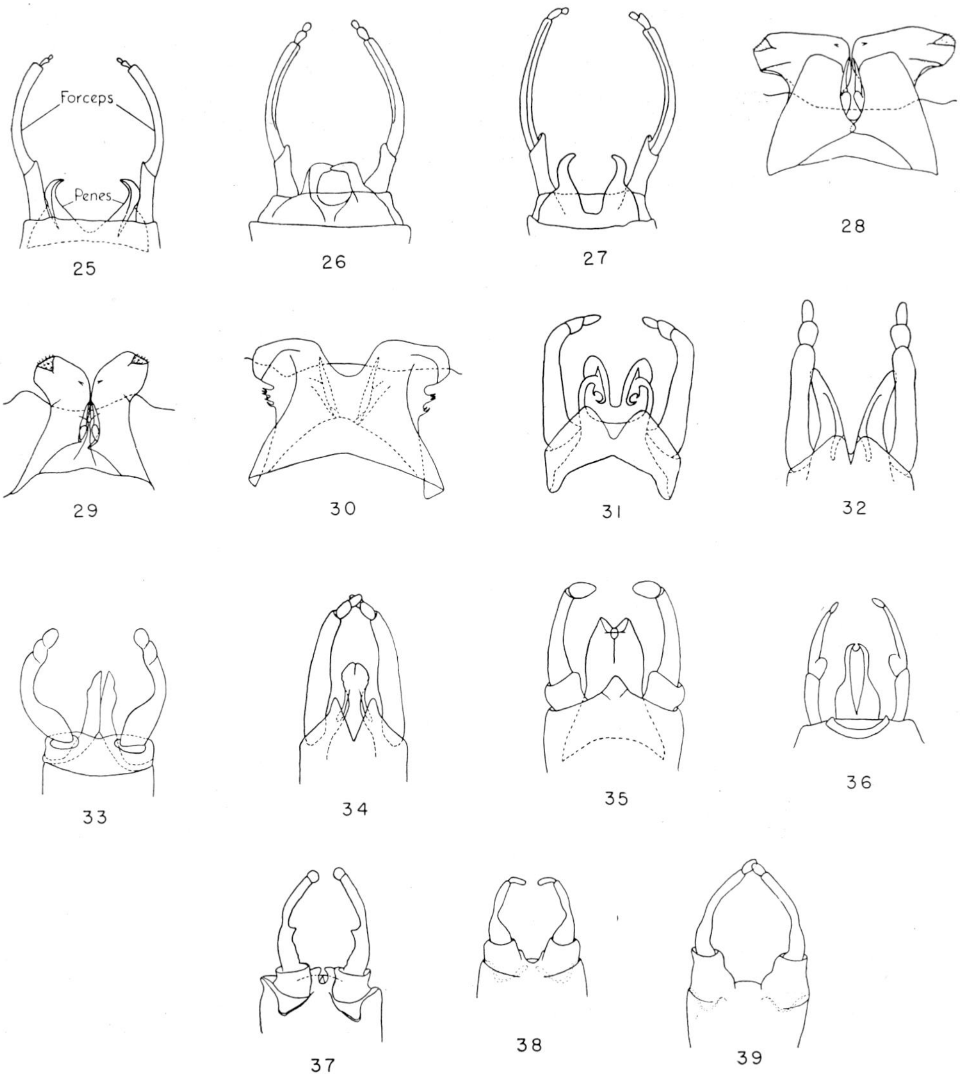

Fig. 25.—*Hexagenia munda orlando,* male genitalia.
Fig. 26.—*Hexagenia bilineata,* male genitalia.
Fig. 27.—*Hexagenia munda marilandica,* male genitalia.
Fig. 28.—*Stenonema smithae,* details of the penes.
Fig. 29.—*Stenonema exiguum,* details of the penes.
Fig. 30.—*Stenonema proximum,* details of the penes.
Fig. 31.—*Paraleptophlebia volitans,* male genitalia.
Fig. 32.—*Paraleptophlebia bradleyi,* male genitalia.
Fig. 33.—*Choroterpes hubbelli,* male genitalia.
Fig. 34.—*Habrophlebiodes brunneipennis,* male genitalia.
Fig. 35.—*Ephemerella trilineata,* male genitalia.
Fig. 36.—*Tricorythodes albilineatus,* male genitalia.
Fig. 37.—*Baetis spinosus,* male genitalia.
Fig. 38.—*Baetis spiethi,* male genitalia
Fig. 39.—*Pseudocloeon alachua,* male genitalia.

15′ A_1 of forewing ends in hind margin; cubital intercalaries present (fig. 16). Eyes of male divided, marked with a distinct band, or simple .... (16)

16 (15′) MA_2 and MP_2 of forewings detached basally from their respective stems (fig. 21). Hind wings much reduced in size and venation, or they may be entirely wanting (fig. 22). Eyes of male turbinate (plate XXI) .... Baetinae .... (29)

16′ MA_2 and MP_2 of forewings normal, attached basally (fig. 16), or MP_2 may be detached basally, but this is paralleled with loss of hind wings. Eyes of male not as above. If hind wings present, venation of metathoracic wings not greatly reduced (fig. 17) .... (17)

17 (16′) Tails 2 .... (26)

17′ Tails 3 .... (18)

18 (17′) Two short intercalaries in forewing between IMP and MP_2 and between MP_2 and CuA (fig. 16). Male forceps with a single terminal segment (fig. 35) .... Ephemerellinae .... (19)

18′ No true intercalaries in forewing in position indicated above (fig. 9). Male forceps with 2 short terminal segments (fig. 32) .... Leptophlebiinae .... (21)

19 (18) Hind wings absent. Forceps with 3 segments; an ovoid swelling at base of second segment on inner margin (fig. 36) .... *Tricorythodes albilineatus*

19′ Hind wings present. Forceps with 3 segments, no swelling on second segment (figs. 16 and 35) .... *Ephemerella* .... (20)

20 (19′) Penes swollen basally, but not apically; abdomen with 3 longitudinal dark stripes dorsally; body light reddish brown in color. See plate XII .... *Ephemerella trilineata*

20′ Penes swollen apically; abdomen without longitudinal dark stripes dorsally; body dark brown .... *Ephemerella hirsuta*

21 (18′) Hind wings reduced in size and venation; distinct costal angulation halfway to apex (fig. 14) .... (22)

21′ Hind wings larger, venation more prominent; no costal angulation; costal margin slightly concave at middle (fig. 12) .... (24)

22 (21) Costal angulation of metathoracic wing obtuse; no sag in fork of MA of forewing (figs. 14 and 13). Penes long, simple, without reflexed spur; forceps base of male undivided (fig. 33). Subanal plate of female slightly emarginate .... *Choroterpes hubbelli*

22′ Costal angulation of hind wings acute; a distinct sag evident in fork of MA of forewing (figs. 9 and 10). Penes with reflexed spurs; forceps base deeply cleft (fig. 34). Apical margin of subanal plate of female deeply cleft .... (23)

23 (22′) Subcosta of hind wing ends in margin at outer side of costal angulation (fig. 10); 2 terminal segments of forceps short, together not 1/4 the length of preceding segment; reflexed spur subclavate at tip (fig. 34). See plate VIII .... *Habrophlebiodes brunneipennis*

23′ Subcosta of hind wing extends almost to apex of wing (fig. 15); 2 end segments of forceps together equal in length to the preceding segment. Reflexed spurs long and slender, not subclavate at tip .... *Habrophlebia vibrans*

24 (21) Middle tail shorter and weaker than laterals .... *Blasturus intermedius*

24′ Middle tail approximately equal to laterals in length and thickness. Penes partially united (fig. 31) *Paraleptophlebia* (25)

25 (24′) Reflexed spur of penes curved, spatulate, reaching to bottom of cleft between penes; penes united for 1/4 their length, long and tapering (fig. 32). Middle abdominal segments of male brown. Blackish rings on apical third of all femora. Wings with a brownish tinge; stigmatic cross veins anastomosed *Paraleptophlebia bradleyi*

25′ Reflexed spur of penes claw-shaped; penes united for 1/2 their length (fig. 31). Middle abdominal segments of male predominantly white. Femora without dark rings. Wings colorless; stigmatic cross veins but little anastomosed (fig. 11) *Paraleptophlebia volitans*

26 (17) CuI of forewing consist of series of forking or sinuate veinlets attaching CuA to hind margin (fig. 7) (Siphlonurinae) *Isonychia* (27)

26′ CuI 2 to 4 in number, free basally, not as above (fig. 8) (Metretopinae) *Siphloplecton speciosum*

27 (26) Venation almost colorless (28)

27′ Venation dark *Isonychia* sp. B

28 (27) Foretibiae wholly dark *Isonychia* sp. A

28′ Foretibiae pale in middle and dark at each end *Isonychia pictipes*

29 (16) Hind wings present (30)

29′ Hind wings absent (40)

30 (29) Forewings with numerous costal cross veins before the bulla; in female, costal margin brownish (fig. 19). Moderate number of cross veins present in hind wings *Callibaetis* (31)

30′ Forewings without costal cross veins before the bulla (fig. 21). No cross veins in hind wings (fig. 22) (33)

31 (30) Male: Intercalaries in mid-posterior margin of forewing occur singly. Abdominal segments 2-6 whitish hyaline with light reddish shadings on dorsum.

Female: Intercalaries in mid-posterior margin of forewing occur singly; vitta brown, margin straight; 20-25 cross veins in forewing; posterior margin of forewing unicolorous (fig. 19) *Callibaetis floridanus* (32)

31′ Male: Intercalaries in mid-posterior margin of forewing occur either in pairs or singly. Abdominal segments 2-6 whitish hyaline with prominent brownish shadings on dorsum. See plate XX.

Female: Intercalaries in mid-posterior margin of forewing occur in pairs; vitta brown, undulatory; 35-40 cross veins in forewings; posterior margin of forewings marked with alternate yellowish and brown areas (fig. 20) *Callibaetis pretiosus*

32 (31) Forewings colorless or only slightly colored; spots covering body brown or reddish brown; body coloration somewhat dulled. Central and north Florida *Callibaetis floridanus* (form A)

32′ Forewings tinged with brown; spots covering body red; body coloration intense. South Florida *Callibaetis floridanus* (form B)

33 (30′) Marginal intercalaries of forewing single *Centroptilum* (34)

33′ Marginal intercalaries paired (fig. 21) (35)

34 (33) Abdominal tergites 2-6 with reddish area *Centroptilum viridocularis*

34′ No extensive reddish areas on abdominal tergites 2-6 *Centroptilum hobbsi*

35 (33′) Costal angulation present on hind wings (fig. 23)............*Baetis*......(36)

35′ No costal angulation on hind wings (fig. 24)..............*Acentrella*......(39)

36 (35) Male: Second segment of forceps with prominent projection on inner margin; excavation and spine between bases of forceps (fig. 37). See plate XXI.
Female: Hind wings with very slight costal angulation; 0.42-0.65 mm. in length. Large U-shaped brown figure on head......*Baetis spinosus*

36′ Second segment of forceps without prominent projection on inner margin ....(37)

37 (36′) Middle abdominal segments of male dark. Small spine in excavation between forceps base. Costal angulation of hind wing much reduced (fig. 22)..............................*Baetis australis*

37′ Middle abdominal segments of male hyaline white. Costal angulation more prominent (fig. 23)..............................(38)

38 (37′) Male: Fourth segment of forceps twice as long as wide (fig. 38). Forewings about 3 mm. or slightly longer; short marginal veinlets present in first interspace; 2 longitudinal veins in metathoracic wings.
Female: Hind wings with acute costal angulation; 0.15-0.28 mm. in length. Posterior margin of head margined with red-brown..............................*Baetis spiethi*

38′ Fourth segment of forceps about as broad as long. Mesothoracic wings 5 mm. in length; 2 long intercalaries in first interspace; 3 longitudinal veins in metathoracic wings..............*Baetis intercalaris*

39 (35′) Reddish markings present on abdominal tergites 2-6..............................*Acentrella ephippiatus*

39′ No reddish markings on abdominal tergites 2-6....*Acentrella propinquus*

40 (29′) Marginal intercalaries occur singly (fig. 21)..............*Cloeon*......(41)

40′ Marginal intercalaries paired..............................*Pseudocloeon*......(42)

41 (40) Abdominal tergites 2-6 of male yellowish white; small submedian red dots on tergites 2-6 near the posterior margin, likewise above the spiracular line near center of each side..............*Cloeon rubropictum*

41′ Characters of adult not known..............................*Cloeon* sp. A

42 (40′) Large ruddy patches on tergites 2-6..............................(43)

42′ Abdominal tergites 2-6 immaculate or with very limited markings....(44)

43 (42) Femora of middle and hind legs with a ruddy dash anteriorly on lower edge and a distinct ruddy spot near the apex and well beyond the middle of the segment. Distal forceps segment much thinner than third..............................*Pseudocloeon parvulum*

43′ Ruddy markings absent from legs. Distal forceps segment as wide as third (fig. 39)..............................*Pseudocloeon alachua*

44 (42′) A mid-ventral row of minute dots on posterior margins of sternites. No reddish markings on tergites 2-6........*Pseudocloeon punctiventris*

44′ Small, paired, red spots on abdominal tergites 2-6..............................*Pseudocloeon bimaculatus*

NYMPHS

1 Mandibles with no tusks visible from above (fig. 60). Gills present on abdominal segments 1-6 only, elytroid and fused on segment 2. Hind wing pads present. Anterolateral angles of prothorax produced forward prominently. See plate V. .... (Neoephemerinae) .... *Oreianthus* sp. No. 1 Traver

1′ Mandible with or without mandibular tusks visible from above (figs. 40 and 60). If gills elytroid and fused on segment 2, hind wing pads absent .... (2)

2 (1′) Mandibles with external tusks projecting forward and visible from above head (fig. 40). Gills present on segments 1-7, rudimentary on segment 1; no protective covering; feathery appearance. See plate IV .... (3)

2′ Mandibles with no tusks are visible from above head (fig. 60). Gills present on segments 1-7, or may be absent from one or more of these segments .... (9)

3 (2) Head with a conspicuous frontal process .... Ephemerinae .... (4)

3′ Front of head rounded, without such a process .... (Campsurinae) .... *Campsurus incertus*

4 (3) Elevated frontal process rounded or truncate (fig. 41) .... *Hexagenia* .... (6)

4′ Elevated frontal process bifid (fig. 42) .... (5)

5 (4′) Mandibular tusks crenate on outer (upper) margin; labial palp with 2 segments .... *Pentagenia vittigera*

5′ Mandibular tusks smooth on margins; labial palp with 3 segments .... *Ephemera simulans*

6 (4) Mandibular tusks strongly upcurved in outer half. Large nymphs—body length of females 26-29 mm.; of males 23-26 mm. Frontal process broadly rounded to truncate. Stream inhabitants .... (7)

6′ Mandibular tusks more gently upcurved in outer half. Smaller nymphs—body length of females 20-26 mm.; of males 15-22 mm. Frontal process conical to narrowly rounded. Lake inhabitants (*bilineata* may occur in streams) .... (8)

7 (6) Found in northwestern Florida in the Apalachicola River region and westward .... *Hexagenia munda marilandica*

7′ Found in peninsular Florida .... *Hexagenia munda elegans*

8 (6′) Found in northwestern Florida .... *Hexagenia bilineata*

8′ Found in peninsular Florida .... *Hexagenia munda orlando*

9 (2) Head strongly depressed; eyes dorsal. Gills present on abdominal segments 1-7; those on 7 reduced to mere filaments (figs. 62 and 63); fimbrillar portion of gills 1-6 covered by protective plate. See plate II .... (Heptageninae) .... *Stenonema* .... (10)

9′ Head not strongly depressed; eyes lateral. See plate XXII .... (12)

10 (9) Gills on abdominal segments 1-6 truncate; seventh gill without tracheae (fig. 62) .... (11)

10′ Gills on abdominal segments 1-6 pointed at apex; seventh gill with one large trachea (fig. 63) .... *Stenonema proximum*

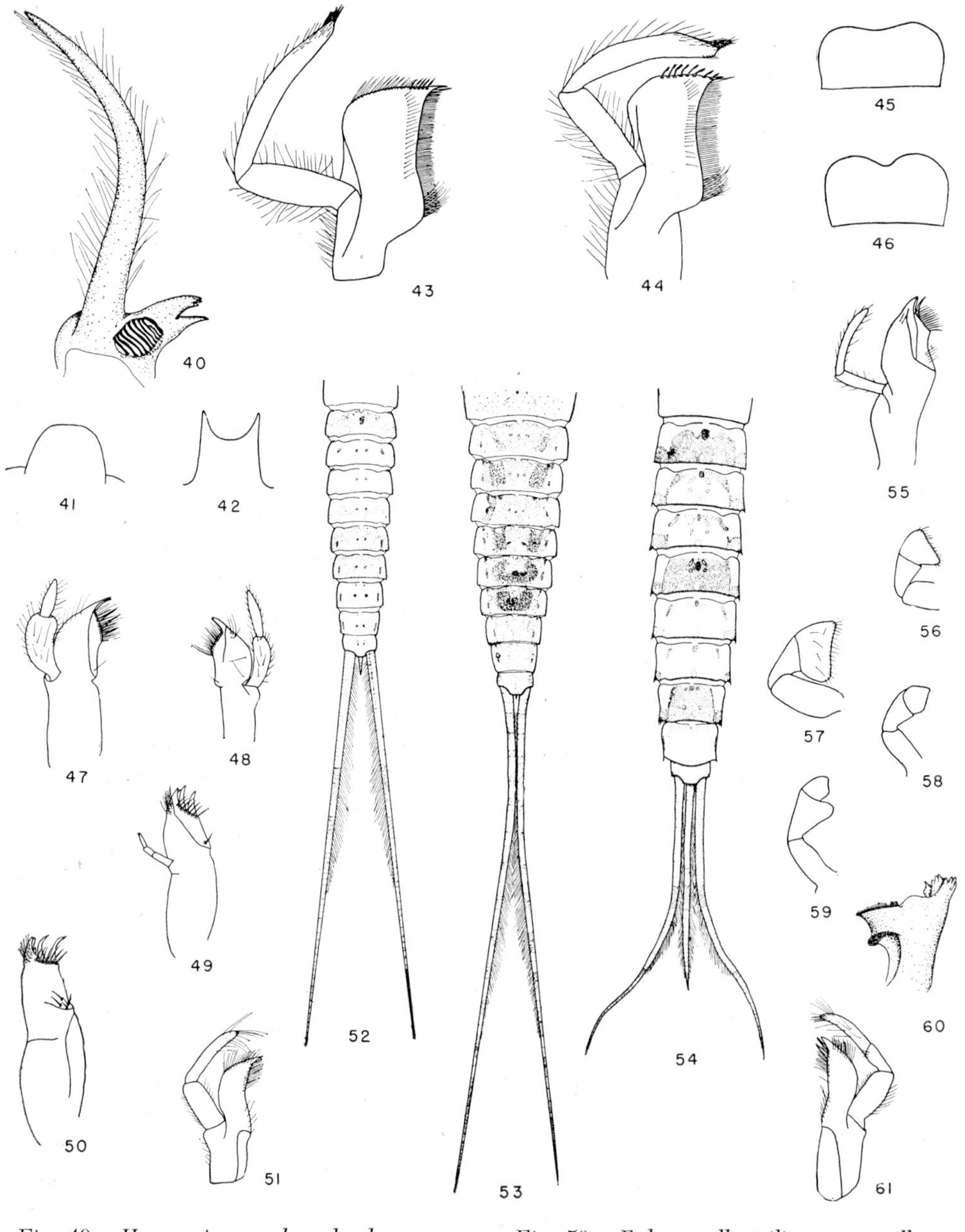

Fig. 40.—*Hexagenia munda orlando,* mandible.
Fig. 41.—*Hexagenia munda marilandica,* frontal process.
Fig. 42.—*Ephemera simulans* (?), frontal process.
Fig. 43.—*Stenonema smithae,* maxilla.
Fig. 44.—*Stenonema exiguum,* maxilla.
Fig. 45.—*Paraleptophlebia volitans,* labrum.
Fig. 46.—*Habrophlebiodes brunneipennis,* labrum.
Fig. 47.—*Callibaetis pretiosus,* maxilla.
Fig. 48.—*Callibaetis floridanus,* maxilla.
Fig. 49.—*Ephemerella choctawhatchee,* maxilla.
Fig. 50.—*Ephemerella trilineata,* maxilla.
Fig. 51.—*Brachycercus maculatus,* maxilla.
Fig. 52.—*Pseudocloeon bimaculatus,* abdomen.
Fig. 53.—*Pseudocloeon parvulum,* abdomen.
Fig. 54.—*Acentrella ephippiatus,* abdomen.
Fig. 55.—*Cloeon rubropictum,* maxilla.
Fig. 56.—*Centroptilum hobbsi,* labial palp.
Fig. 57.—*Centroptilum viridocularis,* labial palp.
Fig. 58.—*Baetis intercalaris,* labial palp.
Fig. 59.—*Baetis spinosus,* labial palp.
Fig. 60.—*Baetis spinosus,* mandible.
Fig. 61.—*Caenis diminuta,* maxilla.

11 (10) No setae on crown of maxillae; 6-8 pectinate spines on anterior border (fig. 44). A yellow band across mesonotum at base of wing pads of last instar nymphs. See plate II .... *Stenonema exiguum*

11′ Setae on crown of maxillae; 4-5 pectinate spines on anterior border (fig. 43). No band on mesonotum .... *Stenonema smithae*

12 (9′) Gills completely concealed beneath greatly enlarged thoracic shield or carapace. Prominent frontal projection and lateral thoracic spines. See plate XI .... (Baetiscinae) .... *Baetisca rogersi*

12′ Gills not concealed beneath the mesonotum. Mesonotum not greatly enlarged; no lateral thoracic spines .... (13)

13 (12′) Outer caudal filaments fringed with hairs on both sides (plate XVIII) .... (14)

13′ Outer tails heavily fringed with hairs on inner side only, although may have a few short hairs on outer side (plate VI) .... (20)

14 (13) Gills present on abdominal segments 1-7 (plate VIII) .... Leptophlebiinae .... (15)

14′ Gills absent from one or more abdominal segments (plate XIII) .... (25)

15 (14) Gills on first abdominal segment differing in shape from others (figs. 67 and 68) .... (16)

15′ Gills similar on all abdominal segments .... (18)

16 (15) Gills lamelliform; each gill with long terminal extensions from the plates (fig. 69) .... (17)

16′ Gills filamentous; each gill on abdominal segments 2-7 consists of 2 clusters of slender filaments (fig. 66) .... *Habrophlebia vibrans*

17 (16) Gills on abdominal segment 1 single, unbranched; terminal extension of upper gill plate of each middle pair rather broad and spatulate (fig. 69). See plate X .... *Choroterpes hubbelli*

17′ Gills on abdominal segment 1 definitely bifid except at base; terminal extension of each gill on middle segments very slender (figs. 67 and 68). See plate VII .... *Blasturus intermedius*

18 (15′) Lateral spines on ninth abdominal segment not more than ¼ the length of that segment; spinules on posterior margins of tergites 1-10. Labrum only shallowly indented on foreborder (figs. 45 and 46) .... *Paraleptophlebia* .... (19)

18′ Lateral spines of ninth segment long and slender, ½ as long as length of that segment; spinules on posterior margins of tergites 6-10 only. Labrum rather deeply indented on foremargin (fig. 46). See plate IX .... *Habrophlebiodes brunneipennis*

19 (18) First gill narrow, tracheae without lateral branches; gills 2-7 broad, lobes separate to base, tracheae with lateral branches (fig. 65). Dorsum brownish with 4 yellowish streaks at anterior border of each tergite .... *Paraleptophlebia bradleyi*

19′ All gills narrow, bifurcate, lobes not separated to base; no lateral branches from tracheae (fig. 64). Dorsum unicolorous .... *Paraleptophlebia volitans*

20 (13′) Claws of middle and hind legs long and slender, about as long as the tibiae; claws of forelegs differ from others in structure in being doubled (fig. 83) .... (Metretopinae) .... *Siphloplecton speciosum*

20′ Claws of all legs similar, sharp pointed, much shorter than tibiae .... (21)

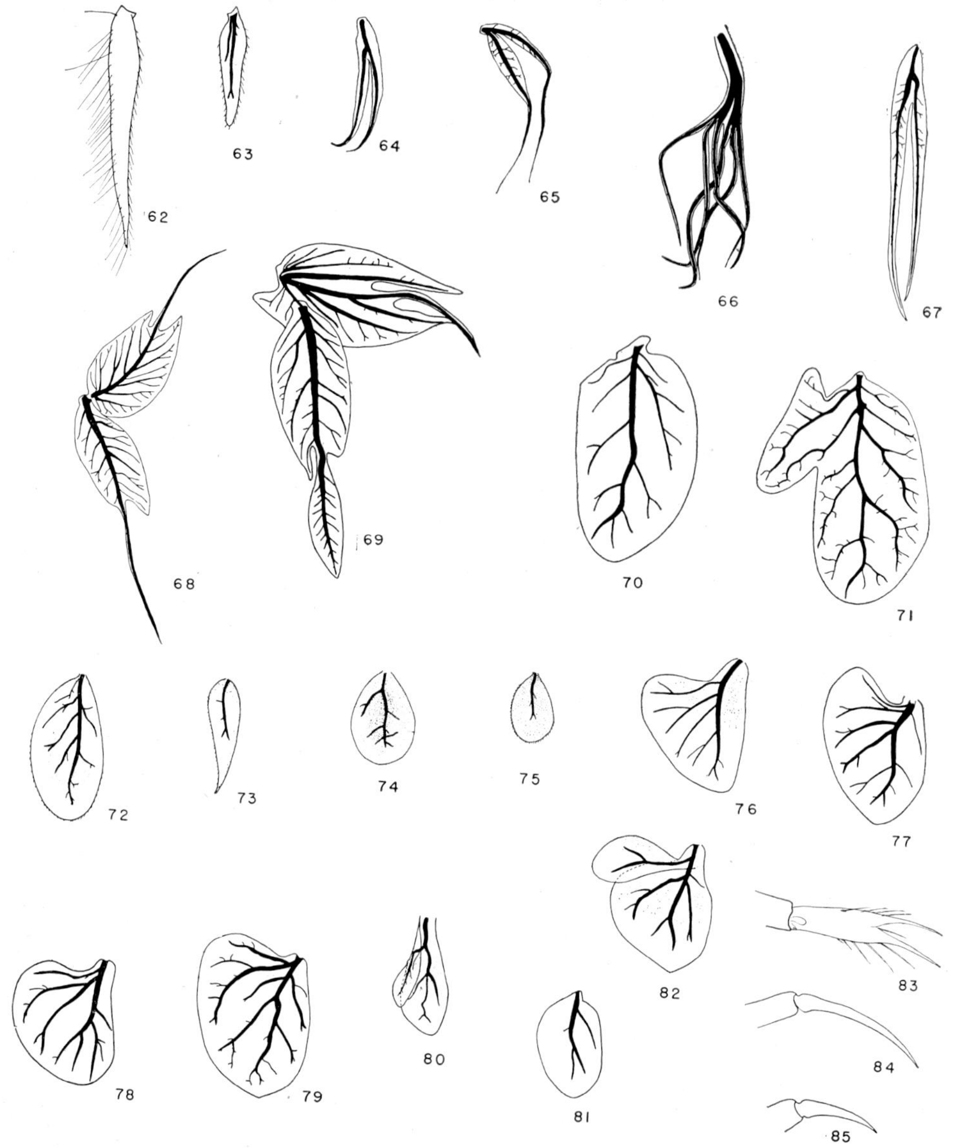

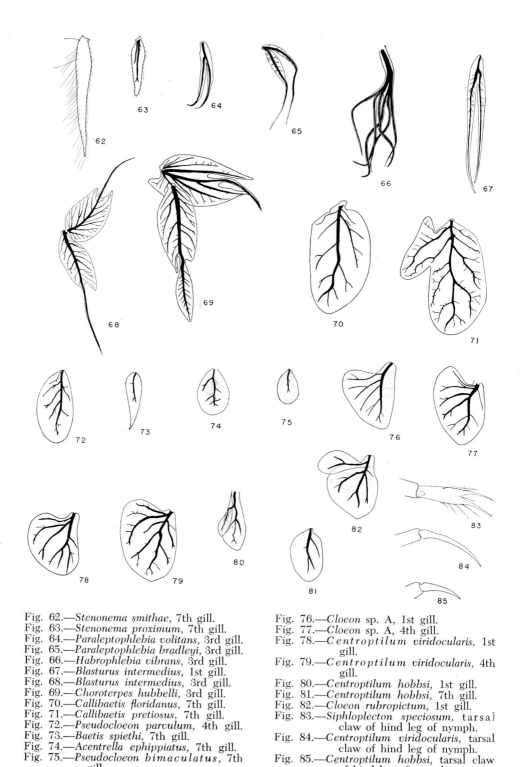

Fig. 62.—*Stenonema smithae*, 7th gill.
Fig. 63.—*Stenonema proximum*, 7th gill.
Fig. 64.—*Paraleptophlebia volitans*, 3rd gill.
Fig. 65.—*Paraleptophlebia bradleyi*, 3rd gill.
Fig. 66.—*Habrophlebia vibrans*, 3rd gill.
Fig. 67.—*Blasturus intermedius*, 1st gill.
Fig. 68.—*Blasturus intermedius*, 3rd gill.
Fig. 69.—*Choroterpes hubbelli*, 3rd gill.
Fig. 70.—*Callibaetis floridanus*, 7th gill.
Fig. 71.—*Callibaetis pretiosus*, 7th gill.
Fig. 72.—*Pseudocloeon parvulum*, 4th gill.
Fig. 73.—*Baetis spiethi*, 7th gill.
Fig. 74.—*Acentrella ephippiatus*, 7th gill.
Fig. 75.—*Pseudocloeon bimaculatus*, 7th gill.
Fig. 76.—*Cloeon* sp. A, 1st gill.
Fig. 77.—*Cloeon* sp. A, 4th gill.
Fig. 78.—*Centroptilum viridocularis*, 1st gill.
Fig. 79.—*Centroptilum viridocularis*, 4th gill.
Fig. 80.—*Centroptilum hobbsi*, 1st gill.
Fig. 81.—*Centroptilum hobbsi*, 7th gill.
Fig. 82.—*Cloeon rubropictum*, 1st gill.
Fig. 83.—*Siphloplecton speciosum*, tarsal claw of hind leg of nymph.
Fig. 84.—*Centroptilum viridocularis*, tarsal claw of hind leg of nymph.
Fig. 85.—*Centroptilum hobbsi*, tarsal claw of hind leg of nymph.

21 (20′)	Posterolateral angles of apical abdominal segments prolonged into thin, flat spines. Forelegs conspicuously fringed with long hairs. Gill tufts present on bases of maxillae and forecoxae. See plate VI ....(Siphlonurinae)........*Isonychia*........(22)
21′	Posterolateral angles of apical abdominal segments hardly more than acute, not prolonged into thin, flat spines. No gill tufts on maxillae and coxae. See plate XXII........Baetinae........(33)
22 (21)	Antennae crossed by a brownish band about ⅓ distant from the base........(23)
22′	Antennae without such bands........(24)
23 (22)	Grayish area on outer margin of each gill plate........*Isonychia* sp. G
23′	No such grayish coloration on the outer margin of the gill plates........*Isonychia* sp. B
24 (22′)	Found in the westernmost part of Florida........*Isonychia* sp. A
24′	Found in the Apalachicola River region and eastward *Isonychia pictipes*
25 (14′)	Gills present on abdominal segments 2-6, 3-7, or 4-7. If present on segments 2-6, those on segment 2 are elytroid, covering all others behind them, and are triangular in shape. Hind wing pads present or absent (plates XIII, XIV, and XV)........Ephemerellinae........(26)
25′	Gills present on abdominal segments 1-6 only, rudimentary on segment 1; gills on segment 2 elytroid, covering all others behind them, and are rectangular in shape. Hind wing pads absent........Caeninae........(30)
26 (25)	Gills present on segments 2-6; elytroid gills triangular. Hind wing pads absent. See plate XV........*Tricorythodes albilineatus*
26′	Gills present on segments 3-7 or 4-7. Hind wing pads present........*Ephemerella*........(27)
27 (26′)	Gills present on abdominal segments 3-7 (plate XIV)........(28)
27′	Gills present on abdominal segments 4-7 (plate XIII)........(29)
28 (27)	A whorl of spines at each tail joining; maxillary palp absent (fig. 50). See plate XIV........(*serrata* group)........*Ephemerella* sp. A
28′	Tails fringed with long hairs in the apical portion, may have whorls of spines at the base; maxillary palp normally developed (fig. 49)........(*needhami* group)........*Ephemerella choctawhatchee*
29 (27′)	Gill rudiments present on segment 1; gills on segment 4 operculate; maxillary palp wanting; segment 9 longer than segment 8 (fig. 50 and plate XIII)........(*bicolor* group)........*Ephemerella trilineata*
29′	Gills on segments 4-7 only; gill on segment 4 semioperculate; segment 9 not longer than segment 8; maxillary palp present........(*simplex* group)........*Ephemerella hirsuta*
30 (25′)	Head with 3 prominent tubercles. Maxillary and labial palpi with 2 segments (fig. 51). Posterolateral spines on middle abdominal segments upcurved. See plate XVIII........*Brachycercus*........(31)
30′	No such tubercles on head. Maxillary and labial palpi with 3 segments (fig. 61). Posterolateral spines on middle abdominal segments not upcurved. See plate XVII........*Caenis*........(32)
31 (30)	Tubercles present on the lateral margins of the pronotum. See plate XVIII........*Brachycercus* sp. A
31′	No such prothoracic tubercles........*Brachycercus maculatus*

32 (30′) Average length of mature nymphs 5.0 mm.—range 3.9-6.7 mm. Numerous yellow spots scattered over pro- and mesonotum. Predominantly pond forms. See plate XVII.......*Caenis diminuta*

32′ Average length of mature nymphs 3.2 mm.—range 2.6-3.8 mm. No yellow spots on pronotum; on mesonotum yellow spots at bases of wing pads and a pair of submedian spots. Predominantly stream forms.......*Caenis hilaris*

33 (21′) Gills on abdominal segments 1 and 2 with 4 lobes, on 3-6 double, and single on 7. Maxillary palp with 2 segments (fig. 47). See plate XIX.......*Callibaetis*.......(34)

33′ Gill lamellae double on first abdominal segment only, or single on all abdominal segments. Maxillary palp with either 2 or 3 segments.......(36)

34 (33) Maxillary palp shorter than body of maxilla; second segment of palp about ½ the length of the first segment; width of first segment about ½ its length (fig. 47). Seventh gill with small recurved flap (fig. 71).......*Callibaetis pretiosus*

34′ Maxillary palp equal in length to body of maxilla; second segment of palp approximately equal in length to first segment; width of first segment less than ⅓ its length (fig. 48). Seventh gill without a recurved flap (fig. 70). See plate XIX.......*Callibaetis floridanus*.......(35)

35 (34′) Found in southern Florida as far north as Highlands County.......*Callibaetis floridanus* (form B)

35′ Found throughout Florida north of Highlands County.......*Callibaetis floridanus* (form A)

36 (33′) Tails 2, middle one represented by a minute rudiment (fig. 53). Metathoracic wing pads absent. See plate XXIV.......*Pseudocloeon*.......(37)

36′ Tails 3 (fig. 54). Metathoracic wing pads present or absent. See plate XXII.......(40)

37 (36) Tails with a brown band at the middle (fig. 52).......(38)

37′ Tails otherwise (fig. 53).......(39)

38 (37) Gills 1-6 unicolorous, tinged with brown; gill 7 with lateral ⅔ red-brown, inner margin clear (fig. 75). Band on tails dark brown (fig. 52).......*Pseudocloeon bimaculatus*

38′ All gills unicolorous, tinged with brown. Band on tails light brown.......*Pseudocloeon punctiventris*

39 (38′) Length of median caudal filament less than width of laterals at base; lateral filaments prominently banded with alternate dark and light annulations (fig. 53). Grayish area in gills (fig. 72).......*Pseudocloeon parvulum*

39′ Length of median caudal filament greater than width of laterals at base; lateral filaments faintly, or not at all, banded. No grayish areas on gills. See plate XXIV.......*Pseudocloeon alachua*

40 (36′) Middle tail shorter and weaker than outer ones. Distal segment of labial palp rounded (fig. 59). Maxillary palp usually not extending beyond tip of galea-lacinia. Hind wing pads present. See plate XXII.......(41)

40′ Middle tail practically as long and stout as outer ones. Distal segment of labial palp dilated and truncate apically (fig. 57). Maxillary palp extending somewhat beyond the tip of the galea-lacinia (fig. 55). Hind wing pads present or absent. See plate XXIII .... (45)

41 (40) Seventh gill reddish brown (figs. 74 and 54) .... *Acentrella ephippiatus*

41′ Seventh gill not deeply colored .... (42)

42 (41′) Seventh pair of gills lanceolate (fig. 73). See plate XXII .... *Baetis spiethi*

42′ Seventh pair of gills rounded (fig. 74) .... (43)

43 (42′) Second segment of labial palp greatly expanded distally (fig. 59) .... (44)

43′ Second segment of labial palp not expanded distally (fig. 58) .... *Baetis intercalaris*

44 (43) Dorsum of last instar male nymphs more or less unicolorous. Lateral caudal filaments fairly long and thin .... *Baetis australis*

44′ Dorsum of last instar male nymphs with a pronounced color pattern. Lateral filaments not much longer than median .... *Baetis spinosus*

45 (40′) Hind wing pads present. See plate XXIII .... *Centroptilum* .... (46)

45′ Hind wing pads absent .... *Cloeon* .... (47)

46 (45) First pair of gills double, others single; branching of tracheae asymmetrical, mostly on inner side (figs. 80 and 81). Spines on lateral borders of abdominal segments 8-10; sternites unmarked. Tarsal claw shortened (fig. 85) .... *Centroptilum hobbsi*

46′ All abdominal gills single (figs. 78 and 79). Spines on lateral borders of abdominal segments 4-10; row of large brown spots on lateral borders of sternites 2-9, becoming large and elongated transversely on 8 and 9. Tarsal claw attenuated (fig. 84) .... *Centroptilum viridocularis*

47 (45′) Gills double on abdominal segment 1 (fig. 82) .... *Cloeon rubropictum*

47′ Gills single on all abdominal segments (figs. 76 and 77) .... *Cloeon* sp. A

SPECIES OF FLORIDA MAYFLIES

EPHEMERIDAE

Ephemerinae

- *Hexagenia munda elegans* Traver
- *Hexagenia munda orlando* Traver
- *Hexagenia munda marilandica* Traver
- *Hexagenia bilineata* (Say)
- *Ephemera simulans* Walker
- *Pentagenia vittigera* (Walsh)

Campsurinae

- *Campsurus incertus* Traver

Neoephemerinae

- *Oreianthus* sp. No. 1 Traver

HEPTAGENIIDAE

Heptageninae

- *Stenonema smithae* Traver
- *Stenonema exiguum* Traver
- *Stenonema proximum* Traver

BAETIDAE

Metretopinae

- *Siphloplecton speciosum* Traver

Siphlonurinae

- *Isonychia pictipes* Traver
- *Isonychia* sp. A
- *Isonychia* sp. B
- *Isonychia* sp. G

Leptophlebiinae

- *Blasturus intermedius* Traver
- *Paraleptophlebia volitans* (McDunnough)
- *Paraleptophlebia bradleyi* (Needham)
- *Habrophlebiodes brunneipennis* Berner
- *Choroterpes hubbelli* Berner
- *Habrophlebia vibrans* Needham

Baetiscinae

- *Baetisca rogersi* Berner

Ephemerellinae

- *Ephemerella trilineata* Berner
- *Ephemerella hirsuta* Berner
- *Ephemerella choctawhatchee* Berner
- *Ephemerella* sp. A
- *Tricorythodes albilineatus* Berner

Caeninae

- *Caenis diminuta* Walker
- *Caenis hilaris* (Say)
- *Brachycercus maculatus* Berner
- *Brachycercus* sp. A

Baetinae

- *Callibaetis floridanus* Banks
- *Callibaetis pretiosus* Banks
- *Acentrella ephippiatus* (Traver)
- *Acentrella propinquus* (Walsh)
- *Baetis spinosus* McDunnough
- *Baetis australis* Traver
- *Baetis intercalaris* McDunnough
- *Baetis spiethi* Berner
- *Centroptilum viridocularis* Berner
- *Centroptilum hobbsi* Berner
- *Cloeon rubropictum* McDunnough
- *Cloeon* sp. A
- *Pseudocloeon alachua* Berner
- *Pseudocloeon parvulum* McDunnough
- *Pseudocloeon punctiventris* McDunnough
- *Pseudocloeon bimaculatus* Berner

ANNOTATED LIST OF MAYFLIES

Genus STENONEMA Traver

In 1933 Traver reviewed the Heptagenine mayflies of North America and erected *Stenonema* to include those species which were formerly placed in *Ecdyonurus* and some of which were formerly considered to be *Heptagenia.* After studying the type species of *Ecdyonurus,* she decided that this genus did not occur in North America, but was confined to the "Old World."

In general, male adults are easily distinguished from the other Heptageniidae, but as far as females are concerned, the difficulties involved in generic placement become much greater because there is much overlapping of characters. Traver's key to the genera of this family ignores the females, and it is only by associating them with males that the genus of this sex may be determined. The thread-like seventh gills of the nymphs easily sets them apart from all other mayfly genera. Specific determinations in the genus *Stenonema* are sometimes rather difficult, particularly in the *interpunctatum* group.[*] Color differences and color patterns are highly variable in some species, and individuals collected from the same place at different times may appear to be different species, although they are, in reality, nothing more than variants of the original form. Keys to the nymphs lead nowhere and serve only to separate the immatures in a broad sense. When more of the nymphal stages are known for the described species of *Stenonema,* perhaps a more comprehensive and useful key can be constructed.

Traver has divided *Stenonema* into two groups on the basis of genitalia and by use of the wing venation. The nymphs of the two groups, *interpunctatum* and *tripunctatum-pulchellum,* are separated by gill structure.

Spieth suggests that *Stenonema* (*Ecdyonurus*) displays two lines of development: one is represented by the *interpunctatum* group; the other, by the *tripunctatum-pulchellum* complex. He also suggests that the *interpunctatum* species are more closely related to *Heptagenia* than is the latter complex of species, a conclusion with which I certainly agree, for examination of the genitalia of the two groups reveals marked parallelism. *Stenonema* is a rather unspecialized genus which is not highly modified from the ancestral type, although it may be fairly high on its particular phylogenetic branch.

Stenonema is one of the largest genera of mayflies in North America. No species are known from western North America, and here *Heptagenia, Rithrogena, Iron,* and others become the predominant forms. The Appalachian Province is the ideal region for the development of *Stenonema,* and in this

[*] Spieth (1947) has attempted to clarify the problem by the erection of subspecific categories.

PLATE I

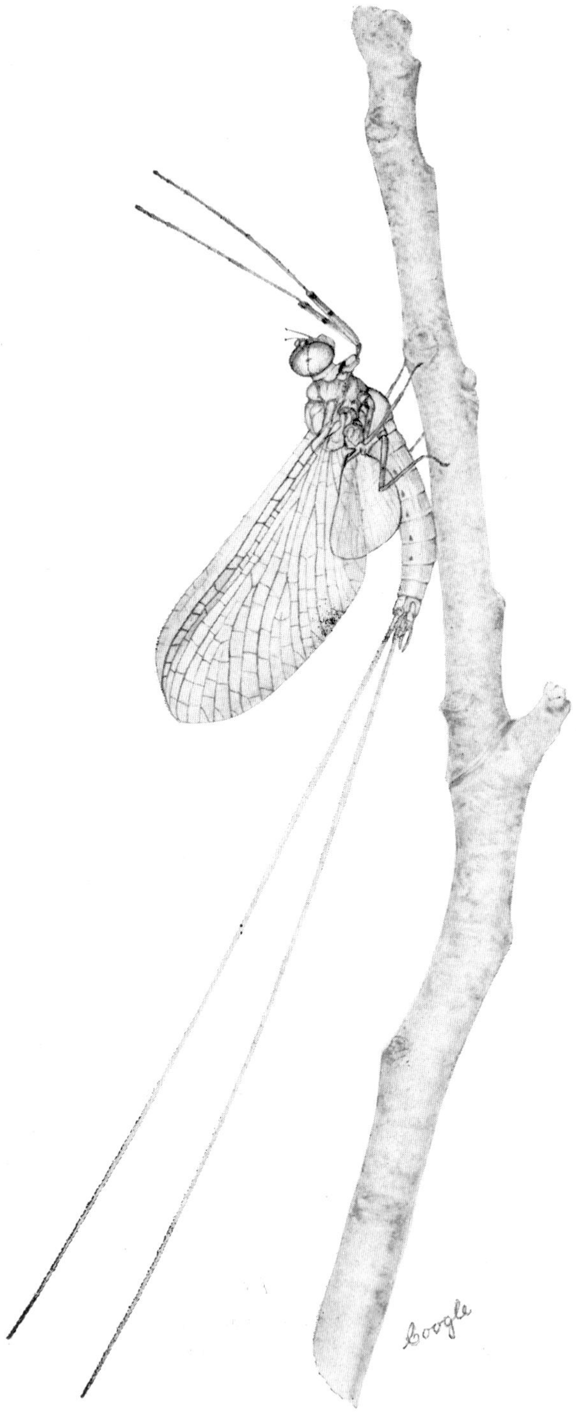

Stenonema smithae Traver, male imago.

area there are many species occupying almost all available situations. Although other Heptagenine mayflies occur in the mountains, *Stenonema* is the one most frequently found. Only a very small proportion of the species of *Stenonema* is known to occupy the Coastal Plain. From their evident success in Florida, however, I would suspect that when various other parts of the province are examined, other species will be found to be just as successful.

Stenonema smithae Traver

TAXONOMY.—*Stenonema smithae* is one of the more recently described species of the genus. During a short collecting trip through several states in the Southeast, Traver discovered S. *smithae* in west-central Alabama near Tuscaloosa. The description included that of the male, female, and nymph. I have found that the Florida insects agree fairly well with the description except that, in the adults, there is an orange coloration on the vertex in the majority of specimens. Traver failed to mention the condition of the mouth parts of the nymphs, but it has been found that the structure of the maxillae is one of the most important characters for separation of S. *smithae* and S. *exiguum* nymphs. Adult S. *smithae* are easily distinguished by the presence of stigmatic spots, by the shape of the male genitalia, by the pattern of cross veins in the forewings, and by the body coloration.

The species falls clearly into the *pulchellum-tripunctatum* group of *Stenonema.* The penes are strongly L-shaped and there is no bar formed in the forewing.

Traver considers *smithae* to be superficially similar to S. *integrum,* and according to published descriptions it also seems to be rather close to *bellum;* however, since I have seen no specimens of the latter species and only few of the former, no further comments can be added. This whole complex of species forms a compact group, differing from each other only in minor structural and ornamental characters.

DISTRIBUTION.—Since the description of *smithae,* the species has not been mentioned in print, and its only published locality record is from Spencers Mill, Tuscaloosa, Alabama. S. *smithae* has been found to be one of the commonest mayflies in Florida, but does not seem to be distributed south of Alachua County. Westward, however, almost every flowing body of water supports populations of nymphs; consequently, *smithae* has been taken as far west in Alabama as collections have been made.

Tuscaloosa is located just beyond the Fall Line on the Coastal Plain. All the Florida records are, of course, from the Coastal Plain. It therefore appears that the species is confined to this physiographic province, although it is possible that it passes beyond this barrier and into the Piedmont. It seems rather doubtful that streams as close to the Piedmont as those at Tuscaloosa should have the species while those just on the other side of the Fall Line should not. Winged insects such as mayflies could easily surmount such an obstacle and enter the streams in the Piedmont, at least in the lower region. It is likely that S. *smithae* is distributed widely over the Coastal Plain wherever there are permanently flowing streams (map 3).

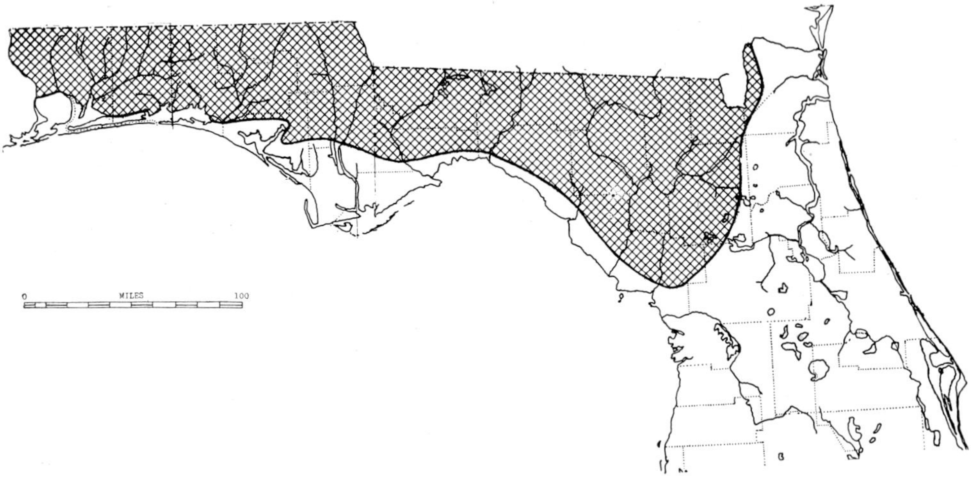

Map 3.—The distribution of *Stenonema smithae* in Florida.

ECOLOGY.—Almost any constantly flowing brooklet, creek, or river will have nymphs of S. *smithae* in it. The insects can be found on various submerged objects which are at least partially anchored in the stream bed, including submerged logs, sticks, leaves, rocks, pebbles, and even tin cans. In the sand-bottomed creeks where there are broad expanses of almost barren sands, no nymphs are found, but if there is, by chance, a log in midstream, the underside is almost certain to have these nymphs on it. A board partially buried in the sand where the current slackens may harbor as many as fifteen to twenty nymphs, the majority being rather young. Stones, likewise, will support a fair number on the darkened, protected undersurfaces. Many immatures can also be found among leaf drift which accumulates in the more slow-flowing water or becomes entangled in submerged branches and vegetation. Frequently, but in no wise so commonly as on more solid supports, nymphs occur on vegetation; however, the immatures are not well suited to such a situation and apparently shun it whenever a better place is offered.

Size of the stream offers no obstacle to *smithae* nymphs as is indicated by their presence in permanently flowing water which was no more than two inches deep and one foot across (at low water), as well as in large rivers such as the Suwannee. Rate of flow is likewise not of particular importance, for in Florida the nymphs occur in waterways which are almost stagnant as well as in the most rapidly flowing streams, of which the Santa Fe River is probably the swiftest. In this latter river the nymphs are very numerous on the undersides of the rocks that are strewn over the stream bed, but they are much less numerous on the *Vallisneria* and other vegetation which form huge mats. Just below the dam at Poe Springs the current of this river is very rapid. Nevertheless, the nymphs are very common even in the swiftest water; at the shore, however, the flow becomes negligible but immatures are almost as numerous there as in the rapids. Depths to which the nymphs descend, as far as I have been able to determine, are somewhat limited by the materials available for support and the amount of flow, but it is likely that the nymphs do go fairly deep in the larger rivers.

It was found that many times the nymphs had occupied areas where the current was rather strong at high water, but as the level of the stream fell and the movement of the water became less vigorous, the nymphs did not migrate into the swifter parts but remained in the almost stagnant portions left behind by the receding flood. In some cases nymphs were even found to be isolated in small pools cut off from the main stream, and here they lived on leaf drift and other debris which accumulated in the bottom of the pool.

The wide limits of toleration of *smithae* nymphs adapt them to many and varied conditions, allowing them to become the most frequently encountered mayfly nymphs in the streams of Florida. The nymphs are not at all, or very little, affected by pH, for I have found them inhabiting a swampy, slow-flowing body of murky water with a pH somewhat below 4.0 (the meter would read no lower). However, nymphs of the same species were found to be quite at home in a clear, sand-bottomed spring run which had a pH of 7.8. Even though this is not as wide a range as that occupied by *Callibaetis floridanus*, it is certainly far beyond that of most mayfly species. Likewise, temperature does not seem of great importance because nymphs inhabit water with a temperature of 30° C.; and the same species may be found in water which at times has ice formed in its quieter portions. Because of the wide ecological valence of *smithae*, it does not appear logical that the species should stop its southward range at Alachua County, but such seems to be the case. One can readily see that stream mayflies could not favorably inhabit the canals of south Florida; however, the clean, sand-bottomed creeks of Hillsborough County are not essentially different from those of north-central Florida. It may be that when this part of Florida is more thoroughly studied, the range of the species will be extended, but until further collecting is done, it must be assumed that *smithae* is absent from this area.

The small pebbly riffles in the sand-bottomed creeks of north-central Florida, and to some extent in northwestern Florida, also afford an excellent collecting place for nymphs of *smithae*. The nymphs tend to congregate in such places because conditions, from the standpoint of food, are excellent, the current is as swift as can be found in the stream, and although predators are rather numerous, the nymphs are still well protected. Young nymphs seem to predominate in the riffles, but this is not always the case and frequently older specimens will be the only ones taken.

At one time or another almost every Florida species of mayfly has been collected along with S. *smithae* except *Hexagenia munda orlando* and *Callibaetis floridanus* (form B). In its favored haunt on submerged logs and boards only few species are present along with S. *smithae*. These include S. *proximum*, S. *exiguum*, *Baetis spinosus*, *B. spiethi*, *Pseudocloeon alachua*, and *Tricorythodes albilineatus.* The *Stenonema* species are the only ones which occupy the undersurfaces of the supports to any extent; the others tend to remain on the sides or on the upper surface. In leaf drift, the Leptophlebine mayflies are found in addition to *Baetis* species.

SEASONS.—September is the only month of the year for which I do not have adult specimens and this is easily explained by lack of collecting during this time. *S. smithae* resembles the Baetidae in its year-round emergence, and there seems to be no definite period of maximum transformation. There is some indication that a greater preponderance of adults may occur during the summer months. Emergence slows down during cold snaps; however, as soon as warm weather returns, the adults begin to appear once more. Nymphs of all ages can be collected from any one stream at any time of the year, and there is no tendency toward brooding. Of course, since temperatures are slightly more uniform and higher in north-central than in northwest Florida, there is a greater amount of emergence in the former section during the winter months because of the longer warm spells. The specimens collected in Alabama by Traver were taken during July and offer no indication of seasonal range in the more northerly part of the geographic area occupied by *smithae.*

HABITS.—*Stenonema* nymphs can be identified at a glance among Florida material—their flattened bodies, broad, flat heads, and spreading legs are characteristics shared by no other species in the state; however, the habitude of the nymphs does not easily separate *smithae* from the other Florida *Stenonema.* When morphology and coloration are employed, the three species are easily differentiated.

If a board, rock, or other object on which the nymphs are living be rapidly lifted from the water and the underside examined, a mad scrambling will be observed as the nymphs of *Stenonema* seek to return to the seclusion of their underwater home. The draining of the water from the support causes the nymphs to cease moving and detection of the immatures then becomes rather difficult. If a little water is dashed over the object, the nymphs again begin to run about. Very few of them actually reach the rim of the support and drop off, although some do successfully escape in this manner.

Swimming actions are very awkward, and it is obvious immediately that *Stenonema* nymphs are certainly not adapted to a free-ranging existence. Though they can swim only forward, the nymphs can walk in almost any direction with equal ease. An undulating motion is the driving force in swimming—the head is first depressed and the wave travels along the abdomen to the caudal filaments, which lash out with little effect because they are almost devoid of long hairs such as those which make the Baetinae efficient swimmers. The swimming act never carries the insect very far, and after a few strokes the nymphs settle on any object which comes within reach, whether it be animate or inanimate. For speed, the swimming of *S. smithae* cannot compare with the rapidity of its crawling movements under water. It is almost impossible to seize the nymphs with forceps without injury when they are attached under water. It is obvious that this latter ability should naturally be greatly accentuated over the ability to swim as an adaptation for a clinging life.

One of the most characteristic behaviors of *Stenonema* nymphs is a strong thigmotactic response to stimuli. The slightest contact with an object in the water will cause the nymphs to fasten their claws immediately and to hold on "for dear life." Of course, such tenacity is necessary when the nymphs are exposed to the strong currents which, at times, must sweep under rocks in midstream. Even though the nymphs may be taken from almost stagnant water, thigmotropism in them is as strongly developed as in nymphs in the swiftest water. The reaction is so pronounced that if several nymphs are placed together in a dish of water with nothing to which they may cling, they will form a clump and remain clumped until some more attractive object is placed with them. The tenacity of the nymphs is shown in their reluctance to release their hold when the object to which they are clinging is lifted from the water. An inexperienced collector frequently injures the nymphs by attempting to pull rather than lift the insects from the support to which they are holding.

Wodsedalek (1912: 35) found in experimenting with nymphs of S. *interpunctatum* that the desire for contact is so strong that they are not fully satisfied until even the dorsum is in contact with some object. To quote him directly: "Two long bricks were placed one over the other in a basin of water and between them small pebbles varying in size so that the space gradually varied in thickness from one end to the other. Then a large number of nymphs were put in the water, and after a short time it was found that nearly all of the specimens were attached to the lower surface of the upper brick with their dorsal side downward, and a large majority of the specimens were in that portion of the wedge-shaped space where their backs came in contact with the brick below." The same author showed that the tropism is so strong that only when the water in which the nymphs live was heated to 45° C. did the insects leave the stones to which they were clinging; and this happened only after the nymphs had been overcome by the heat.

Reaction to light is also quite strong in *Stenonema* nymphs, and they will readily react to changes in intensity. The tropism is definitely negative. A stone was placed on a white background in a dish of water; in the same dish, a stone was also introduced which had a dark background. A series of nymphs were then released on the former stone; after a few hours, nearly all nymphs had migrated to the darkened area where they remained without any apparent desire to leave.

Wodsedalek (1912b: 260), during his research on Palmen's organ in mayflies, found that the supposed reaction to light is in part a result of orientation and concluded:

In their natural habitat they are always found clinging to the under surfaces of small rocks, and this same position is regularly assumed by all normal ones in the aquaria of the laboratory. When a stone, to which the specimens are attached is inverted in the water, the insects soon make for its under side, many of them doing this as the stone is being turned over. This is also true of normal specimens in the dark-room, and hence it is obvious that this tendency of the nymphs to cling to the lower surfaces of rocks, with their dorsal side downward, is not due entirely to their negative reaction to light. It is unquestionably due, in part, to a definite power of orientation independent of phototaxis.

The imagoes are slightly phototropic, more so in the subimaginal stage. The amount of reaction to light seems also to be linked with temperature, for on cool nights none were attracted to my lighting sheet, although adults were known to be on the wing. However, on mild nights in midsummer when the temperature was high and the adults were active, a number were drawn to the sheet. The Northern *Stenonema* species seem to be much more reactive to light than are those of Florida if the number of specimens brought to me by friends who have visited Lake Erie during the period of emergence is any indication.

As might be expected, the food of nymphs consists almost entirely of multicellular plants. There are a few diatoms scattered among the materials in the alimentary tract, but filamentous algae, epidermis of living and dead plants, and various unidentifiable substances predominate. Wodsedalek found that in order for the nymphs to be stimulated to feed, the particle of food, unless attached to a large object, must be large enough to afford comfortable attachment of the nymphs. I have been able to keep nymphs alive for some time and apparently in good health in aquaria in which *Vallisneria* leaves were placed. These were covered with diatoms and small algae; on the other hand, aquaria, in which only decaying leaves were put, furnished an adequate food supply for the developing immatures, and they reached adulthood in very good shape.

Dodds and Hisaw (1924) discussed the adaptations of a relative of *Stenonema, Cinygmula ramaleyi,* in the waters of Colorado. They consider that the wide distribution of this type of mayfly is probably due to the constant conditions which exist under stones, both in streams and lakes, and assert that "there is very little or no current beneath the rocks, and the oxygen content here has been found to be about the same in the lakes as in the streams. A species living under these mild current conditions would be expected to have conservative current resisting adaptations." The flattening of the body and the femora, and the spreading of the legs all reduce resistance to a minimum in the rheocolous genus, *Stenonema.* *S. smithae* is no more, nor any less, flattened than other species of *Stenonema* and could probably withstand just as strong current as the strongest in which its kindred species dwell.

LIFE HISTORY.—I have attempted to rear this species from the egg several times, but in every case the nymphs died a few days after hatching. To get the eggs to hatch was simple; to go much beyond this stage was another matter. The crux of the situation is probably in the food supply. Introduction of food invariably introduces other organisms, and once these are present, they seem to multiply at such a rapid rate that the competition offered by them is too great for the delicate mayfly nymphs.

Nymphal development in the genus *Stenonema* has been studied intensively for only two species, *S. interpunctatum* and *S. canadense.* The former species was examined from the first through the eleventh instars and the changes noted; however, no data were presented on the actual number of instars, nor was there an estimate of the length of life of the nymph. Ide (1935b), studying *S. canadense,* described the first through the sixth, the eighth, tenth, eleventh, twelfth, fourteenth, and fifteenth instars, and estimated the total number of molts to be between forty and forty-five. But he, too, failed to give the actual length of life of the individuals. Skipping the intermediate instars, he then described the octult, septult, quintult, quartult, tertiult, penult, and last instars of the species, explaining that the intermediate instars were not studied because of the difficulty encountered in estimating the instar from the material collected. The morphological differences between *interpunctatum* and *canadense* are very slight and development is essentially the same in both.

The surest method of securing *S. smithae* eggs, which are reasonably certain to hatch, is to capture ovipositing females and force them to continue oviposition in dishes containing small amounts of water. The eggs are liberated from the female in small masses as she dips her abdomen and lightly touches the surface of the stream. After liberation, the ova settle to the bottom where they absorb water which causes a jelly-like substance to swell about each ovum, forming a transparent protective covering which is also adhesive. It may be of some significance that unfertilized eggs take anywhere from a few minutes to several hours to form this jelly-like covering, but if the eggs are fertilized, the covering forms within five minutes of deposition of the ova.

The eggs, either aerated or not aerated, but kept in shallow dishes of water, hatched in the laboratory over a period varying from eleven to fifteen days.* These figures include the earliest and latest hatchings observed in the eggs of three females. Because the eggs were accidentally mixed when collected, I was unable to determine to which female the earlier and later hatching eggs belonged. My first attempt at hatching this species was fairly successful, although the females which I secured were nearly spent and contained only a few eggs in the anterior portion of the body; however, these were fertilized and a number of nymphs hatched within approximately fourteen days. A group of females, which were collected later and which had just begun oviposition, were completely filled with eggs, and hundreds of nymphs hatched from this batch.

These last eggs were deposited on October 5. They were examined on October 15 and it was found that nearly every one contained an almost fully developed nymph coiled within. The first hatching was noted exactly 270 hours and 13 minutes after oviposition, and from this time, hatching continued

*Needham (1935: 90) gave the time of incubation of the eggs of *S. tripunctatum* as eleven to twenty-three days. This is the only published record of the actual time required for embryonic development of a species of this genus.

for four more days, with a predominance at about twelve to thirteen days after laying.

The movement of the nymph within the chorion as it tries to escape is very clear and easily observed. There is a forward, backward, and sideward motion of the head as though the insect were scraping it against the egg shell; however, very little action occurs in the more posterior portions of the body which are tightly coiled. I observed this movement within individual eggs lasting as long as two days, with the final liberation of the immature after this period.

Finally, the egg ruptures along its longitudinal axis and the nymph emerges, but the exit is rather slow and arduous. The first part of the body to pass from the shell is, of course, the head. As it is pulled, or rather pushed, the chitin of the insect yields in the median region much as cellophane does when it is stretched. As the head is pushed out, the thorax gradually squeezes through, giving here and there to accommodate the narrow opening in the chorion. It is rather surprising that the opening is so small, because it provides quite an obstacle to the rapid hatching of the nymphs, which would seem to be a prerequisite of the habitat in which the insects dwell. However, this slow hatching may be a condition peculiar to laboratory-reared individuals, but information gleaned from other authors indicates that this slowness is not confined to *smithae*. Soon after the thorax is freed, or at least the anterior two-thirds, the antennae pop out and are immediately extended. The wriggling motion increases and the body begins to swing from side to side as sufficient mass is freed to give some force to the movement. At the same time, a series of waves that are created by intense muscular exertions pass along the body. The abdomen is soon entirely free of the shell and the caudal filaments are drawn out, but just before this happens the middle legs are suddenly released. The other two pairs are then withdrawn. The legs appear to be held together by some adhesive and it is only by the greatest effort on the part of the nymph that they are separated from each other. They are extended straight out from the body, but the insect is still unable to bend them; soon a tremor passes through the nymph and all the legs are flexed simultaneously. The nymph immediately becomes active and scrambles away from the shell, pulling the remainder of its tails free. The time required for hatching in the specimens observed varied from three to ten minutes from the time the head appeared through the slit in the chorion until the caudal filaments were entirely free. Hatching occurred just as frequently during any part of the day or night.

The newly hatched nymph is essentially similar to that of S. *canadense* illustrated by Ide (1935b). The anterior part of the alimentary tract is filled with yolk, and the nymph can survive on this material for as long as two days. The most noticeable part of the insect is its enormous head with the five eyes, the only really dark part of the body.

Thigmotactic and phototropic responses are present when the nymphs are hatched, and the immatures can be seen running about seeking darkened places and clinging tightly to the objects on which they are situated. The young nymphs, like the more mature insects, move readily backward, sideways, or forward, according to the direction of the stimulus.

Mayfly nymphs, immediately after molting, are very pale, almost white. Molting was observed in *smithae* nymphs at 1:45 P.M., and at 4:00 P.M. melanization had occurred to such a great extent that the nymph was almost back to its normal color. By 5:30 P.M. the normal coloration had been reached.

I have observed the mating flight of this species several times, and each time it was essentially the same. Just before dusk, a small band of approximately ten to twenty males gather directly over a stream—whether it be small or large makes no difference. The characteristic flight is then begun about twelve to fifteen feet above the water. The horizontal flight is slight; the vertical rise and fall is about two feet. The flight was observed in May at 7:00 P.M. and in early July at 8:00 P.M.; in the early spring, it occurred at 6:15 P.M. There seems to be an intimate correlation between amount of light and the time at which the swarming takes place. The length of time that the flight lasts was not determined because the insects were still flying when darkness fell; however, since the females begin to oviposit while it is still light, it would seem that mating ceases either very shortly before dark or soon thereafter.

The forward flight of the males was not much over eight to ten feet with a rapid return to the original position. Occasionally a female would fly into the swarm and be immediately seized by a male. These two then separated from the swarm and were usually lost to sight, but some were seen settling to the ground where they soon separated, the male returning to the swarm, the female flying off to oviposit. Cooke (1940) made an interesting observation on the copulatory approach of the sexes of *S. vicarium* during flight:

> Females showed very little tendency to take part in the flights. The captures of two complete companies of imagoes and of the greater portion of a third yielded only one female, which was taken with the second group. It is probable that this female had just entered the swarm from beneath, because when a single female was seen passing below a company of males it was seldom disturbed; on the other hand, when it passed a few feet above or directly through the swarm, it was instantly attacked by them. The large eyes of the males are situated on the dorsolateral regions of the head . . . , a location which perhaps enables this sex to see females above them better than below. All attempts to mate were made by the male flying up beneath the female placing his forelegs over her prothorax and head. With upcurved abdomen he grasped the body of the female with forceps near her seventh abdominal segment . . . and mating thus became effective.

Observations on the oviposition of the females demonstrated an interesting behavior. The insect flies low over the water, about six to twelve inches above the surface. At intervals, the abdomen is touched to the surface where the current is rather noticeable, the action resembling that of a dragonfly or damselfly. The ovipositing flight, for the most part up and down but occasionally cross stream, is not composed of pronounced risings and fallings such as take place during the mating flights, but is in a straightforward direction. The eggs are released, a few at a time, until the female is spent, when she flies to some low-hanging bush. The horizontal flight may cover a distance varying between twenty and fifty feet, oviposition occurring both on the downstream and return flight. I did not determine the entire time required for release of the ova, but it must not take much more than five to ten minutes, because spent females were collected before darkness had fallen completely.

The urge to oviposit is very great. During a heavy downpour I observed, and later captured, a female in oviposition flight. Seemingly she dodged between the raindrops. Certainly, if one had struck her, the flight would have immediately ended.

An interesting note concerning the manner of oviposition of other mayflies in the same taxonomic grouping as *smithae (tripunctatum-pulchellum)* is the parallelism in the method exhibited by *smithae.* The stream forms, observed by Dr. Smith (*in* Needham, 1935: 72), have an oviposition flight which is not greatly different from the mating flight; however, in the lake species "there was no up and down dance, but there were occasional pauses and then sudden starts, somewhat after the manner of a dragonfly. Once in a while a female would slow up a bit and dip her abdomen in the water without actually alighting, and sometimes she would alight on the water for just a moment." From this habit there would seem to be a close relationship between the lake-dwelling *Stenonema* species and the stream forms of the Coastal Plain, but if one refers to the morphological characters, such a relationship is not so evident.

Emergence of *Stenonema smithae* subimagoes occurs in late afternoon, at approximately the same time as the mating flight. I have frequently taken subimagoes rising from the stream at the same time that mating flights were in progress. The nymph, when preparing for transformation, floats freely at the surface of the water. The thoracic skin then splits suddenly and the subimago appears. The actual emergence takes only a few seconds. With almost no rest period, the insect arises and flies upward until out of sight (unless some support is nearby), the flight never being horizontal but always upward at a moderate incline. After twenty to twenty-four hours, the imaginal molt occurs. Female imagoes may live for about two and one-half days.

An interesting point concerning the eyes was noted in *Stenonema smithae.* In mating males, that is, those taken during a mating flight, the eyes are black. Laboratory-reared males kept in the dark will have black eyes, yet when they are exposed to the light for a short while, the eyes become pale, but this change is gradual and, in laboratory-reared insects, may take as long as forty minutes. Lyman observed a similar color change in *Stenonema* in 1943.

LOCALITY RECORDS.—Alachua County: Santa Fe River at Poe Springs, May 21, 1934, nymphs; February 24, 1937, nymphs; March 12, 1938, nymphs; March 25, 1939, nymphs and adults; April 6, 1940, nymphs and adults. Hatchet Creek, March 22, 1937, nymphs; February 8, 1938, nymphs; March 23, 1938, nymphs and adults; April 2, 1938, nymphs and adults; April 18, 1938, adults; May 5, 1938, nymphs and adults; November 13, 1938, adults; March 22, 1939, nymphs and adults; April 1, 1939, adults; April 13, 1939, nymphs and adults; May 6,

1939, nymphs and adults; June 24, 1939, adults; October 11, 1939, nymphs and adults; October 28, 1939, nymphs; April 24, 1948, adults. 2½ miles west of Gainesville, January 16, 1938, nymphs and adults; January 29, 1938, nymphs and adults; February 3, 1938, nymphs and adults; March 5, 1938, nymphs; January 7, 1939, nymphs; January 28, 1939, adults; March 10, 1939, nymphs and adults; April 21, 1939, adults; October 6, 1939, nymphs and adults; January 6, 1940, nymphs and adults; March 18, 1940, nymphs and adults; May 7, 1940, nymphs and adults. Hogtown Creek, February 2, 1932, nymphs; March, 1933, nymphs; February 19, 1934, nymphs; February 15, 1937, nymphs; April 3, 1937, nymphs; April 26, 1937, nymphs and adults; May 10, 1937, nymphs; September 25, 1937, nymphs; October 30, 1937, nymphs; July 17, 1938, adults; April 30, 1940, nymphs and adults. Near Worthington Springs, February 5, 1939, nymphs. 3 miles north of Paradise, February 12, 1938, nymphs. 1 mile west of Newnan's Lake, May 11, 1937, nymphs; January 8, 1938, nymphs; January 25, 1938, nymphs and adults; August 13, 1938, adults. Rocky Creek, February 1, 1938, nymphs. Campus, University of Florida, March 5, 1938, nymphs. Five miles northwest of Gainesville, November 8, 1937, nymphs. Bay County: 5.6 miles north of Panama City, May 30, 1940, nymphs. 16.8 miles north of Panama City, June 8, 1938, nymphs. 26 miles north of Panama City, June 8, 1938, nymphs. 28.7 miles north of Panama City, June 8, 1938, nymphs. 32 miles north of Panama City, June 8, 1938, nymphs. Calhoun County: 4.2 miles east of county line, April 7, 1938, nymphs. Columbia County: Falling Creek, February 4, 1938, nymphs. 11.5 miles north of Lake City, October 27, 1938, nymphs. Escambia County: Carpenter's Creek, April 4, 1938, nymphs. Flomatin Road, April 4, 1938, nymphs. 14.3 miles south of Flomatin, April 5, 1938, nymphs. Bayou Marquis, June 1, 1940, nymphs. Gadsden County: 15.1 miles east of Chattahoochee, April 1, 1938, nymphs. River Junction, March 17, 1939, nymphs. 4½ miles south of River Junction, March 17, 1939, nymphs; June 30, 1939, nymphs and adults; June 8, 1940, adults. Gilchrist County: Suwannee River at Old Town, April 5, 1938, nymphs. Hamilton County: 6 miles north of Live Oak road at U. S. Highway 41, February 4, 1938, nymphs. 8.3 miles south of Jasper, February 4, 1938, nymphs. White Springs, February 4, 1938, nymphs. Holmes County: Sandy Creek, December 11, 1937, nymphs; July 2, 1939, nymphs and adults; December 14, 1939, nymphs; May 1, 1946, nymphs. Jackson County: Blue Springs Creek near Marianna, May 5, 1933, adults; December 11, 1937, nymphs; June 6, 1940, adults. 3.6 miles north of Altha, December 10, 1937, nymphs; June 9, 1938, nymphs and adults; December 1, 1939, nymphs and adults. Jefferson County: April 1, 1938, nymphs. Leon County: 15.6 miles west of Tallahassee, June 5, 1938, nymphs; March 17, 1939, nymphs; November 30, 1939, nymphs. 11.2 miles west of Tallahassee, March 16, 1939, nymphs and adults. 1 mile south of Florida Highway 20 on Sopchoppy road, June 5, 1938, nymphs and adults. 7 miles south of Florida Highway 20 on Sopchoppy road, June 5, 1938, nymphs. Levy County: Gulf Hammock, December 21, 1938, nymphs. Liberty County: Sweetwater Creek, May 7, 1933, nymphs; June 10, 1938, nymphs; November 4, 1938, nymphs; July 1, 1939, nymphs; December 1, 1939, nymphs and adults; May 2, 1941, adults; April 29, 1946, adults. Kelly Branch, December 10, 1937, nymphs. Little Sweetwater Creek, December 10, 1937, nymphs; June 10, 1938, nymphs. Hosford, November 30, 1939, nymphs. 10 miles south of River Junction, March 17, 1939, nymphs. 4.5 miles north of turnoff to Torreya State Park, June 10, 1938, nymphs. Madison County: 8 miles east of Jefferson County line, February 5, 1938, nymphs. 4.3 miles east of Jefferson County line, February 5, 1938, nymphs. 3 miles west of Greeneville, February 5, 1938, nymphs. Nassau

County: 19.1 miles north of Duval County line, August 28, 1938, nymphs. Okaloosa County: Shoal River, December 11, 1937, nymphs. Crestview, December 12, 1937, nymphs. 5.1 miles west of Walton County line, May 31, 1940, nymphs. 2.3 miles east of Niceville, April 3, 1938, nymphs. Niceville, June 7, 1938, adults. 3.6 miles north of Niceville, April 3, 1938, nymphs. Santa Rosa County: 7.1 miles west of Milton, April 4, 1938, nymphs and adults. Pace, June 1, 1940, adults. Taylor County: Fenholloway River, May 29, 1940, nymphs and adults. Wakulla County: Smith Creek, June 5, 1938, nymphs and adults. Walton County: 1 mile north of Walton County line, June 7, 1938, nymphs. 2.1 miles west of Walton County line, May 31, 1940, nymphs. 7.3 miles northwest of Ebro, June 7, 1938, nymphs and adults. 15.8 miles west of Ebro, June 7, 1938, adults. 9.6 miles west of Portland, May 31, 1940, nymphs and adults. Portland, April 3, 1938, adults. 10 miles east of Freeport, April 2, 1938, nymphs. 5.4 miles east of Freeport, April 2, 1938, nymphs and adults. 2 miles west of Freeport, April 3, 1938, nymphs. 2.6 miles west of Freeport, June 7, 1938, nymphs. 13.8 miles west of Freeport, June 7, 1938, nymphs. 15.6 miles west of Freeport, June 7, 1938, nymphs. 12.5 miles east of Niceville, April 5, 1938, adults. 11.4 miles east of Niceville, April 3, 1938, nymphs and adults. Washington County: 5.8 miles south of Vernon, April 2, 1938, nymphs. Holmes Creek, December 11, 1937, nymphs; April 2, 1938, nymphs and adults; June 9, 1938, nymphs; July 2, 1939, nymphs and adults.

Stenonema exiguum Traver

TAXONOMY.—*Stenonema exiguum* is another Florida species which falls into the *pulchellum-tripunctatum* complex. It is rather close to S. *smithae* but can easily be distinguished by the characters set forth in the key. In her description, Dr. Traver suggests that *exiguum* is also close to S. *integrum*. The original description included only the male and the female, and the nymph was described by Daggy in 1945. The type specimens were collected near Woodlawn, North Carolina, at the Chattahoochee River, Atlanta, Georgia, and the Etowah River at Rome, Georgia. Subsequently, Traver has recorded *exiguum* from Alabama somewhere between Birmingham and Tuscaloosa.

DISTRIBUTION.—In Florida S. *exiguum* is known to occur from Hillsborough County in the south and as far north and west as the state line; eastward the species does not go much beyond the eastern limits of Alachua County. Specimens from Mobile County, Alabama, and Baker County, Georgia, are also at hand. The distribution of *exiguum* parallels that of S. *smithae* and is almost identical with it. The close resemblance of the nymphs of the two species makes determination of very young specimens rather difficult, and for this reason some of the individuals which I am calling *exiguum* may possibly be *smithae*, and vice versa. There is a fairly large series of adult *Stenonema* from Hillsborough County in my collection, but not a single one is identifiable as S. *smithae* —a circumstance that further leads me to believe that *exiguum* is the only representative of the *pulchellum* complex in this southern area.

The distribution of *exiguum* is interesting because the species is known from four physiographic provinces—the Coastal Plain, the Piedmont, the Valley and Ridge, and the Central Lowlands. It is not known in which of these provinces

PLATE II

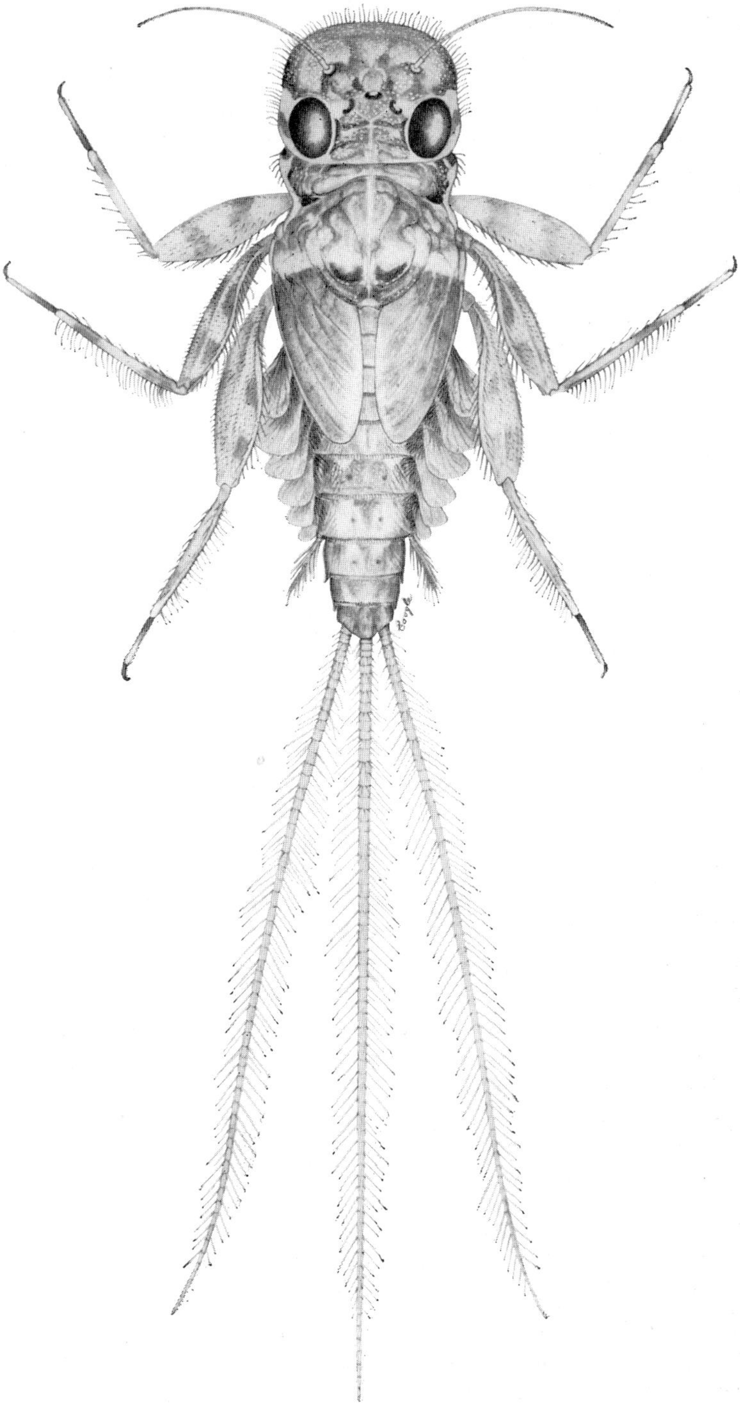

Stenonema exiguum Traver, nymph.

exiguum predominates, but the species is certainly well distributed in the first-mentioned region (map 4).

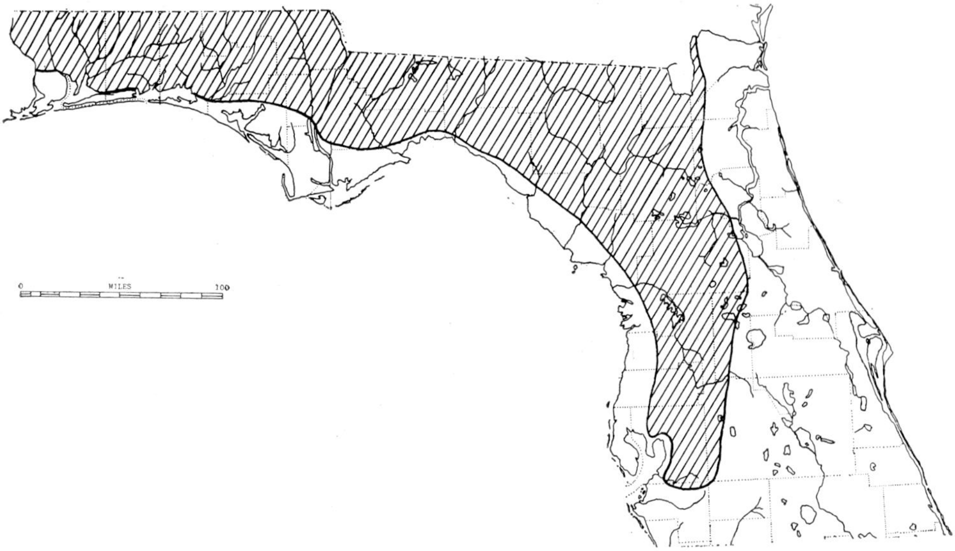

Map 4.—The distribution of *Stenonema exiguum*, *S. proximum*, and *Baetis spinosus* in Florida.

ECOLOGY.—Ecologically, the species is very similar to S. *smithae*, except in two respects. In the first place, nymphs in west-central and north-central Florida are very seldom found in smaller streams, but are very numerous, probably outnumbering *smithae*, in the larger creeks and rivers. The sites occupied within the inhabited streams are, however, identical with those of *smithae*. Secondly, a single nymph has been collected from a sand-bottomed lake, a habitat in which *smithae* has never been found. This specimen was brought from a depth of fifteen feet by means of an Ekmann dredge, but it was dead when collected. However, it was in a very good state of preservation and death must have occurred either just before collection or during collection because of an injury suffered in the dredge, although there were no external manifestations of such injury. It is surprising that only a single nymph should have been collected, for I examined rather thoroughly the shore materials on which the nymphs would be expected to be found. There is a small, slow-flowing stream entering the lake at one side and it is barely possible that the nymph was washed in from here shortly after death.

SEASONS.—Emergence is year-round like that of S. *smithae*. The fact that no adults are available for January, May, September, and November does not indicate failure of the species to emerge, but only that I was unable to get to streams that support populations of this species. However, there are nymphs in my collection which certainly would have emerged during these months. The type specimens were collected in June and July and indicate nothing of the seasonal life of the species farther north.

HABITS.—The habits of *smithae* and *exiguum* are identical. In the field, mature *exiguum* nymphs may be differentiated from *smithae* nymphs by the fact that across the thorax of the former species there is a rather broad yellowish-brown band.

LIFE HISTORY.—The life history of the species does not differ essentially from that of S. *smithae* except in that the subimaginal period is generally somewhat shorter, lasting from seventeen to twenty hours. I have observed no mating flights of S. *exiguum*.

LOCALITY RECORDS.—Alachua County: Santa Fe River at Poe Springs, May 14, 1934, nymphs; May 21, 1934, nymphs; March 19, 1935, nymphs; April 15, 1935, nymphs; March 24, 1937, nymphs; March 12, 1938, nymphs and adults; March 18, 1938, adults; February 11, 1939, nymphs and adults; February 18, 1939, nymphs and adults; February 28, 1939, nymphs and adults; March 25, 1939, nymphs and adults; October 3, 1939, nymphs and adults; April 6, 1940, nymphs and adults. Santa Fe River at U. S. Highway 41, May 14, 1938, nymphs. Hatchet Creek, October 28, 1939, nymphs. Bay County: 16.8 miles north of Panama City, June 8, 1938, nymphs. Pine Log Creek, May 31, 1940, nymphs. Citrus County: Withlacoochee River, April 2, 1937, nymphs. Escambia County: Perdido Creek, April 5, 1938, nymphs. Bayou Marquis, June 1, 1940, nymphs. Gadsden County: 4.5 miles south of River Junction, March 17, 1939, nymphs. Gilchrist County: Suwannee River at Old Town, April 1, 1938, nymphs and adults; April 19, 1938, nymphs and adults. Hillsborough County: Bell Creek, March 26, 1938, nymphs. 2 miles west of Alafia, March 26, 1938, nymphs. Hillsborough River State Park, August 18, 1938, adults. Hillsborough River, February 11, 1939, nymphs; October 21, 1940, adults. Holmes County: Sandy Creek, May 1, 1946, nymphs. Jackson County: Blue Springs Creek near Marianna, December 11, 1937, nymphs; June 9, 1938, nymphs; July 1, 1939, adults; June 5, 1940, adults; May 3, 1941, adults. 3.6 miles north of Altha, June 9, 1938, adults. 12.2 miles southeast of Marianna, June 9, 1938, nymphs. Jefferson County: April 1, 1938, nymphs. Liberty County: Sweetwater Creek, June 10, 1938, nymphs; November 4, 1938, nymphs; December 1, 1939, nymphs and adults. Madison County: At Jefferson County line, Aucilla River, June 4, 1938, nymphs. Marion County: Ocala National Forest, July 23, 1938, adults. Silver Springs, May 7, 1934, nymphs. 9.6 miles southwest of Salt Springs, April 10, 1948, nymphs. Okaloosa County: 5 miles west of county line, June 7, 1938, nymphs. Baggett Creek, April 4, 1938, nymphs. Niceville, June 7, 1938, nymphs. Putnam County: Red Water Lake, March 26, 1939, nymphs. Johnson, April 10, 1948, nymphs. Santa Rosa County: 2 miles west of Milton, April 4, 1938, nymphs. Suwannee County: Suwannee River at Branford, June 6, 1940, adults.

Stenonema proximum Traver

TAXONOMY.—It is questionable as to whether a series of specimens in my collection should be called *Stenonema proximum*, S. *pallidum*, or S. *interpunctatum*. Adding all characteristics together, I have finally decided to consider the species to be *proximum*, but if they are, then *proximum* must be so defined that the variations observable in my specimens will be included. Among the specimens at hand are individuals which could certainly be considered as *pallidum;* likewise, there are specimens which could easily be identified as *interpunctatum*.

However, taken as a whole, the description of *proximum* seems to include the various degrees of variability shown by the group of Florida specimens better than do the other two. According to Spieth's recent revision of the *interpunctatum* group, the species would probably be S. *interpunctatum interpunctatum.*

Although Traver had females and nymphs when she described the species in 1935, she failed to include these forms, except for a figure of the abdominal color pattern of the nymph. Subsequent to the original description, the species has not been recorded in print, though Traver did record S. *interpunctatum* from Alabama, which may be identical with the Florida form.

S. *proximum* is the only representative of the *interpunctatum* group of *Stenonema* found in Florida. Both in the nymphal and adult stages, the insect is quite distinct from the members of the *tripunctatum-pulchellum* group as it occurs in this area. This distinction has been pointed out in the key.

DISTRIBUTION.—Traver described S. *proximum* from a series of adults and nymphs collected near White Church, New York. At the same time she mentioned having specimens, which seemed to be of this species from Ithaca and Honeoye Falls, New York, and Lake Forest, Illinois. S. *proximum* has been found to be generally distributed throughout Florida wherever there is permanently flowing water. Adults and nymphs have been taken in Hillsborough County in the west-central part of Florida and northward to Burke County, Georgia. The species is known to extend from Nassau County, Florida, to Mobile County, Alabama.

If the Florida species is *proximum,* the distribution certainly seems unusual. If, on the other hand, the Florida specimens are *interpunctatum* or *pallidum,* the distribution is much more easily explained, for both these species have been previously recorded from the Southeast. Such a great break in distribution does not seem consistent, and doubtless it will later be proved that all three of these species can be included under the one name, *interpunctatum.**

Only twice before has a species of *Stenonema* belonging to the *interpunctatum* complex been reported from the eastern Coastal Plain. In 1933, Traver mentioned the occurrence of S. *interpunctatum* (?) in Goshen Swamp, Duplin County and in Chicod Swamp in Pitt County, North Carolina. The last record was published by the same author in 1937 when she recorded S. *interpunctatum* from Wilson Lake, and from North River near Tuscaloosa, Alabama (map 4).

ECOLOGY.—In many respects, S. *proximum* resembles S. *smithae* ecologically, but its ecological valence seems to be of a lesser degree than that of the latter species. One of the chief inhabitants of the larger streams of north-central Florida, it is very rarely found in the smaller ones which support large populations of *smithae.* Its habitat in these streams is identical with that of *smithae* —submerged logs, rocks, sticks, and boards in almost any part of the stream. The species seems to flourish to a much greater extent in alkaline creeks which

*Spieth (1947) considers *pallidum* to be a synonym of S. *interpunctatum interpunctatum* and *proximum* to be a synonym of S. *interpunctatum frontale.*

are of moderate dimensions, varying from the size of Hatchet Creek near Gainesville to that of the Santa Fe River.

Nymphs of S. *proximum* are also inhabitants of sand-bottomed lakes, but are confined to the shore region where they can be found on the undersides of boards, logs, sticks, or any other permanently submerged object which might give the nymph a good foothold and furnish a continuous food supply. In the lakes, the insects confine themselves to that part entirely free of any silt and characterized by a cleanly washed sand shore and beach. There is little wave action or deposition of silt in this region. Vegetation is almost lacking, and suitable situations for the nymphs are not plentiful. Consequently, when a board is located on which there are nymphs, it will usually have quite a number of individuals on it.

The stream associates of S. *proximum* are the same as those of S. *smithae;* the lake associates include two species of mayflies definitely, and a third questionably. These are *Ephemerella trilineata, Choroterpes hubbelli,* and the questionable one, *Stenonema exiguum.* In the lake from which *proximum* was taken most commonly, other animals present were indicative of a stream association. The mayflies listed above are typical stream forms, and Dr. A. F. Carr reports that the reptilian and amphibian faunae are likewise of a rheocolous nature. I have found such normal stream inhabitants as sponges, caddisflies, and parnid beetles living with S. *proximum.* This condition has been observed in only two of the sand-bottomed lakes of Florida although it is probably much more widespread. The finding of a lake-shore fauna which so closely resembles a stream fauna is reminiscent of the conditions found along some of the rocky beaches of Northern lakes where the mayfly fauna is quite abundant. However, it seems apparent that only very tolerant forms can withstand the conditions prevailing along the lake margins in Florida. This situation is somewhat understandable when one considers the slight wave action, the scant debris along the shore, the complete absence of rocks or pebbles, and the small amount of vegetation.

SEASONS.—Adults are known for every month of the year except September, and this is due only to lack of collecting during this month. Nymphs which would certainly have emerged during September were collected in August. The species is known only for June (published dates) in its Northern environs. S. *interpunctatum* has been reported from North Carolina and Alabama in April and July; *pallidum* was collected in May and June.

HABITS.—The habits of this species are much the same as those of *smithae.* The nymphs are very tolerant and have been kept alive in aquaria in the laboratory for more than two months, during which time many specimens emerged. There is some difference, however, in the size of the adults which emerged when the nymphs were first brought into the laboratory and those which emerged after one or two months under these conditions. The latter are somewhat smaller, having been dwarfed by confinement or by food which was not conducive to a full growth.

LIFE HISTORY.—The subimago emerges in late afternoon after sunset, flies to a nearby support where, after eighteen to twenty-two hours, the final molt takes place, and the adult is soon ready for mating.

The mating flight begins just after the sun has set and continues until darkness has fallen. I lost sight of the individuals in the swarm as darkness fell and, therefore, could not determine how long the flight lasted. The following description of swarming is taken from my notes of April 1, 1939:

. . . At 7:00 P.M. while standing on the bridge over Hatchet Creek, I noted several S. *smithae* females flying up and down the stream in the act of oviposition, but could see no mating flight of this species. . . . While the S. *smithae* females were ovipositing, I suddenly noticed just below me a bit of activity, which rapidly began to increase. It was becoming rather dark, but by shifting my position, I could see that this activity was caused by a group of S. *proximum* males in their mating flight. The swarm was small, composed of not more than twenty males. These insects flew up and down continuously; the upward flight was rapid, the downward slower, as though the insects were drifting. I paced off the horizontal distance covered by the swarm and found it to be almost negligible, not more than five feet, and this was not varied during the fifteen minutes which I observed the insects. At 7:15 it became so dark that further observation was impossible.

Now and then, I saw a female that approached the swarm, entered it and coupled with a male, after which the two seemed to drift from the swarm and settle to the ground where I lost sight of them. The female could easily be distinguished by the brilliant orange eggs which showed clearly through the translucent abdomen. I swept my net through the swarm once and captured seven males and two females. The latter were probably in copulation with two of the males when they were interrupted, for as I lifted the females from the net, there remained behind two clumps of orange-colored eggs dropped by them as they came in contact with the cloth. This dropping of eggs is very likely correlated with mating, because females that have not mated will retain the ova. Yet if copulation is in progress or has been completed, the eggs are released on contact. All during the flight, the strong breeze which was blowing did not seem to disturb the flight of the imagoes.

The minimum height of the flight of the swarm was between six and eight feet above the stream. The rise and fall of the males was between five and six feet. The females seldom rose as high as the males.

The most pronounced difference in the flight of *proximum* and *smithae* is in the great rise and fall of the males of the former species. The flight of other species of the *interpunctatum* group has not been described; however, Smith (*in* Needham, 1935: 72-73) has described the method of oviposition of one of the species, and since this has not been observed in *proximum*, I repeat his observations:

This individual, when I first saw her, was flying straight upstream, about five feet above the water, and was barely making headway against the evening breeze. On the underside of the abdomen she was carrying a fairly large mass of eggs, just as *Ephemerella* does. The eggs were of an orange or reddish color. Finally she settled down to the water and was carried downstream about a foot. She then flew up again, but without the eggs. This individual was captured. She was spent, and the tip of the abdomen was still curved down, where it had fitted over the egg mass.

Unfortunately no other females carrying egg masses were captured; so I could not find out what happens to the mass when it is laid in water, but the eggs are equipped with two very long coiled threads, which probably adhere to the bottom and to other eggs.

Mayflies belonging to this group are seldom seen in the early evening, but may often be found later at lights. It therefore seems likely that oviposition usually occurs after dark.

On April 19, while living material that had been collected on April 6 was being examined, I found several young *Stenonema* nymphs which were in either the first or second instar. Although I am by no means certain of the identification, I believe that they were the young of S. *proximum*. Because so few specimens were available, I will not attempt to describe the development, but there are certain definite differences between these and the nymphs of S. *canadense* as described by Ide (1935). The eggs from which these nymphs hatched were probably brought into the laboratory attached to vegetation which was used as aquarium plants. The vegetation was not examined when first brought in, and it was not until two weeks later that I discovered young mayfly nymphs on a leaf. Even after such a short period of time in the laboratory, the aquarium was literally teeming with organisms, especially dipterous larvae, a few caddisfly larvae, and young snails, most of which had apparently hatched in the laboratory. The mayflies had been hatched long enough so that some had entered the second instar as demonstrated by the presence of gills. These nymphs differed from the first instar of *smithae* in that they were reddish brown, while the *smithae* nymphs were white. For three more weeks some of the nymphs remained alive, but very shortly after this, all had disappeared. Growth during these three weeks was rapid: the nymphs developed additional gills and more spines on the femora, the caudal filaments elongated, and the thorax became much more robust.

The postembryonic development of S. *interpunctatum* has been studied to some extent and the results published in *The Biology of Mayflies;* only the first eleven instars are described, and the statement is made that the remaining nymphal instars show only increase in size of the various body parts. No mention is made of the actual number of instars through which the nymph must go.

LOCALITY RECORDS.—Alachua County: Lake Santa Fe, April 7, 1937, nymphs and adults; January 30, 1940, nymphs and adults; December 15, 1940, nymphs. Hatchet Creek, March 23, 1938, nymphs and adults; April 2, 1938, nymphs and adults; April 18, 1938, nymphs and adults; May 5, 1938, nymphs and adults; November 13, 1938, adults; April 1, 1939, adults; June 24, 1939, adults. Santa Fe River at Poe Springs, May 14, 1934, nymphs; February 12, 1938, nymphs; March 12, 1938, nymphs and adults; March 18, 1938, nymphs and adults; February 18, 1939, nymphs and adults; October 3, 1939, nymphs and adults; April 6, 1940, nymphs and adults. Santa Fe River at Highway 41, May 14, 1938, nymphs.

Near Worthington Springs, February 5, 1939, nymphs. Bay County: 26 miles north of Panama City, June 8, 1938, nymphs. 28.7 miles north of Panama City, June 8, 1938, nymphs. Calhoun County: Chipola River, November 5, 1938, nymphs. Duval County: 11 miles north of Jacksonville, August 28, 1938, nymphs. Gadsden County: 15.1 miles east of Chattahoochee, April 1, 1938, nymphs. Gilchrist County: Suwannee River at Old Town, April 5, 1938, nymphs and adults; April 26, 1938, adults. Hillsborough County: 2 miles west of Alafia, March 26, 1938, nymphs and adults. Hillsborough River State Park, August 16-18, 1938, adults. Dug Creek, August 15, 1938, adults. Hillsborough River, February 11, 1939, nymphs and adults; February 25, 1939, adults; October 21, 1940, adults. Jackson County: 3.6 miles north of Altha, June 9, 1938, nymphs and adults. 12.2 miles southeast of Marianna, June 9, 1938, nymphs. Blue Springs Creek near Marianna, May 5, 1933, adults; December 11, 1937, nymphs; June 9, 1938, nymphs; July 1, 1939, adults; June 5, 1940, adults. Leon County: 15.6 miles west of Tallahassee, June 5, 1938, nymphs and adults. 7 miles south of Highway 20 on Sopchoppy road, June 5, 1938, nymphs. 13 miles west of Tallahassee, November 30, 1939, nymphs. Levy County: Gulf Hammock, December 21, 1938, nymphs. Liberty County: 4.5 miles north of turnoff to Torreya State Park, June 10, 1938, nymphs. Sweetwater Creek, April 29, 1946, adults. Madison County: Aucilla River, June 4, 1938, nymphs. 4.3 miles from Jefferson County line, February 5, 1938, nymphs. Marion County: Rainbow Springs, February 26, 1939, nymphs; April 15, 1939, adults; March 9, 1940, nymphs and adults. Ocala National Forest, July 23, 1938, adults. Oklawaha River at Eureka, February 12, 1938, nymphs and adults. Withlacoochee River, March 25, 1938, nymphs. Silver Springs run, December 22, 1937, adults. Nassau County: 19.1 miles north of Duval County line, August 28, 1938, nymphs. Okaloosa County: 5 miles west of county line, June 7, 1938, nymphs. Niceville, June 7, 1938, nymphs. Putnam County: Red Water Lake, March 26, 1939, nymphs and adults. Santa Rosa County: 2 miles west of Milton, April 4, 1938, nymphs and adults. 7.1 miles west of Milton, April 4, 1938, nymphs. Pace, June 1, 1940, adults. Wakulla County: Smith Creek, June 5, 1938, nymphs. Wakulla Springs run, May 29, 1940, nymphs and adults. Walton County: 5.4 miles west of Washington County line, May 31, 1940, nymphs. 10.6 miles west of Washington County, May 31, 1940, nymphs. 15.8 miles west of Ebro, June 7, 1938, nymphs and adults. Portland, April 3, 1938, adults. 9.5 miles west of Portland, May 31, 1940, nymphs and adults. 5.4 miles east of Freeport, April 2, 1938, nymphs. 13.8 miles west of Freeport, June 7, 1938, nymphs and adults. 11.4 miles east of Niceville, April 3, 1938, nymphs. Washington County: Holmes Creek, April 2, 1938, nymphs; April 30, 1946, adults.

Genus HEXAGENIA Walsh

Hexagenia, described by Benjamin Walsh in 1863, was one of the first North American genera to be considered new. Needham, as late as 1920, recognized only two "good and distinct species" in the eastern United States—a lowland form, *H. bilineata,* from lakes and rivers, and an upland bog-stream species, *H. recurvata.* McDunnough (1927: 117) commented as follows concerning Needham's paper:

. . . With the above conclusions I must emphatically disagree; from a study of a large number of dried specimens and further from personal observations on living material (both subimagos and imagos) during the annual 'swarming' period at Sparrow Lake, Ont., in the latter half of June, 1925, I am convinced that there are a number of good species in this genus, closely related, it is true, but well separable, partially on male genital characters and also on color pattern of the abdomen, size of the eyes, etc.; none of these features varies to any appreciable extent in any given species and Needham's so-called intergradients are in reality good species.

Traver described seven additional species in 1931, another in 1935, and a ninth in 1937. After studying Walker's types in the British Museum, Spieth (1940: 327-30) reduced *occulta* and *viridescens* to the status of subspecies of *limbata*. In 1941, Spieth (1941: 233-80) revised the genus and listed fourteen species and subspecies of *Hexagenia* occurring in North America north of Mexico.

The genus is generally distributed over North America but is poorly known on the Pacific Coast, as only a single species has been recorded from that region (Upholt, 1937). *Hexagenia* is also Neotropical and African, but its greatest development seems to be in the Nearctic.

Four species and subspecies of *Hexagenia* are known to occur in Florida, and their combined ranges cover much of the state. None of them is known to occur in the Everglades, and this may possibly be explained by the absence there of suitable streams or sand-bottomed lakes.

The classification of the species occurring in Florida follows Spieth's revision; however, additional study of the *munda* complex is necessary before a satisfactory taxonomic grouping of the genus *Hexagenia* in Florida can be accomplished.

Hexagenia and *Ephemera* are almost certainly close relatives, and they were probably only recently separated from each other. Spieth (1933: 350) has suggested that there are three distinct stocks in the family Ephemeridae as it is known in North America and that *Hexagenia* and *Ephemera,* which are closely related, represent one. He asserts:

. . . *Campsurus* represents another stock. *Pentagenia* and *Polymitarcys [Ephoron]* are close relatives and represent still another stock, although they are more distinct from each other than *Ephemera* and *Hexagenia* are from one another. *Potamanthus* stands as an intermediate between the other Ephemeridae and the Leptophlebiidae. . . .

Hexagenia munda orlando Traver

TAXONOMY.—Spieth (1941: 264-66), after studying a series of mayflies from Florida, concluded that *H. orlando* as described by Traver should be considered a subspecies of *H. munda.* Specimens from High Springs (obviously the Santa Fe River) and from Rock Bluff (on the Apalachicola River) were recorded as intergrades of *orlando* and *elegans.* Other specimens which were before him at the time of his work were apparently typical *orlando* from lakes. My specimens from localities near those from which he records intergrades are not intergrades, but true *elegans.* I have no records of the species other than from lakes.

H. munda orlando can be separated from other Florida species of *Hexagenia* by its color pattern, its smaller size, and the fact that it is confined to the sand-bottomed lakes of the Central Highlands of Florida.

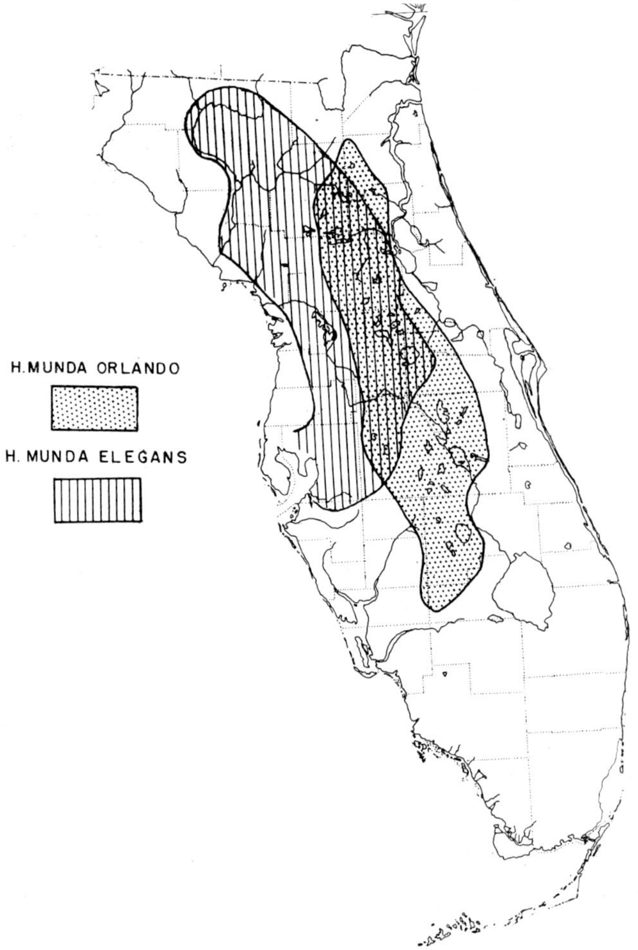

Map 5.—The distribution of *Hexagenia munda orlando* and *H. munda elegans* in Florida.

DISTRIBUTION.—*H. munda orlando*, which is endemic to the Central Highlands of Florida, is confined to lakes. Lake Harney, from which Eaton's specimens of "*variabilis*"* came, is on the edge of the lake region, and is an enlargement of the St. Johns River not far from its headwaters. I have been unable to visit this lake and have no specimens from it; however, circumstantial evidence leads me to believe that the *Hexagenia* occurring in Lake Harney must certainly be the species here named *H. munda orlando* (map 5).

* Eaton described *H. variabilis* from adults taken at Utah Lake; Texas; Florida; St. Louis, Mo.; Galena and Rock Island, Ill.; Detroit, Mich.; New England; Philadelphia, Pa.; and Washington, D. C.

ECOLOGY.—*H. munda orlando* is one of the most ecologically limited species found in Florida. The nymphs burrow in the sand and silt of the lake bottoms. The only method by which the nymphs have ever been taken has been through the use of an Ekmann dredge, because they live in water ranging from nine to thirty feet in depth. The heaviest concentration of the immatures apparently lies in the region from eighteen to thirty feet deep, but a few nymphs have been collected from the deepest parts of most of the lakes which were sampled. One exceptional specimen was even taken at a depth of forty-five feet, the deepest spot that could be located in Kingsley Lake. There was some evidence of thermal stratification in the deeper parts of this lake during August, and no nymphs could be found below this zone; however, immatures were taken closer to shore. That part of the lake bottom in which *Hexagenia* nymphs are most common is sandy, but is usually overlaid with a very thin layer of silt and has *Utricularia* scattered over it. In the deeper portions of some of the shallow, sand-bottomed lakes, there is a deposit of rather slimy, but firm, black mud covering the sand bottom. Nymphs are frequently collected from this mud, and along with them there may be brought up many *Corethra* larvae, a few annelids, and an occasional amphipod (*Eucrangonyx gracilis* Smith).

Although such large lakes as Newnan's and Orange are very interesting in themselves, they are notable for the absence of burrowing mayflies. Only the more tolerant of pond forms thrive in them. Newnan's Lake, near Gainesville, has a bottom deposit of silt which is thick, loose, and fluffy, and in which very few organisms exist. The tremendous amount of silt apparently makes it impossible for those bottom organisms requiring a moderate supply of oxygen to remain alive. The bottom receives a continuous and large supply of dead vegetation, since lakes of this type have many floating plants—water hyacinths, and to a lesser extent water lettuce, which drift about at the mercy of the winds—and sometimes an abundance of spatterdock extending out to a depth of as much as seven feet.

Immatures have not been collected from the lakes of the southern part of the Central Highlands, but adults from these lakes are not different from those of the northern part of this region. A superficial examination of the southern lakes indicates that they are identical with the sand-bottomed type of the more northern area.

The lake bottoms must constitute an ideal habitat, because enormous numbers of individuals live in them as burrowers. Various people have reported that when *H. munda orlando* is emerging, the adults become pests, and that they pile up under street lights and have to be swept from the streets. I have seen them congregating in such numbers that the sides of a bathhouse were almost hidden by the insects—nearly every blade of grass supported several individuals. The mayflies, when disturbed, fluttered upward in great masses to settle on the nearest support, and even clung to body and clothing. In every depression on the sandy beach, dead mayflies were accumulated several layers deep. Along the edge of the water, there was a row, five to six inches wide,

PLATE III

Hexagenia munda orlando Traver, male imago.

of dead adults and nymphal exuviae, which had been washed up. Many nymphal skins, as well as dead adults, were found floating on the surface of the lake. Persons who have observed the emergences of the Florida species and those of the Great Lakes (*Hexagenia limbata occulta* and *H. rigida)* say that the actual number of individuals emerging from the Florida lakes is considerably smaller in proportion.

Ephemera simulans (?) is the only species of mayfly associated with *H. munda orlando* as a burrower in the lakes of Florida. Other mayflies found along the margins of some of the sand-bottomed lakes are *Ephemerella trilineata, Choroterpes hubbelli, Stenonema proximum,* and, questionably, S. *exiguum.*

SEASONS.—The greater part of the emergence takes place during the summer months, particularly in June, July, and August; however, I have adults taken from April through November and a single mature nymph, collected on February 26, which would have emerged in March. Mr. O. E. Frye has informed me that he saw three individuals at Winter Haven on March 5. It seems likely that the species emerges during the spring, summer, and autumn, but that the winter temperatures are sufficiently low to depress the metabolic rates to such an extent that emergence does not occur during the cold months. There is some likelihood that in the southern part of its range, the species transforms during the entire nine months mentioned above. The sexual stage of *H. munda orlando* thus differs from that of many other mayflies in Florida in being seasonal, and in having the season spread over a very long period.

HABITS.—The burrowing habit is very strongly developed in the nymphs, and all bodily structure is adapted to suit the immatures to a hypogeic existence. The gills present an enormous aerating surface, the legs are modified for digging and casting out the soil as it is passed back, and the incisors of the mandibles are elongated into tusks as accessory digging organs and for lifting the roof of the burrow.

The gill movements of *H. munda orlando* appear to be the same as those of *H. recurvata,* which Morgan and Grierson (1932: 233) described as follows:

> Waves of motion pass backward over the gills sometimes too rapidly for the eye to follow but at other times in listless ripples. The gills may cease moving altogether, usually for a few seconds or for 1 or 2 minutes, though at times they may be quiet for half an hour. The motions vary with those of the nymph's body and also with the temperature and chemical content of the water. When resting, the gills are held almost upright or bent slightly backward. When waving, each pair bends backward, the filaments of the opposite gills touch each other, instantly separating as the gills are pulled upward, the whole performance resembling the rapid opening and closing of a V. So far as observed, the gills move whenever the nymph is active, especially when it burrows. While digging, the nymph holds its front legs closely against its wedgeshaped head, then lunges forward, at the same time pushing these legs outward. Almost synchronously with this, its middle legs are pushed outward and backward and the hind legs backward. As its body wedges forward, the nymph fans the silt with its gills, opening and closing them in the V-shaped angle. Thus, their surfaces are cleared of the mud thrown back by the hind legs, and the water is continually circulated about them.

Hexagenia munda orlando nymphs are rather adept swimmers. When they are placed in a dish of water in which there is no silt or sand for burrowing, they will swim around vigorously seeking to escape and continually butting their heads against the sides of the container. The burrowing habit is so strongly developed that, under these conditions, small nymphs will seek larger individuals and attempt to burrow under them. Swimming is accomplished by wave-like motions of the abdomen and up and down sweeps of the caudal filaments. The gills seem to play little or no part in this activity. Morgan and Grierson (1932: 235), experimenting with *Hexagenia* nymphs from which the gills had been removed, found that the gill-less individuals swam almost as well as, but more slowly than, those with gills.

The method of feeding is similar to that of an earthworm, in that the nymphs feed as they burrow. Examination of the digestive tract of immatures showed the presence of a few diatoms and some cells which seemed to be algae, but for the most part the contents were not determinable. Surprisingly, there were no sand grains in any of the digestive tracts examined.

LIFE HISTORY.—I have not determined the length of the nymphal period, but from emergence data it would seem to be approximately one year. On the basis of statistical measurements, Neave (1932) estimated that *H. limbata occulta* took two years to go through its life history in Lake Winnipeg. After studying *Ephemera simulans* and *H. limbata occulta* in Lake Wawasee, Indiana, Spieth (1936 and 1938) concluded, also on the basis of measurements, that both species take only one year for development. He subsequently revised his opinion, and now (verbal communication) believes a two-year period more likely. He notes (1938a: 32) that there is "no reason why a species in the northern part of its range might not take two years to mature, while in the southern part of its range one year would be sufficient."

Since temperatures in the lakes of Florida seldom, or never, become low enough to stop growth completely, it would not be surprising to find that *H. munda orlando* completes its development within one year. Collections made in early spring include rather small, half-grown, and almost mature nymphs, as well as intermediates. Since growth rate is a function of temperature, and since Spieth's work (1938a) at Lake Wawasee shows that some nymphs of *H. limbata occulta* grew as much as 16 millimeters in eighty-eight days during the summer and early fall, it would certainly be reasonable to expect that all of these nymphs, in the warm Florida waters, would have emerged sometime during the same year. Nymphs collected in autumn ranged in age from very young to two-thirds grown individuals. No mature specimens were taken. The larger individuals of autumn would probably have emerged in spring or early summer of the following year; the half-grown specimens would appear in summer, and the very young nymphs would have emerged in late summer or early autumn of the next year.

There is a definite brooding of *H. munda orlando.* During the summer, the broods may sometimes be separated by not more than a week or ten days; at other times, the intervening period may be much longer. In the spring and early summer the emerging broods are smaller, on an average, than they are during July and August, and they decrease in size in September and early fall until emergence stops completely. There may be a few isolated individuals on the wing during interemergence periods but they are scarce and difficult to find.

Just prior to emergence, the nymph, ordinarily negatively phototropic and positively geotropic, reverses these tropisms and swims from the lake bottom to the surface, where it breaks through the surface film and the adult immediately bursts free. The winged insect rests a moment, and, if not taken by a fish or bird, flies to shore where it alights on the nearest available support. At the time of emergence, the predators become extremely active and gorge themselves on nymphs and subimagoes. I have observed subimagoes "popping" out of their nymphal skins at the surface, and many bass and other fish voraciously striking at them before they had an opportunity to fly away. At the same time, birds were sitting on the trees and bushes along the lake shore eagerly awaiting the advent of the insects, and whenever one approached a bird immediately seized it.

Subimagoes taken into the laboratory molted by the next evening after a resting period of about twenty-four hours. A single female reared in the laboratory emerged in the late afternoon on August 20, and underwent its subimaginal molt between 9:00 P.M. and 10:00 P.M., August 21. During the afternoon of August 23, the insect died after a life in the winged stage of about seventy hours.

I have not observed the mating flight of *H. munda orlando,* but on one occasion I did notice restlessness on the part of a large group of these insects about thirty minutes before sunset. I have been told that the species forms large swarms which fly out over the lakes just about sunset, all at the same height (approximately thirty feet).

Duration of the egg stage has not been determined for *H. munda orlando.* According to Clemens (1922: 78), eggs of *H. recurvata* hatched in the laboratory fourteen days after oviposition; Wiebe (1926) found that artificially inseminated eggs hatched in the laboratory in nine days. Spieth (1938a) hatched eggs of *H. limbata occulta* by placing them in jars which were immersed in a small stream. These eggs took twenty days to develop. At the same time, another group of eggs which was kept in the laboratory hatched in only fifteen days.

LOCALITY RECORDS.—Alachua County: Santa Fe Lake, April 2, 1936, nymphs; April 7, 1937, nymphs. Clay County: Kingsley Lake, May 19, 1933, nymphs; April 6, 1934, nymphs; July 1, 1938, adults; July 2, 1938, adults; August 7, 1938, nymphs; August 10, 1938, adults; August 20, 1938, adults. Lake Geneva, June 8, 1938, adults; June 25, 1938, adults; July 2, 1938, adults; August 20, 1938, adults. Highlands County: Child's Crossing, August 11, 1938, adults; Sebring, May 12, 1939, adults. Avon Park, May 4, 1940, adults. Lake County: Beekman Lake, May 25, 1933, nymphs; Sellar's Lake, May 25, 1933, nymphs; Umatilla, October

1, 1938, adults. Marion County: Swim Pond in Ocala National Forest, February 26, 1938, nymphs; Lake Bryant, April 16, 1938, nymphs. Near Eureka, September 7, 1938, adults. Orange County: Winter Park, June 5, 1939, adults. Osceola County: East Tohopekaliga Lake, August 7, 1939, adults. Polk County: Auburndale, June 8, 1939, adults. Lakeland, July 26, 1938, adults; September 8, 1939, adults. Winter Haven, July 25, 1938, adults; April 12, 1939, adults; April 15, 1939, adults; September 1, 1940, adults; November 2, 1940, adults. Putnam County: Sunnyside Beach, July 23, 1938, adults. Volusia County: Lake Dias, June 12, 1947, adults.

Hexagenia munda marilandica Traver

TAXONOMY.—In his revision of the genus *Hexagenia*, Spieth (1941: 261-63) recorded *H. munda marilandica* from Chipola Lake (probably the Dead Lakes of the Chipola River) and from Rock Bluff (on the Apalachicola River). Prior to this work, I had considered the species to be *H. weewa*, which Spieth synonymized with *H. munda elegans*. *Marilandica* can be more easily differentiated from *elegans* by the color pattern of the females than by that of the males. The middorsal line of *marilandica* is not so prominent as that of *elegans* and may be obsolescent or absent on the anterior abdominal segments; the oblique lateral lines on the dorsum of the abdomen of *marilandica* are weaker than the corresponding ones on *elegans;* and the body of *marilandica* is a lighter yellow than that of *elegans*. *Marilandica* specimens are, on the average, somewhat larger than those of *elegans*.

DISTRIBUTION.—Spieth (1941: 262-63) has recorded *marilandica* from Alabama, District of Columbia, Florida, Georgia, Maryland, North and South Carolina, New Jersey, and Virginia. In regard to the subspecies, he states that it "occupies a narrow and elongated range along the eastern side of the southeastern highlands, extending in small numbers down into the lower elevations." He examined intergrades of *marilandica* and *elegans* from North Carolina, from the District of Columbia, and from several places in Georgia. One specimen from Dalton, Georgia, appeared to be an intergrade between *munda* and *marilandica*.

Nymphs which I am placing, with reservations, as *marilandica* have been collected from the streams of the Tallahassee Hills region. *H. munda marilandica* immatures are known from the tributaries of the Apalachicola River, both on the eastern and western sides of this stream; nymphs and adults have also been collected as far west as the drainage of the Yellow River in Okaloosa County. The subspecies is known with certainty only from the western part of the state, and, in this region, is confined principally to the Apalachicola River system (map 6).

ECOLOGY.—The immatures are burrowers in the silt of stream margins. It is easy to locate the nymphs because they leave small round openings leading into their burrows. If there is a rather large deposit of silt in quiet water, there may be many of these openings over a comparatively small area. Holmes Creek, in western Florida, offers ideal conditions for the development of the nymphs, and it was at this stream that I found them more common than at any

PLATE IV

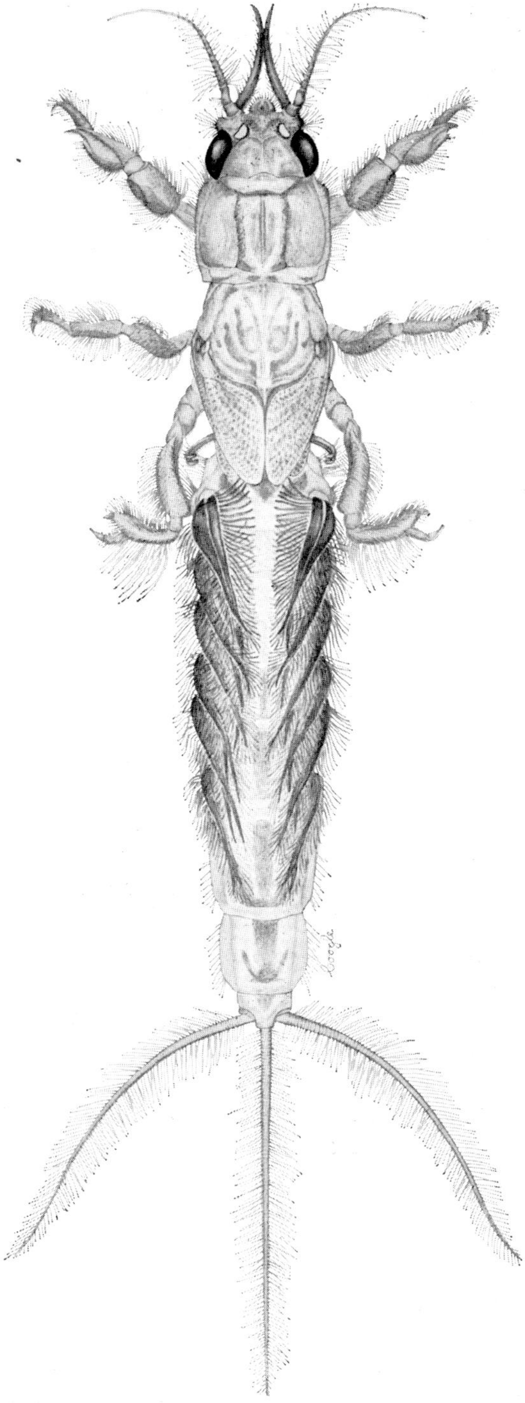

Hexagenia munda marilandica Traver, nymph.

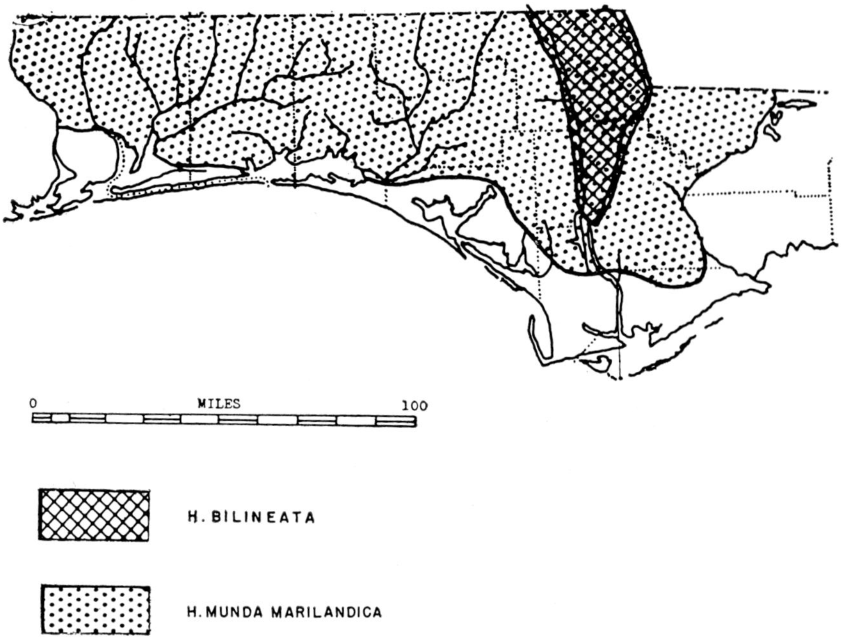

Map 6.—The distribution of *Hexagenia bilineata* and *H. munda marilandica* in northwest Florida.

other locality from which the insects have been collected. Here the water flows slowly over a very silty bottom. Vegetation grows profusely in the stream bed, but in places where the current is slightly more rapid, the floor is bare except for the silt. At the time of collecting, the water was quite low, not more than a few inches in depth, and burrows were located without trouble from the bridge over the stream. The nymphs build burrows which descend only a very short distance so that a shallow scoop of the Needham apron net easily removed the insects.

The nymphs were found burrowing in large numbers in the banks of sand-bottomed streams in Gadsden and Liberty counties. The beds of these streams are composed mostly of clean-swept, yellow sand, but deposits of reddish clay mixed with silt are formed at the edges and the nymphs burrow in this material.

Stream size is probably not a limiting factor because in the Tallahassee Hills region nymphs were taken from a small stream not more than three feet across; yet the immatures must be present in prodigious numbers in the Apalachicola and Chipola rivers if the reports of emergences which have reached me bear any semblance of truth.

I have been told that when the "willow flies" come out, they cover everything, forming huge piles under lights and making pests of themselves. When the adults are out in numbers along the Apalachicola, they cover the willow trees which line the banks of the river; hence the name "willow fly." I have been told that the bass lie in wait under the willows, and as the insects accidentally fall from their perch onto the water surface, the fish immediately seize them. At this time all that one needs to do in order to catch a "mess of bass" is cast under the willows; the bait is sure to be taken. The people of this region eagerly look forward to the arrival of the flies for their best fishing.

Professor Needham's description of these insects is well worth repeating. Traver (1931: 607) quotes from Dr. Needham's field notes:

> . . . The capture of *Hexagenia weewa [H. munda marilandica]* occurred on this wise. Prof. C. R. Crosby and I were passengers on the Apalachicola River steamboat *John C. Callahan, Jr.* On the evening of the eighth of April, this big boat ventured up the Chipola River above Dead Lake to Cotton Bluff to take on some barrels of turpentine. Near Cotton Bluff the boat got stuck in one of the bends of that crooked little stream just at nightfall. The searchlight was turned on, and in the beam of light that it cast upstream the mayflies rose above the black waters by thousands. Along with multitudes of midges and a few other mayflies of the genus Ephemerella, the big yellow female Hexagenias came fluttering up to the light. They settled all over the front of the boat, two or three layers deep wherever there was support. They flew in our faces and clung to our clothing. One could gather them up by barrelsful. But among the many thousands of females, only two or three males were seen.

The nymphs are very hardy. I brought twelve of them a distance of three hundred miles in two small vials filled with moist silt, and these individuals were not removed from the bottles during a period of more than twenty-four hours. Upon my return to Gainesville, every nymph was found to be alive and apparently in good health. Moreover, while I was away from Gainesville for two weeks, the water evaporated from the pans containing the nymphs until the surface of the mud was completely dry. When water was added to the mud and it was stirred, out swam the nymphs none the worse for the desiccation.

Some of the burrowing dragonfly larvae are associated with *marilandica* in the silt of the stream bed, but the only other mayflies are *Caenis hilaris, C. diminuta,* and *Brachycercus* sp. A, which do not burrow but frequently live on top of the silt or silty sand.

SEASONS.—The specimens Needham collected were taken on April 8, 1927. I have adults captured in March, April, June, July, and August, as well as some specimens which emerged in the laboratory during September. A single nymph, which was rather close to maturity, was collected in late November and would probably have emerged early the following spring. I have little doubt that *marilandica* is a seasonal species emerging over a period of about eight months. There is definite brooding, but I have been unable to obtain additional data about this characteristic. Nymphs collected at various times of the year fall into distinct size groups, and there is much overlapping of young with old nymphs.

LIFE HISTORY.—It is likely that the developmental period of *marilandica* is one year, although there is little definite information to support this statement. There is fairly good evidence to indicate that *elegans* needs but one year to complete its life history, and because of the close similarity between *marilandica* and *elegans,* both structurally and ecologically, it would seem probable that *marilandica* also can develop within a year.

Subimagoes, in the laboratory and in the field, normally emerge in late afternoon, just about dusk, but I have observed a single female emerge in the field about 10:00 A.M. After a resting period of fourteen to twenty hours, the final molt occurs, and this is followed by mating if conditions are propitious.

I have observed several mating flights of this species, and they are usually composed of from thirty to seventy-five males. The flight takes place just at dusk and on succeeding nights can be timed within five minutes of its occurrence. The males fly at a height of about twenty to twenty-five feet above the ground and usually over vegetation consisting chiefly of willow trees. As the flight begins, one or two males will begin moving up and down in the typical manner, and almost immediately they are joined by others until the swarm soon reaches its maximum size. The up-and-down flight varies considerably, but as a rule appears to be about five or six feet. The upsweep is accomplished with the caudal filaments depressed and held close together and with the wings beating rapidly; on the downsweep, the filaments are raised and separated, the wings, which are outspread, do not beat, and the insect merely floats downward on opened wings. Each male appears to choose its place in the swarm and maintain its position until it leaves the flight to mate. I have never seen females enter the group of swarming males, but I have observed the coupled pairs flying at about the same height as the swarm and off to one side. In every case where I could follow the flight, the pair separated after a few seconds: the male returned to the swarm, and the female disappeared into the vegetation. The flight of males lasts about fifteen minutes and gradually disperses just about dark.

At the same time that the males are flying, females can be observed flying in swarms of a size about equal to those of the males. Instead of being an up-and-down flight, theirs is horizontal back and forth, and they fly about ten feet below, and away from, the males. The flight of females continues until after dark. It was interesting to note that on two successive evenings, while I stood on a bridge and watched flights of females below me, three or four at a time would leave the swarm and fly over the bridge at precisely the same spot. For some reason, they chose the southwest corner of the bridge, just over a projecting willow branch. I was able to stand at this point and collect a large number of the females as they left the swarm. The flight of the females was crosswise to the flow of the stream, while that of the males was parallel to the flow. Females leaving the swarm had evidently not oviposited, for they frequently deposited their ova into the net when captured. Sometimes two large clusters of orange-colored eggs were protruding from the abdomen also indicating that oviposition had not occurred.

LOCALITY RECORDS.—Calhoun County: Chipola River, November 5, 1938, nymphs. Gadsden County: 15.1 miles east of Chattahoochee, April 1, 1938, nymphs. 4.5 miles south of River Junction, March 17, 1939, nymphs; June 30, 1939, adults. 10 miles south of River Junction, July 1, 1939, nymphs. Holmes

County: Sandy Creek, July 2, 1939, nymphs. Jackson County: 3.6 miles north of Altha, July 1, 1939, nymphs. Blue Springs Creek, May 3, 1941, adults. Liberty County: Hosford, March 17, 1939, nymphs; November 30, 1939, nymphs. Sweetwater Creek, May 7, 1933, adults; June 10, 1938, nymphs; November 4, 1938, nymphs; December 1, 1939, nymphs; May 2, 1941, nymphs; April 29, 1946, adults. Okaloosa County: Near the Shoal River, December 12, 1937, nymphs. 1 mile east of Crestview, December 12, 1937, nymphs. Walton County: 2 miles west of Freeport, April 3, 1938, nymphs. 15.8 miles west of Ebro, June 7, 1938, adults. 2.1 miles west of Walton County line, May 31, 1940, nymphs and adults. Washington County: Ebro, May 30, 1940, adults. Holmes Creek, April 2, 1938, nymphs and adults; June 9, 1938, adults; July 2, 1939, nymphs and adults; adults reared July 7 - September 11, 1939; April 30, 1946, adults.

Hexagenia munda elegans Traver

TAXONOMY.—*H. munda elegans* was placed as a subspecies of *munda* by Spieth (1941: 259-61), and *H. weewa* Traver and *H. kanuga* Traver were considered as synonyms of *elegans*. *Elegans* can be separated from *marilandica*, which it resembles more closely than it does *orlando*, by the characters described under *marilandica*. Spieth, after examining specimens taken at High Springs, Florida (at the Santa Fe River), stated that *elegans* and *orlando* intergraded there. I have never taken any specimens of *orlando* other than from lakes and my specimens from the Santa Fe River are definitely of the *elegans* type.

DISTRIBUTION.—*H. munda elegans*, in Florida, has been taken from as far south as Hillsborough County and as far west as Chipola Lake (Dead Lakes on the Chipola River). The western record is that of Spieth (1941: 260). My collections indicate that *marilandica* replaces *elegans* in western Florida. *Elegans* ranges from Maryland to Texas and Oklahoma, chiefly occurring along the coastal plains and inland to the highlands (map 5).

ECOLOGY.—The ecological relationships of *elegans* are the same as those of *marilandica*. In January, 1940, I found nymphs of a burrowing species in Hogtown Creek, about one mile west of Gainesville. Previously, I had not taken burrowing nymphs nearer than the Santa Fe River. Only forty-five of the insects were collected in about three hours, for they were quite scarce. These were brought into the laboratory where they were put into a pan with a layer of silt about one-half inch in depth. A few days later, I returned to the creek and collected twenty more immatures, which were placed in cages submerged in the stream; however, Hogtown Creek became heavily contaminated by the wastes from a chemical plant, and nearly all the insects, as well as fish and other animals, were killed. Two males, which are typical *elegans*, were reared from the insects kept in the laboratory.

The nymphs proved to be very hardy under laboratory conditions. In the early stages of rearing, there were forty-five nymphs confined in a rather small pan; at the end of one month the mortality was found to be very low, only five specimens having died. This low mortality rate continued, with scattered emergences, until summer, when nearly all of the remaining nymphs reached the last instar. For some unknown reason, these were unable to transform.

All of the nymphs seemed undersized and the two imagoes that were reared were certainly dwarfed, probably because of limited quarters and insufficient food.

Associates of *elegans* are necessarily less numerous than those of *marilandica,* since not so many mayfly species occur in the region where the former species is present. In Hogtown Creek, *Caenis diminuta* was very frequently encountered among the dead leaves covering the silt in which *Hexagenia* nymphs burrowed. That part of the Santa Fe River inhabited by the burrowers has relatively few mayflies, and these are predominantly *Baetis spinosus, Stenonema smithae,* and *Tricorythodes albilineatus.*

SEASONS.—Adults have been either collected or reared in March, April, May, June, and August. Among the nymphs that I attempted to rear, two reached their last instar in August, but did not transform to subimagoes. The data from the Hogtown Creek specimens point to an early summer emergence as the predominating period for the species. On May 1, many of the nymphs were in the tertiultimate, penultimate, or last instar, and by the first part of July most of them had died.

HABITS.—The habits of *elegans* are the same as those of *orlando* except for choice of habitat.

LIFE HISTORY.—Nymphal development probably takes one year. There is no definite evidence to support this statement, but nymphal growth rate seems to point to such a conclusion.

Some indication of the rate of growth under rigorous conditions is shown in Table 3 which is based on measurements of laboratory-reared nymphs collected from Hogtown Creek. An attempt was made to rear nymphs in the stream as a control, but because of the contamination mentioned above, the check failed.

TABLE 3

GROWTH OF NYMPHS OF *Hexagenia munda elegans*

Date Measured	No. of Specimens Measured	Smallest Nymph	Largest Nymph	Average
February 2	32	6.5 mm.	25 mm.	13.6 mm.
May 1	31	9.5 mm.	23 mm.	16.3 mm.
June 19	19	9.0 mm.	22 mm.	17.5 mm.

The male nymphs, just before transformation, ranged from 17.5-19 millimeters in length. The larger nymphs were females.

Emergence in the laboratory occurs in late afternoon, and the subimaginal molt takes place from fifteen to twenty hours later.

I have not observed the mating flight of *elegans,* nor has it been described; I believe, however, that it would be identical with that of *marilandica.*

LOCALITY RECORDS.—Alachua County: Hogtown Creek, January 17, 1940, nymphs; adults reared June 9 and 10, 1940; May 16, 1947, nymphs. Santa Fe River, February 18, 1939, nymphs; February 28, 1939, nymphs; adults reared April 25, 1939; April 6, 1940, nymphs; adults reared August 11, 1940. Gilchrist County: Suwannee River, May 7, 1939, adults. Hillsborough County: Hillsborough River, June 18, 1939, adults. Holmes County: Sandy Creek, December 14, 1939, nymphs. Levy County: Waccasassa River, May 11, 1947, nymphs. Marion County: Withlacoochee River, May 13, 1938, adults; June 6, 1939, adults. Oklawaha River, April 15, 1939, adults; April 10, 1948, adults. Rainbow Springs, July 29, 1940, adults. Putnam County: St. Johns River at Welaka, August 26, 1941, adults.

Hexagenia bilineata (Say)

TAXONOMY.—The taxonomy of *H. bilineata* is probably in better shape at present than that of the majority of other species of *Hexagenia.* In 1920, Needham decided to apply this name to "all variants of the species that occupies the beds of our larger lakes and streams." McDunnough (1927: 117) strongly disagreed with Needham and restored the species which the latter had synonymized. Although many of the present species of *Hexagenia* are considered by Spieth to be subspecies of *limbata* and *munda, bilineata* is so distinct that there can be no doubt as to its specific status.

While collecting at a light in front of an ice house facing the dammed-up part of Blue Springs Creek near Marianna, I noted a large mayfly circling about the lights. With the aid of Dr. Horton Hobbs, I captured the insect and immediately recognized it as a species new for Florida. Before the evening was over, nine individuals, all females, were taken. Thus the identification of the species, while almost certain, still lacks the final verification of the male. These females fit the description of *bilineata* fairly well, but differ in minor details.

DISTRIBUTION.—The species was originally described from Minnesota by Say. Walsh recorded *bilineata* from Illinois; Eaton, from Louisiana and Texas; and McDunnough mentioned that it seems to be confined to the Mississippi River and its tributaries. Other localities given by Traver include: Fairport, Iowa; Lucedale, Mississippi; Waco, Texas; Knoxville, Tennessee; St. Cloud, Minnesota; Gloverport and Lexington, Kentucky; and Rome, Marietta, Atlanta, and the Tombigbee River, Georgia. Her records for Alabama are from Gadsden, Tuscaloosa, Wilson Lake at Muscle Shoals, and from a river located between Birmingham and Decatur. Spieth also records the species from Washington, D. C., Indiana, Maryland, Missouri, New Mexico, Ohio, Oklahoma, and Virginia.

The Florida specimens were taken from a tributary of the Chipola River, which in turn flows into the Apalachicola River. The Chattahoochee River, which flows through Atlanta, is listed as a locality for *H. bilineata,* and therefore it does not seem at all remarkable that the species should occur in other tributaries of the Apalachicola River drainage system (map 6).

ECOLOGY.—The nymphs are known to be burrowers in the silt of the larger lakes and streams. No immatures have been taken in Florida which I can recognize as the nymph of *bilineata.* I rather suspect that, at the place where the adults were collected, the nymphs live in the silt of the artificial lake formed by the damming of Blue Springs Creek. Spieth (1941: 244) states that "the nymph apparently dwells in the deeper waters and consequently is rare in collections."

SEASONS.—Nothing is known of the seasonal habits of the species in Florida, except that the adults were taken in early June. In other parts of its range, Spieth reports that the imagoes emerge from the middle of June to the middle of September, and the peak of emergence seems to occur in August. Eaton suggests that it emerges from June (in the South from May) to September.

HABITS AND LIFE HISTORY.—Needham (1920: 278-81) has discussed the habits, life history, and mating flight of *H. bilineata* in his paper on the burrowing mayflies of larger lakes and streams.

LOCALITY RECORDS.—Jackson County: Blue Springs Creek near Marianna, June 5, 1940, adults.

Genus EPHEMERA Linnaeus

Ephemera simulans Walker (?)

TAXONOMY.—Approximately sixty nymphs of a species of *Ephemera* were collected by Professor J. S. Rogers from Kingsley Lake in May, 1935; subsequent efforts to obtain additional material have been unsuccessful. Some of the nymphs sent to Dr. Spieth were tentatively identified by him as *Ephemera simulans* Walker. Until Florida males are known, this identification must stand, but it is quite likely that an undescribed species is represented.

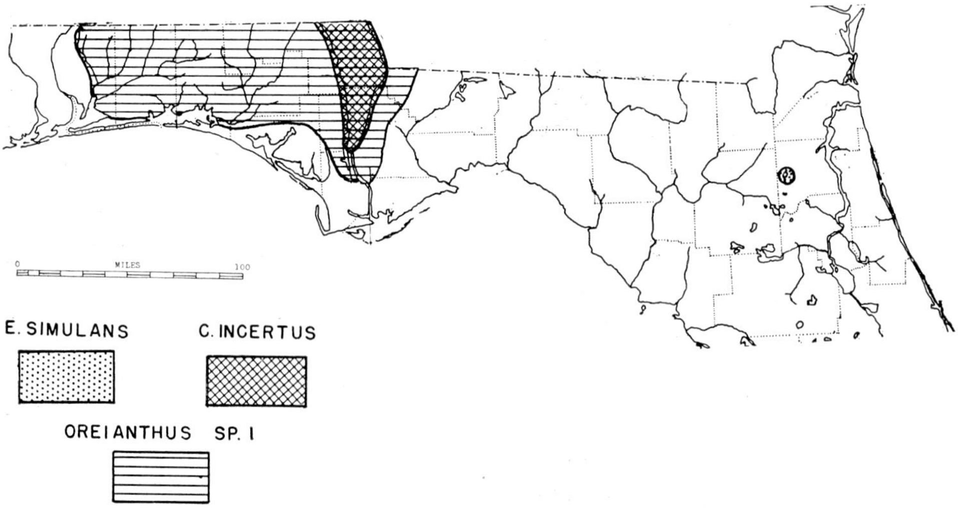

Map 7.—The distribution of *Ephemera simulans* (?), *Campsurus incertus*, and *Oreianthus* sp. No. 1 in Florida.

DISTRIBUTION.—No species of *Ephemera* has previously been recorded from the Coastal Plain or even the Piedmont, although there is a single record of *Ephemera* sp. taken in an airplane towing at Tallulah, Louisiana, at a height of two thousand feet. The genus is rather common in the Appalachians, where the species mostly inhabit streams. *E. simulans* is known from Ontario, several of the Northwestern states, the Midwest, and throughout the Northeast. Traver states that the species is usually found in lakes and the larger rivers (map 7).

ECOLOGY.—Although little is known of the ecology of the Florida *Ephemera*, there has been much published concerning *E. simulans* as it occurs elsewhere in North America. Needham (1920: 283) summarized its ecological habits as follows: "The 'mackerels' are lacustrine rather than fluviatile in habitat, especially *E. simulans*." The Florida nymphs were taken in an Ekmann dredge from the lake bottom, but whether from sand or silt is uncertain. It has been said that the nymphs are burrowers in the silt, but judging from the habits of the related *Hexagenia munda orlando* in Kingsley Lake, I believe that *E. simulans* probably inhabits the more sandy areas of the bottom.

H. munda orlando, a burrowing mayfly, is the only other species found in the same habitat with *E. simulans* in Kingsley Lake.

SEASONS.—The Florida nymphs are nearly full grown, approximately in the tertiultimate instar. Emergence is probably limited to a short period during the summer, for these nearly mature nymphs, collected during May, are nearly all of the same size.

LIFE HISTORY.—Spieth (1936) has studied the life history of *E. simulans* in Lake Wawasee, Indiana. He found that in this lake it is a burrowing form restricted to the littoral area, and he concluded that it takes only one year to complete its life cycle. The Florida *Ephemera* is not likely to require a longer period than the Northern form to reach maturity; on the other hand, like the remaining burrowers found in this state, it probably lives one year as a nymph instead of having a shortened developmental period such as characterizes many other Florida mayflies. Ide (1935: 436-46) described the first thirteen and the last seven instars in Canadian material of *E. simulans*, but he did not determine the total number of instars through which the species goes.

Mating flights were well described by Needham in 1908 and 1920. In 1913 Clemens described the flight and the copulatory act (pp. 332-33), which does not differ essentially from that of other mayflies. He did, however, notice an interesting peculiarity, namely, "that the male *Ephemera* frequently attempted copulation with the male *Hexagenia*, apparently being deceived by color."

LOCALITY RECORDS.—Clay County: Kingsley Lake, May, 1935, nymphs.

Genus PENTAGENIA Walsh

Pentagenia vittigera (Walsh)

TAXONOMY.—A single male adult of *Pentagenia vittigera* was collected in May, 1941, at Alum Bluff on the Apalachicola River. This single representative of the species in my collection fits the description of the species reasonably well.

DISTRIBUTION.—In common with *Hexagenia bilineata,* this species is widely distributed and has been recorded from Illinois, Kansas, Iowa, Texas, Tennessee, Minnesota, Arkansas, and northwestern Alabama. The Florida record from the Apalachicola River is not unexpected in view of the facts that Traver records the species from Sheffield, Alabama (1937), on the Tennessee River, and that the tributaries of the Apalachicola come rather close to those of the Tennessee, thus making the dispersal of the species relatively easy.

ECOLOGY.—Little is known of the ecology of this species, except that it appears to be confined to the larger and deeper rivers, where the nymph burrows in the silt.

SEASONS.—The Florida adult was collected on May 3, 1941. The Mississippi specimens described by Needham (1920: 282) emerged from June 16 to September 7, with maximum occurrence in late June. Traver's specimen from Alabama was collected on July 4.

LIFE HISTORY.—There are no data available on the life history of *P. vittigera.*

LOCALITY RECORDS.—Liberty County: Alum Bluff on Apalachicola River, May 3, 1941, adults.

Genus CAMPSURUS Eaton

Campsurus is a rather unusual mayfly in that the legs (except the forelegs of the males) have become weakened and aborted so that they are no longer of any use to the adult insect as ambulatory organs. This and other distinctive features make *Campsurus* taxonomically easily separable from all other genera of North American ephemerids.

In South America, *Campsurus* is one of the predominant mayfly genera. Needham and Murphy (1924: 13) state that the genus includes nearly a fifth of the described species of the entire Neotropical region. In North America, on the other hand, only five species of *Campsurus* are known, nearly all being limited in their distribution to the Coastal Plain. *C. primus* was described from Grand Tower, Illinois, at the edge of this province; *C. decoloratus* was described from Matamoros, Mexico, just across the border from Brownsville, Texas, and its range extends northward to College Station, Texas, which is very close to the Fall Line; *C. puella* is known only from New Orleans; and *C. incertus* occurs in southern Alabama, southern Georgia, and northern Florida. *C. manitobensis* was recorded at Cartier, Manitoba, by Ide (1941).

Spieth (1933: 347-48) considers that, because of the stump-legged condition, the genus stands distinct from the other burrowing mayflies of North America and represents a separate phylogenetic branch. His arguments are clear, and seem in all respects conclusive.

The mating flight of *Campsurus segnis* in British Guiana has been picturesquely described by Dr. Ann Morgan (1929: 62-63). The anatomy and taxonomy of this species were also treated by her and by Needham and Murphy (1924: 19-20).

Campsurus incertus Traver

TAXONOMY.—For two years, I searched for nymphs or adults of *Campsurus* in Florida, because the distribution of the North American species clearly indicated that some member of the genus should certainly occur in Florida. Naturally, since *incertus* was known from the drainage of the Apalachicola River in Georgia and Alabama, it was logical to expect that it would also be found in the same drainage system in Florida, which proved to be the case. The only Florida specimens secured were taken by Mrs. Long, the proprietress of a store and filling station near the Apalachicola River at Bristol. She had previously informed me that during the summer the "willow flies" formed huge piles under the lights outside the store. I asked if she would be kind enough to drop a few of these "willow flies" into the bottles which I left with her, and departed, expecting to hear no more about the matter. Much to my surprise, two months later the bottles arrived filled with females of *C. incertus,* the same insect which all my search in the region had failed to reveal. No other Florida specimens have been seen.

When Traver described *C. incertus,* she stated that it might be synonymous with *puella,* described by Pictet in 1843 from New Orleans, and not since taken. This question is still unsettled. I have no males of *incertus* and cannot say without question that the Florida specimens are of this species; however, the color pattern is very close to that described by Traver and the distributional evidence supports this determination.

DISTRIBUTION.—*C. incertus* is known from Spring Creek and Albany, Georgia; Eufaula, Alabama; and Bristol, Florida—all in the drainage of the Apalachicola River. It is also known from Macon, Georgia, which lies in the drainage of the Ocmulgee River, at the edge of the Piedmont. The first stream drains into the Gulf of Mexico; the second, into the Atlantic Ocean (map 7).

ECOLOGY.—Aside from the fact that the nymphs of *Campsurus* are believed to be burrowers, and appear to be confined to fairly large rivers, nothing is known concerning the ecology of any species of this genus.

SEASONS.—*Campsurus incertus* emerges during the summer, according to information gathered from various people living along the Apalachicola River. The Florida specimens here recorded were taken about July 29. The insects from Georgia which Traver used in her description were collected in August and on September 4.

LIFE HISTORY.—The developmental stages are not known for any North American species. Because of the vestigial condition of the legs and the consequent fact that the winged form is entirely aerial, the subimaginal stage must necessarily be brief. Morgan (1929: 63) says:

. . . The subimago stage of *Campsurus segnis* (South American) is probably of brief duration, far different from that which is characteristic of its related *Hexagenia bilineata* and *Ephemera simulans* in which this stage lasts twenty-four hours or more. In these and in many other species this subimago period is the one in which there is a rapid elongation of the legs and caudal setae. These changes are most obvious in the males, the lengthening slenderness of whose front legs is zigzagged across the encasing subimago skin as if in a stocking much too short. No growth of the legs seemed to have occurred in these *Campsurus* subimagos.

LOCALITY RECORDS.—Liberty County: Bristol, July 29, 1939, adults.

Genus OREIANTHUS Traver

The unique mayflies belonging to this genus were described by Traver in 1931, 1935, and 1937 as *Oreianthus purpureus* and *Oreianthus* species No. 1. A third species, described by Joly as *Caenis maximus* in 1871, on the basis of nymphs from the Garonne River in France, has been assigned tentatively to *Oreianthus* by Traver (1931: 108). The relationships of the genus are particularly interesting, for while the characteristics of the adult definitely ally it with the Ephemeridae, the nymphal characters very closely parallel those of the Caeninae. However, it has been considered that the nymphal characters are secondary adaptations, and that the wing venation, which is fundamentally of the Ephemeridae type, indicates the true relationships. McDunnough's genus, *Neoephemera,* has been placed by Traver in a separate subfamily, Neoephemerinae, along with *Oreianthus.*

In the United States, this genus is known to occur in North Carolina, Tennessee, Georgia, and Florida—in the Valley and Ridge, Blue Ridge, and Piedmont provinces, and the Coastal Plain, respectively.

Phylogenetically, *Oreianthus* is rather difficult to place. In 1932, Traver suggested that "the general appearance of the imago, other than the venation, is as close to certain members of the subfamily Baetinae as to the Ephemerinae, and I think it more closely allied to the former than to the latter group." However, in subsequent publications, she has definitely placed *Oreianthus* with the Ephemeridae. Even though the nymph is very highly specialized, the genus may possibly occupy the same type of intermediate position with regard to the Ephemeridae and Baetidae that *Isonychia* occupies with regard to the Baetidae and Heptageniidae. More likely, it represents still a fourth, and entirely separate, family stock, and cannot rightfully be linked with any of the other extant families.

Oreianthus sp. No. 1 Traver

TAXONOMY.—The nymphs of *Oreianthus* sp. No. 1 were well described by Traver in her paper (1937: 83-84), entitled "Notes on the Mayflies of the Southeastern States." Her specimens were taken in the identical locality from which I have subsequently collected nymphs corresponding very closely with the original description. Traver suggested that the differences between *Oreianthus* sp. No. 1 and *O. purpureus* might be generic rather than specific. The validity of this suggestion must be tested through a study of the adults, and the adult of species No. 1 is still unknown.

At first glance, one is likely to confuse young nymphs with those of *Caenis*, but closer examination immediately distinguishes the two, for *Oreianthus* has prominent metathoracic wing pads. In the later instars *Oreianthus* nymphs are much larger than are those of *Caenis*.

DISTRIBUTION.—*Oreianthus* sp. No. 1 has been recorded from the Piedmont in Georgia, Columbus, Georgia (on the Fall Line), and from Liberty County in the Coastal Plain of Florida. The species has subsequently been collected in various streams draining into the Apalachicola River, and from Okaloosa County in the Yellow River drainage. The Florida distribution of *Oreianthus* is very limited, but the reason for this limitation is not at all clear as the species is quite tolerant (map 7).

ECOLOGY.—Although many streams in northern Florida were examined and collections made in all types of habitats in them, *Oreianthus* sp. No. 1 was seldom discovered. In the streams where it was found (those draining the clay hills along the Apalachicola), the nymphs were confined to those parts of the creeks where the flow is continuous although not necessarily rapid. Most frequently, the nymphs were collected from the roots of terrestrial plants which projected into the water. Debris and silt accumulates among the tangle of the roots, and this trash, along with the roots, furnishes an apparently ideal situation for the development of the nymphs. Only rarely do the insects appear in collections made from the cleanly washed roots in the swifter waters where *Isonychia* is so common.

Although not as numerous as in the root situation, the nymphs also live in leaf drift and among branches which become anchored rather permanently in the stream bed. However, they seem to be marginal forms and do not appear to be adapted to swifter waters.

The habitat of the Florida species is apparently different from that of *Oreianthus purpureus*. Traver (1931: 103) found the nymphs of the latter species only in swift water, at a depth of from one and one-half to two feet, where they lived beneath large flat rocks. Her observations on the same species in 1937 indicated that "the nymphs were usually found beneath small, isolated boulders or irregularly shaped sedimentary rocks in rapid water."

The spindly legs of the immatures and the presence of gill covers indicates that the Florida nymphs find their natural habitat in the silty places where they have been taken, rather than on the undersides of cleanly swept rocks in mid-

PLATE V

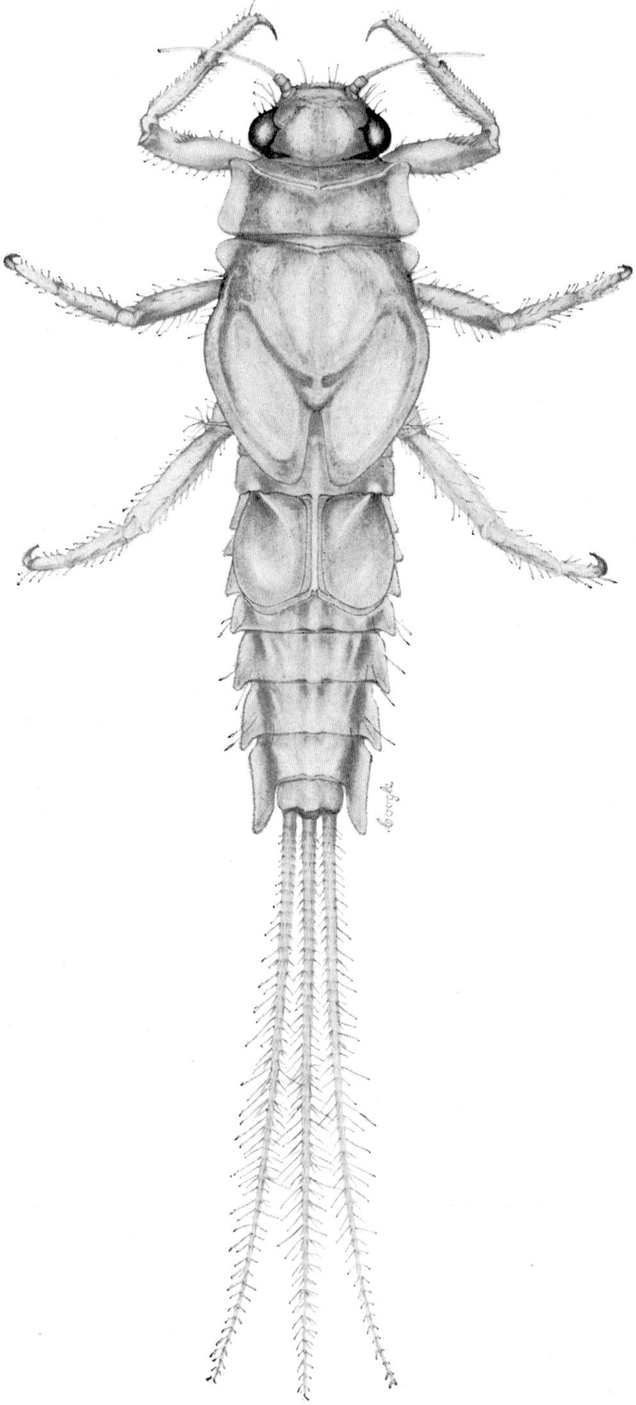

Oreianthus sp. No. 1 Traver, nymph.

stream. Frequently, large, submerged, and well-anchored logs were found in the stream bed, yet no *Oreianthus* nymphs were on them. The other situations, then, must be more acceptable to the immatures, although even here they are not particularly numerous. I have collected for several hours from the roots described above, but the catch was small, approximately twenty nymphs in all.

The nymphs are rather tolerant insects. Traver had one nymph (*O. purpureus*) which survived an automobile trip of three days, during which time the water was changed infrequently each day. Several nymphs of *Oreianthus* sp. No. 1, confined in a half-pint jar in which the water was not changed, were brought to Gainesville from Sweetwater Creek, a distance of about 250 miles. They were removed overnight to an open pan, but the following day were replaced in the jar for the remainder of the trip. In Gainesville they were kept in a pan of shallow water where they lived for a time, but, for some undetermined reason, none lived longer than twenty days in captivity. Two nymphs reached the emergence point, but died before transformation was accomplished.

The most frequently encountered associates of *Oreianthus* nymphs are *Isonychia pictipes* and *Isonychia* sp. A. Also very common in the root association are the immatures of *Caenis hilaris, Tricorythodes albilineatus, Baetis spinosus, B. australis, B. spiethi, Pseudocloeon punctiventris, Paraleptophlebia volitans, Habrophlebiodes brunneipennis,* and occasionally *Blasturus intermedius. Hexagenia munda marilandica, Baetisca rogersi,* and *Brachycercus* sp. A are likewise found in the streams which *Oreianthus* inhabits.

SEASONS.—*Oreianthus* sp. No. 1 is strictly seasonal. Emergence takes place during the spring in late March and April, and, to some extent, in May. This is stated so positively because nymphs were taken from Sweetwater Creek in November of 1938, yet in June, in the identical situation, not a single specimen could be found. In March, 1939, I again collected a number of nymphs from this creek, but in July of the same year no nymphs were taken; in early December of 1939 half-grown nymphs were found to be as plentiful as on any other occasion. In June, 1940, no specimens could be found in a stream in Gadsden County where, in March of the previous year, immatures had been collected. Again on April 29 and 30, 1946, when collections were made in Sweetwater Creek only very small nymphs were collected, and in February, 1949, nymphs, which were two-thirds grown, were the only ones found.

A study of some fifty nymphs from Florida streams substantiates the evidence offered above. Those taken in November and December are of a size which shows that subimagoes would certainly not have emerged sooner than the following March or April. Immatures taken in March were well grown or would have matured within a month. *Oreianthus* sp. No. 1 thus presents one of the most restricted seasonal ranges of any mayfly found in Florida. This fact, as has been noted previously, is in accord with other evidence suggesting a Southern origin for the species. As was there indicated, Northern species are nonseasonal in Florida, whereas Southern forms tend to have definite emergence periods.

HABITS.—The nymphs swim but little, and their motions are extremely awkward when they do. The tails are bent over the abdomen and suddenly lashed so that the movement of the abdomen, assisted by the beating of the almost bare caudal filaments, drives the insect slowly and with much difficulty through the water.

The attitude of the nymphs when taken from the water is much the same as that characterizing all those species of mayflies which have operculate gill covers. A slow, deliberate movement is coupled with an occasional flicking motion of the caudal filaments as the insect brings them completely back over its dorsum until they point anteriorly, and then lashes out with them as though it were using the filaments for propulsion. The nymphs, which walk slowly and laboriously, are ready prey for most carnivores. Their coloration and choice of habitat doubtless act as protection against the many predators which infest the streams.

Having so few specimens of *Oreianthus* sp. No. 1, I examined the alimentary tract of only one nymph. The food materials, although thoroughly macerated, appeared to consist mostly of plant epidermis with an occasional algal cell interspersed in the mass.

LIFE HISTORY.—The above comments on the seasonal distribution of *Oreianthus* sp. No. 1 indicate that the life history of the species occupies approximately one year. I was unable to rear any adults in the laboratory or to secure any in the field. Traver succeeded in rearing a few individuals of *O. purpureus*. Concerning the life history of this species, she wrote (1937: 37): "As to the time of transformation from nymph to subimago, I record my observations on those nymphs kept in the rearing cages at Penrose. Some subimagoes emerged in the forenoon, between 8 and 9:30 A.M. One male emerged about 2:45 P.M., one female at 11 A.M. Is it characteristic of the species, that the individuals emerge at such varying times during the day?" She did not observe mating flights of the species.

LOCALITY RECORDS.—Gadsden County: 4.5 miles south of River Junction, March 17, 1939, nymphs. Liberty County: 10 miles south of River Junction, March 17, 1939, nymphs; Sweetwater Creek, November 4, 1938, nymphs; December 1, 1939, nymphs. Okaloosa County: Shoal River, December 11, 1937, nymphs.

Genus SIPHLOPLECTON Clemens

Siphlurus flexus was described by Clemens in 1913, but by 1915 he concluded that the species was sufficiently distinctive to be placed in a separate genus, *Siphloplecton,* which he erected for its reception. McDunnough (1923: 47), after examining material of the *Siphlonurus* group, suggested that *flexus* was synonymous with *basale,* and the latter species therefore became the type for the genus. He again mentioned the synonymy of the above two species in 1924, and in both papers considered *Siphloplecton* to be one of the Heptageninae. Ide (1930: 227-28) followed McDunnough's placement of the genus in this subfamily. Ulmer (1933: 209-11) treated the *Siphlonurus* group as a family, Siphlonuridae, and included this family under the suborder Heptagenioidea; however, he completely ignored *Siphloplecton* in his key. Spieth (1933: 327) reversed the order and considered Siphlonuroidea as the superfamily and Siphlonuridae as a family, and under the superfamily he also included the Heptageniidae. Still further complexities as regards the higher categories were introduced when Traver (1932: 111) published her paper on the "Mayflies of North Carolina" and considered *Siphloplecton* to be one of the Baetinae. She retained this viewpoint in 1935 when she placed *Siphloplecton* as a Baetidae, but put the genus in the subfamily Metretopinae. Her reasons for the placement are as follows: "This subfamily might well be placed in the Heptageniidae, if venation and genitalia only are considered. Since, however, the basal joint of the tarsus is not wholly free, but more or less fused with the tibia as in the Siphlonurinae, and the known nymphs are distinctly of the Siphlonurine type, it may equally well be placed in the Baetidae. It would seem to be a connecting link between the Heptageniidae on the one hand and the Siphlonurinae on the other."

Since this last treatment of *Siphloplecton* and the higher categories, there have been no further discussions of its relationships, and there seems to be a tendency on the part of the current workers to accept Traver's classification, at least in a modified form.

Siphloplecton is a small genus, including only five known species, the fifth recently described by Spieth as *S. costalense.*

The genus is distributed throughout the eastern part of North America from Manitoba to Florida. Spieth (1938: 3) discussed briefly the distribution of his new species in relation to the distribution of closely related species. He finds that *S. costalense* is geographically separated from its nearest relative by more remotely related species—a unique sort of distribution among mayflies.

Phylogenetically, *Siphloplecton* is little known, but it is certainly a primitive genus when considered from the standpoints of venation and genitalia, and it may well be a connecting link between the Heptageniidae and the Siphlonurinae.

Siphloplecton speciosum Traver

TAXONOMY.—The very distinctive nymph of *S. speciosum* was described by Traver in 1932, along with the adult, from streams near Macon, Georgia, and others near by. The species has not since been mentioned, except for its redescription in *The Biology of Mayflies* and a statement of its distribution by Spieth (1938: 3) in his discussion of the distribution of *S. costalense.* Because only the nymph stage is represented in my collection, the specific determination must retain a shade of doubt, but my specimens agree very closely with Traver's description of the nymphs from Georgia.

There is a single nymph among the Florida nymphs which resembles *S. speciosum* in all respects, except in that there is a broad, black, median line running from the posterior margin of the head to the middle of tergite ten.

This is probably only an individual variation, for the nymph was taken in company with typical forms of S. *speciosum*.

DISTRIBUTION.—The locality records that are available hardly justify generalizations concerning the distribution of a species. Nymphs have been taken from only three localities, two in Florida and one just beyond the Florida line in southern Alabama. The type locality of S. *speciosum* is very close to the Fall Line and the other localities given by Traver also abut this physiographic dividing line; whether or not the species enters the Piedmont I have been unable to determine. From the evidence at hand it would seem that S. *speciosum* is a Coastal Plain species which only occasionally enters the Piedmont. In Florida, the species is confined to the westernmost portion extending from just east of DeFuniak Springs to the Styx River in Alabama. Though I have searched particularly for *Siphloplecton* nymphs in the more easterly and southern regions, I have found none. Even those localities from which nymphs had previously been taken produced not a single specimen on return trips. Probably the species is confined to an area extending eastward from Augusta, along the Fall Line, into the Alabama River drainage which it follows southward to its limits in this direction (map 8).

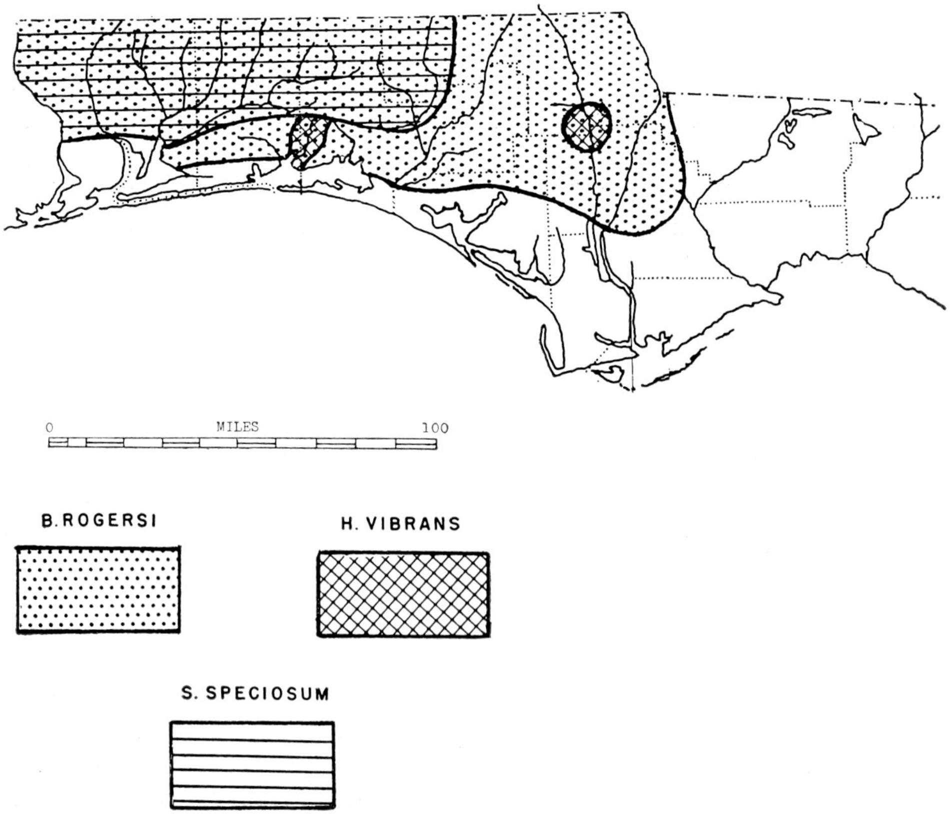

Map 8.—The distribution of *Baetisca rogersi*, *Habrophlebia vibrans*, and *Siphloplecton speciosum* in northwest Florida.

ECOLOGY.—In the Styx River, Alabama, from which nearly all of my specimens were collected, the insects were found at the shore among the debris which had accumulated in a mass of *Myriophyllum* growing at the water's edge. The Styx River is a fairly large, deep, almost stagnant body of water which drains rather flat country and finally empties into the Perdido River near the Gulf of Mexico. I found no *Baetis* nymphs, although other stream mayflies were present; however, the conditions existent at the time of collection would seem to preclude all except the most tolerant forms, and this was found to be the case. The bed of the stream consists of soft, red clay and the vegetation is confined to the shore zone, but even here it is abundant only in scattered areas. Where the nymphs of *Siphloplecton* were collected, there was no perceptible flow, and the plants, predominantly submergent, formed dense mats.

A single specimen taken at Sandy Creek was likewise found in vegetation in the most slow-flowing part of the creek. This is a fairly large stream with a noticeable current and very little vegetation. The plants grow in protected places where there is little movement of the water, and the situations in this creek which are favorable for the development of mayfly nymphs are extremely scarce. This paucity of habitats is reflected in the number of specimens of ephemerids taken from the stream. Broad, sandy stretches clearly indicate the origin of the name of the stream. The bed is loose, coarse-grained sand which constantly shifts with the current, allowing but little material to become anchored.

Another specimen was taken from the Shoal River, a stream which does not differ essentially from Sandy Creek—a slow-flowing, deep stream with a shifting sand bottom. The *Siphloplecton* nymph was found clinging to vegetation which grew around a piling driven deep into the sands to serve as a bridge support.

Found with *Siphloplecton* in the Styx River plant association were nymphs of *Stenonema proximum, Paraleptophlebia bradleyi, Blasturus* sp., and *Callibaetis* sp. In the Shoal River a specimen of the rare *Oreianthus* sp. No. 1 was found living on the same piling from which the *Siphloplecton* nymph was collected. *Baetis spinosus, Blasturus intermedius, Isonychia* sp. A, and *Stenonema smithae* nymphs were taken along with the *Siphloplecton* nymph from Sandy Creek.

SEASONS.—All specimens of *S. speciosum* in my collection were taken in December and range in size from less than one-third grown to penultimate instar. This would seem to indicate a winter, spring, and probably summer emergence. I have little doubt but that emergence is year-round, since almost mature nymphs were collected as late in the year as December. It so happened that at the time these nymphs were collected the temperature was near freezing. The nymphal metabolism is most likely slowed during a cold snap, but as soon as the temperature rises growth is resumed and emergence ensues. There is a direct correlation between temperature and emergence; if there is a prolonged period of warmth during the winter, there may be a fairly large number of adults emerging. Of course, as far as this species is concerned, this is entirely hypothetical, as my statements are based on observations of other species which also have a high ecological valence. Traver's adult specimens from Georgia were collected April 14 through 17.

HABITS.—Very little has been written of the habits of *Siphloplecton*. According to Traver (1932: 95), the North Carolina species are "very similar in habits and habitat to *Siphlonurus*. Often like the former, found among aquatic plants or trash near the shore, where there is a slight but constant movement of the water."

LIFE HISTORY.—Clemens (1915) gave a little life history data for *S. basale (flexus)*, but for the other species of the genus there is almost no information. Traver does give the time of subimaginal life of her Georgia specimens as approximately forty-eight hours. Nothing further is known concerning the life history of the species.

LOCALITY RECORDS.—Holmes County: Sandy Creek, December 11, 1937, nymphs. Okaloosa County: Shoal River, December 11, 1937, nymphs.

Genus ISONYCHIA Eaton

Eaton proposed the name *Isonychia* for *manca* Eaton and *ignota* Walker, but in 1881, under the impression that the generic name was preoccupied by *Isonychus* Mannh., he suggested the name *Chirotenetes* to replace *Isonychia*. McDunnough (1923: 47) re-examined the group and concluded that according to the International Rules of Nomenclature *Isonychia* Eaton is valid. In his paper, McDunnough also wrote a key for the separation of the genera of the Siphlonurinae as known at that time.

The genus in North America contains twenty-seven described species. Sixteen of the species have been named by Traver, three by McDunnough, and eight by various other authors. Although identification is tentative, it seems that there are at least two new species represented in my collection from Florida and southern Alabama.

Taxonomically, the nymphs are still poorly known, for only eleven have been associated with their adults. Identifications of immatures based on the literature must therefore retain a considerable element of doubt, especially since there is such close resemblance between the nymphs of the various species. The length of the tibial spur of the foreleg as compared to the length of the tarsus of this leg is a character frequently used in separating nymphs, and one which I had also employed. I have now come to believe that it is merely an individual variation. One nymph in the penultimate instar was kept in the laboratory for several days, and during this period it molted once. An examination of the exuviae showed the tibial spur to be long and sturdy, but comparison with the spur on the actual nymph revealed that the spur was now quite short and insignificant. This evidence would seem to indicate that the size of this structure varies from instar to instar. In addition, it has been found that nymphs in the last instar invariably have short tibial spurs, whereas the spurs of those in earlier instars are variable in length, ranging in size from one-half to more than three-fourths the length of the tarsus.

By use of the structure of the forceps or styliger plate, the genus can be divided into two distinct groups. A further subdivision into four groups has been made on the basis of the peculiarities in the structure of the penes. The Florida species seem to fall into two different groups, but since material has not been available for study which shows the various genitalia patterns, they will not be classified according to this plan. Apparently one species belongs to the *sicca* group, and a second to the *albomanicata.* The other two species, which are known only from nymphs or females, cannot be placed in a specific group.

The nymphs of *Isonychia,* likewise, may be grouped into those with a brown band across the antennae and those lacking the band. Both of these forms occur in Florida but further subdivision must remain in question until adults are reared.

Isonychia is a Holarctic genus and is rather widely distributed in the Nearctic region. Traver's work in North Carolina showed that in the Southeast the genus has undergone considerable speciation, for, of the twenty-seven described species, seventeen occur in this region. The two other species from Florida bring this total to nineteen.

In Florida, *Isonychia* is confined to the northern part of the state, and does not appear east of the Suwannee River drainage system. Westward it is found all the way to the state boundary and collections from southern Alabama, that is, Baldwin and Mobile counties, include a few nymphs of this genus. *Isonychia* is inadequately known from the Coastal Plain, the Florida species and the Mississippi species *(I. rufa)* being the only ones recorded from that province.

Much work has been done on mayflies in the southeastern part of Canada, yet *Isonychia,* unlike the species of the mountainous section of the Southeast, has not been found to be a predominant element of the stream fauna. I collected mayfly nymphs in a stream in the northern part of the Coastal Plain in Alabama during June, and found that *Isonychia* nymphs were the most conspicuous element in the stream. The opposite becomes true as one travels southward, and in the Florida streams *Isonychia* is much less common than are the various Baetine species.

The familial placement of *Isonychia* is debatable. Spieth (1933: 329) considers that all of the affinities of this genus lie with the Heptageniidae—the shape and number of the segments of the genital forceps, the structure of the maxillae, and the structure of the gills. Traver (1935) takes another view and places *Isonychia* in the Baetidae, allying it with the Siphlonurinae, but she gives no evidence in support of her opinion.

In this paper, the genus will be considered as one of the Baetidae, although characters are shared in common with both the Baetidae and Heptageniidae. It is doubtful that *Isonychia* is an intermediate between these two families, but probably it represents an offshoot somewhere near the base of the stock. According to Spieth, the shapes of the genitalia, and of the forceps base, as well as those of the mandible and of the labium, are indicative of modifications that are peculiar to the genus and that distinctly set it apart from other existing genera. Of unknown significance are the maxillary gills and those found at the bases of the forecoxae, but the structure of the abdominal gills may, in the future, shed some light on the phylogenetic relationships of *Isonychia*. The wings of the genus are very similar to those of *Siphlonurus*, which Spieth places in a distinct family, the Siphlonuridae, but which Traver includes in her family Baetidae.

Isonychia sp. A

TAXONOMY.—Very likely this is a new species; but without specimens of closely related species for comparison, and with relatively few adults of the Florida form (two males), this species will not be described until additional information is gathered. In Traver's key to the imagoes of *Isonychia*, the male adults fall to caption 3, which reads as follows:

Venation almost colorless; foretibia pale in the middle and black at each end .... *pictipes*
Venation dark; tibia wholly dark 4

The males of *Isonychia* sp. A, although the venation is almost colorless, do have wholly dark foretibiae.

Traver's key to the nymphs leaves one in a hopeless tangle, for the Florida specimens might fit any one of several species; and her use of the tibial spine of the foreleg as a differentiating character has proved to be of absolutely no value in separating species. In every stream from which nymphs were collected, long- and short-spined forms have been found existing side by side, and since this spining is the only visible difference (except for the number of teeth on the claws, which to some extent parallels the length of the spines on the forelegs), it would seem that these two groups of immatures are merely variants of the same species. These characters taken as a unit, that is, relative length of spine and number of teeth on claws, may be of specific value, but until adults are reared, I prefer to retain these nymphs under a single name.

DISTRIBUTION.—*Isonychia* sp. A appears to be limited in distribution to the northwestern portion of Florida. Because of the indefiniteness of characters for separation of nymphs, it is uncertain whether the specimens from the Suwannee and Santa Fe rivers are *Isonychia* sp. A or *I. pictipes;* it is likely, however, that they are the nymphs of the latter species since one adult male was collected from a bridge over the Suwannee River. Disregarding these more easterly localities, *Isonychia* sp. A is distributed from the eastern tributaries of the Apalachicola River to the western boundaries of Florida. A few specimens from Mobile County, Alabama, likewise seem to fall into the same general group and are probably *Isonychia* sp. A. Large areas in this northern part of

Florida seem to lack these insects, and their absence is probably due to ecological causes, many of the streams of the region being unsuitable for the nymphs (map 9).

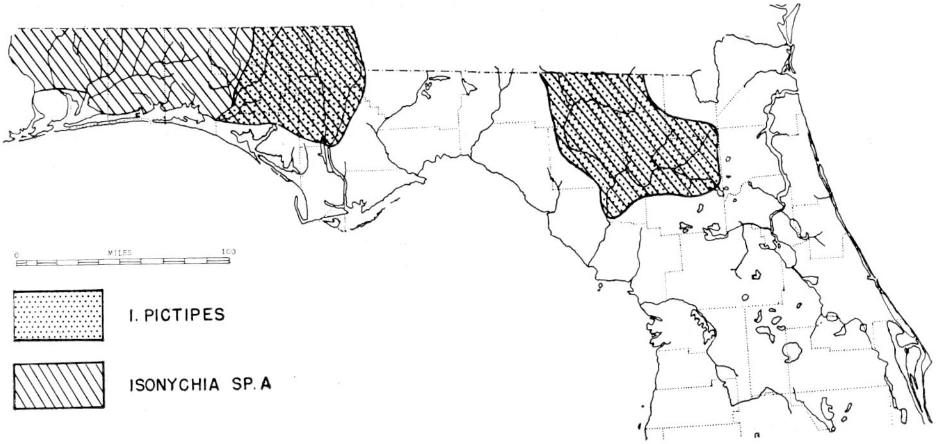

Map 9.—The distribution of *Isonychia pictipes* and *Isonychia* sp. A in Florida.

ECOLOGY.—*Isonychia* sp. A nymphs are found only in flowing water where the current is most rapid. The majority of streams which are inhabited by this species are of the sand-bottomed type. Rarely were nymphs collected from the beds of *Vallisneria* and *Potamogeton,* which form such an important element of many of the streams; however, they were very frequently found on submerged logs, rocks, and boards which were solidly anchored in the stream. Collections from the few riffles encountered in these west Florida creeks included no specimens of *Isonychia.* The most productive situations in the Florida streams are masses of debris caught in the swiftest parts of the current. This mass is usually made up of branches, leaves, small sticks, and other trash, which may have collected on some permanently fixed obstruction in midstream. Here the nymphs generally cling to the sticks and larger objects, and as soon as these are disturbed, the immatures release their hold; thus they are easily collected by means of a Needham apron net.

By far the most productive habitat in the sand-bottomed creek is the tangle of roots of terrestrial plants which are exposed at the banks by the continual washing of the rather swift waters. Many leaves and sticks, as well as much silt and other debris, collect among the exposed roots. Even though there is so much extraneous material present, the nymphs seem to find the conditions here excellent, for they have been found more plentifully in a situation of this type than on submerged logs where they would seem to be more naturally situated. Where the roots are exposed, but the flow of water is reduced, the nymphs are not so frequently encountered as they are in those places where the roots are exposed to the full force of the current. Often masses of Spanish moss become enmeshed among the roots where the current is strong and *Isonychia* nymphs have been found particularly common among them. Since, as Clemens (1917) has shown, the nymphs depend entirely on the flow of water for their supply of food, they would be expected to occupy the swiftest parts of the stream.

When the nymphs occur on submerged logs or rocks, they do not occupy the upper surfaces of these objects but select the undersides or more protected areas. Although the full force of the current does not strike them, they are assured a constant, swift flow of water. Clemens found (1917: 21-23) from his experiments on *Isonychia albomanicata* that the nymphs cling to the under surface of stones to avoid the full force of the current. Under the rocks the water moves much less rapidly than nearer the surface, but it still supplies adequate amounts of food and oxygen to the nymphs.

The acidity or alkalinity of the stream appears to have little influence on the presence of the nymphs, as is indicated by the fact that specimens have been collected just as frequently from acidic as from basic waters. The current must be permanent and fairly strong, there being a direct ratio between the flow and presence of *Isonychia* sp. A nymphs.

Typical associates of *Isonychia* sp. A nymphs in the root and debris associations are *Caenis hilaris, Baetis spiethi, B. spinosus, B. intercalaris, Tricorythodes albilineatus, Paraleptophlebia volitans, Habrophlebiodes brunneipennis, Ephemerella* sp. A, *Stenonema smithae, S. exiguum, Oreianthus* sp. No. 1, *Acentrella ephippiatus,* and *Pseudocloeon bimaculatus.*

SEASONS.—That emergence occurs throughout the year is proved by the presence of nymphs in all stages of development in March, April, June, July, November, and December. The adults were collected in June. *Isonychia,* like the majority of Florida mayflies, ignores seasons, but in its Northern ranges the species are sharply restricted.

HABITS.—*Isonychia* nymphs are probably the strongest swimmers of all mayfly immatures, as is evident when a specimen is liberated in very rapid water. The nymphs can swim against strong currents and actually move upstream—an act which other North American mayflies are apparently incapable of performing. The powerful caudal filaments, with their heavy growths of hair, make excellent paddles for driving the insects rapidly through the water. These are probably of greatest use to the nymphs in moving from one place to another in rapid water, and, although it has not been previously recorded, the movement under normal conditions is probably in the main by swimming rather than by crawling. The latter is the chief means of locomotion of the majority of stream-inhabiting species.

The rate of regeneration of caudal filaments is rather rapid. A nymph brought into the laboratory at Gainesville was somewhat damaged in that it lacked half of its right and two-thirds of its middle tail. This condition was noted on March 12 and the nymph was in its penultimate instar. After the molt to the last instar on March 18, the complete tails were again present and were of normal length.

That the subimagoes are rather sensitive to humidity was discovered when attempts were made to secure imagoes by keeping the subadults in a paper sack. Out of several subimagoes collected, not a single specimen molted completely, and most of them died before even beginning the imaginal molt. After emergence the adults remain directly over the stream. All specimens were collected either at light at the shore of creeks or from trees or bridges over them. The subimagoes and imagoes are distinctly phototropic, although the latter are less so. At the lighted sheet, the subimagoes tend to alight in a poorly illuminated area near the ground. Those specimens which were attracted did not all come at once, but came intermittently over a period of approximately two hours.

LIFE HISTORY.—The fact that the subimagoes arrived at the lighting sheet over a two-hour period would seem to indicate that emergence probably occurred soon after dark. The light was kept burning from seven to nine in the evening, beginning just as darkness fell and continuing until no more ephemerids were attracted. The subimaginal period of those specimens which partially molted lasted twenty to twenty-two hours. The length of this period does not differ much from that of *Isonychia albomanicata* and *I. aurea* in North Carolina which Traver found to be twenty-two to twenty-six hours and thirty-one hours, respectively.

I have not observed the mating flights of *Isonychia.* As far as I have been able to ascertain, there are only a few scattered notes describing the mating flights of some of the species of this genus. Concerning *I. sadleri,* Traver wrote (1937: 59): "Often they danced high in the air; again, individuals as well as mating couples drifted down almost to stream level. They were strong, tireless dancers, the first to appear and the last to leave the scene of festivity."

Clemens (1917) has discussed the emergence, the subimaginal period, the imaginal stage, and the mating flight of *I. albomanicata.* Although the nymphal emergence is probably different in the Florida species because of differences in habitat, the mating flight is most likely very similar.

LOCALITY RECORDS.—Bay County: 16.8 miles north of Panama City, June 8, 1938, nymphs. Gadsden County: 4.5 miles south of River Junction, March 17, 1939, nymphs. Jackson County: 12.2 miles southeast of Marianna, June 9, 1938, nymphs. Holmes County: Sandy Creek, December 11, 1939, nymphs. Liberty County: Sweetwater Creek, June 10, 1938, nymphs; November 4, 1938, nymphs; July 1, 1939, nymphs; December 1, 1939, nymphs; April 8, 1941, adults. 10 miles south of River Junction, March 17, 1939, nymphs. Okaloosa County: Shoal River, December 11, 1937, nymphs. Baggett Creek, April 4, 1938, nymphs. Niceville, June 7, 1938, nymphs. Santa Rosa County: 2 miles west of Milton, April 4, 1938, nymphs.

Isonychia pictipes Traver

TAXONOMY.—Traver described *Isonychia pictipes* from specimens collected at the Apalachee and Alcova rivers in Georgia, basing her species on imaginal characters such as the bicolored foretibiae, pale venation, and small size. The

Florida specimens agree very well with her description, but since she did not figure the male genitalia, there is still a modicum of doubt as to the correctness of the determination. She did state that the penes are very similar to those of *I. sicca,* but the drawings in *The Biology of Mayflies* of the genitalia of this species are not particularly clear. However, McDunnough figured the genitalia of *I. sicca* in 1931, and the Florida mayflies which I am calling *pictipes* agree fairly well with his drawing. Traver did not rear this species and the nymph remains undescribed. I have been unable to differentiate nymphs which can be definitely classified as *pictipes* rather than as *Isonychia* sp. A. Specimens at hand seem to indicate that there are no really good diagnostic characters which will differentiate the nymphs of the two species, although it is reasonably certain that both species are represented in the immature stages. Because a single adult male was collected at the Suwannee River and several adult and subimago specimens were taken in Liberty and Gadsden counties, the nymphs from these localities are here included under *I. pictipes.* Those nymphs taken west of the Apalachicola River are being placed tentatively as *Isonychia* sp. A, although both species are probably represented in the collection from this region.

Published references to *pictipes* since its description include only its re-description by Traver in her taxonomic treatment of North American may-flies (1935).

DISTRIBUTION.—The species is found in the Coastal Plain and Piedmont provinces. The published locality records have been given above. In Florida, the species has been found to range from Alachua County to the Apalachicola River. *I. pictipes* probably is dispersed throughout western Florida into Alabama and thence into the Piedmont of Georgia, from which area the allotype and paratypes were taken (map 9).

ECOLOGY.—The ecological relationships of the nymphs are identical with those of *Isonychia* sp. A.

SEASONS.—Adults in my collection were taken in May, June, and July, and dried adults were found in December stuck in tar covering the underside of the bridge over Sweetwater Creek. Traver's specimens from Georgia were taken in late May and in August. Nymphs in all stages of development were collected from the Suwannee River in April; the Santa Fe River specimen was mature by March 18. It appears from the above data that emergence occurs throughout the year.

HABITS AND LIFE HISTORY.—Habits and life history of the species are no different from those of *Isonychia* sp. A.

LOCALITY RECORDS.—Alachua County: Santa Fe River at Poe Springs, March, 1935, nymphs; March 12, 1938, nymphs. Gadsden County: 4.5 miles south of River Junction, March 17, 1939, nymphs; June 30, 1939, adults; June 6, 1940, nymphs and adults. Gilchrist County: Suwannee River at Old Town, April, 1938, nymphs. Liberty County: 10 miles south of River Junction, March 17, 1939, nymphs. Sweetwater Creek, July 1, 1939, nymphs; December 1, 1939, nymphs and adults; May 2, 1941, nymphs and adults; April 28, 1946, adults.

PLATE VI

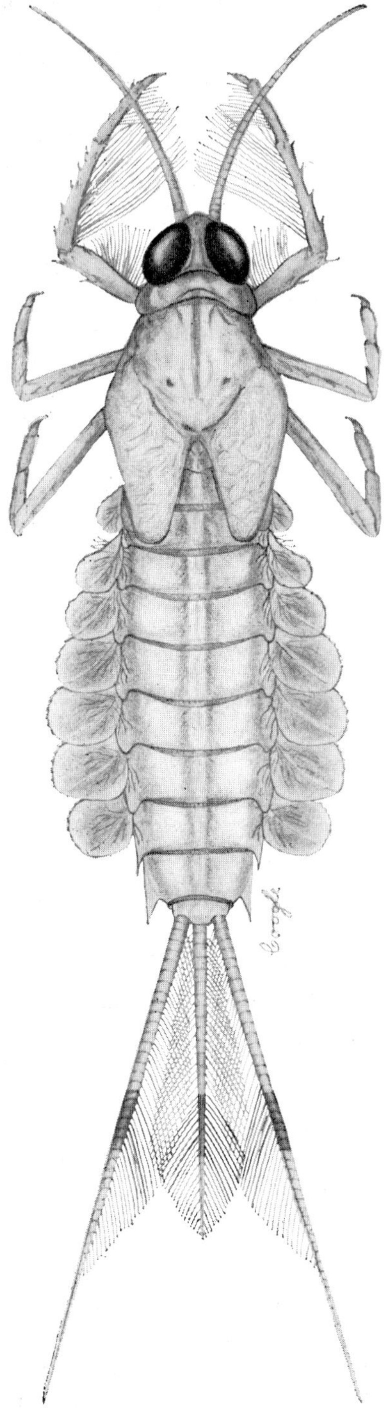

Isonychia pictipes Traver, nymph.

Suwannee County: Suwannee River at Branford, May 29, 1940, adults. Washington County: Holmes Creek, July 29, 1939, adults.

Isonychia sp. B

TAXONOMY.—Determination of a species in this genus makes it almost imperative that males be at hand. As I have been unable to secure this sex for *Isonychia* sp. B, and as only a single female and a few nymphs are known from Florida, it does not seem advisable to give these a name. The adult is quite different from adults of other species in having a shallowly excavated subanal plate, dark venation, and an abdomen with a much more distinctive color pattern than have the other species. Until couplet 18 is reached, this female fits well into Traver's key, which distinguishes the species on the basis of pale tails which are ringed at each joining with red or brown, or tails which are dark brown or black. The Florida specimen has pale tails but they are entirely unbanded. The nymphs are distinct from the nymphs of the *pictipes*-sp. A group in that the antennae are crossed by a brown band about one-third the distance from the base; the color pattern of the venter also distinguishes these nymphs. Corresponding to the condition in the other group of *Isonychia* nymphs in Florida, *Isonychia* sp. B might be separated into two distinct species by virtue of the length of the tibial spur, but as pointed out previously the character is completely linked with the instar of the nymph. This character has been checked in specimens from the Coastal Plain, the Piedmont, and the mountains in Georgia, and in all cases the variation of the tibial spur agrees with my finding in the Florida specimens.

DISTRIBUTION.—*Isonychia* sp. B is known only from Holmes Creek at the Washington-Holmes county line. All specimens were collected on the same date.

ECOLOGY.—I did not collect the few specimens at hand, but I am rather familiar with the stream from which they were taken. Drs. Hobbs and Marchand, who collected the insects, suggested that they were probably found on submerged logs lifted from the stream bed. Holmes Creek is broad, marshy, densely overgrown with *Vallisneria, Persicaria,* and *Isnardia,* and has a steady but gentle current. At low water, a definite, sand-bottomed channel, which is entirely free of the vegetation mats, can be detected. This part of the stream is small, not more than a few feet in width, and on the bed are many logs, sticks, and other debris. Later I made collections at Holmes Creek, but took no other *Isonychia* sp. B nymphs or adults.

Holmes Creek is a typical acid-swamp stream of the sort found in northwest Florida, and the description of this type of creek in the section dealing with aquatic habits includes a fuller discussion of this waterway.

Several other species of mayflies have been collected from the vegetation and submerged logs in Holmes Creek. These include *Baetis spinosus, B. spiethi, Cloeon rubropictum, Cloeon* sp. A, *Paraleptophlebia volitans, Caenis hilaris, Hexagenia munda marilandica, Ephemerella trilineata,* and *Tricorythodes albilineatus.*

SEASONS.—Only mature and half-grown nymphs and a single adult are represented in my collection. There appears to be no reason to doubt that emergence in this species is identical with that of the other Florida species of *Isonychia,* and occurs throughout the year.

HABITS AND LIFE HISTORY.—Habits and life history are unknown, but probably differ very little from other Florida species of *Isonychia.*

LOCALITY RECORDS.—Washington County: Holmes Creek, April 2, 1938, nymphs and adults.

Isonychia sp. G

TAXONOMY.—Perhaps the two nymphs which I am calling *Isonychia* sp. G should be included under *Isonychia* sp. B. The two species differ in the color pattern of the venter and in the presence of grayish areas in the outer margins of each gill of *Isonychia* sp. G. These may be only local variants of the Holmes Creek species, but until so proved, I believe that the differences warrant a separation. These nymphs, in common with *Isonychia* sp. B, have antennae which are crossed by a brown band; the tibial spurs are also long as would be expected since neither specimen is mature.

DISTRIBUTION.—The single stream from which *Isonychia* sp. G is known is located in Gadsden County and is one of the tributaries of the Apalachicola, draining into the river about two or three miles distant from the place where the nymphs were caught.

ECOLOGY, SEASONS, HABITS, AND LIFE HISTORY.—All other data are probably the same as those for the other Florida species of *Isonychia.*

LOCALITY RECORDS.—Liberty County: 10 miles south of River Junction, March 17, 1939, nymphs.

Genus BLASTURUS Eaton

The genus *Blasturus* was described by Eaton in 1881 and redefined in his monograph on the Ephemeridae in 1884. In 1932, Traver discussed the genus as it occurs in North Carolina and described five new species. In her paper, she included the description of *Blasturus gracilis,* but subsequently removed this species to the genus *Leptophlebia,* after first shifting all the North American species formerly placed in that genus to *Paraleptophlebia.* Ide (1935: 124) suggested that *Leptophlebia johnsoni* might be intermediate between *Blasturus* and other *Leptophlebia* species, and that it would be well to drop entirely the generic name *Blasturus.* Spieth (1938: 214) followed Ide's suggestion, and in his discussion of coloration in relation to seasonal emergence used the generic name *Leptophlebia* to refer to species placed in *Blasturus* and *Paraleptophlebia* by Traver.

In 1932, Traver wrote, "Both as to nymphs and imagoes, the genus *Blasturus* is a difficult one to separate into its component species. Structural differences are minor and difficult to recognize until some time is spent in studying the group as a whole. Color differences exist, but are likewise minor and relative, and are an unsatisfactory basis for the separation of species." To date, additional information has not changed the situation.

Even though there is justification for synonymizing *Blasturus,* I am retaining the name for the Florida species, since it does serve as a convenient category, and is useful in presenting a picture of the Leptophlebines of Florida. This action is, furthermore, apparently in line with Spieth's most recent views (1940a: 325).

The genus (or subgenus) *Blasturus* is entirely Nearctic, and is generally distributed throughout this region. Its phylogenetic position is stated in the discussion of the relationships of *Blasturus, Leptophlebia,* and *Paraleptophlebia* given under the last-named genus. Spieth (1933: 345), before he considered *Blasturus* to be a part of the genus *Leptophlebia,* thought that it was probably the most primitive genus of the family Leptophlebidae (considered here as the Leptophlebiinae), and that *Blasturus* and *Leptophlebia* presented a rather close affinity to each other. He stated: "Indications that they all (including *Choroterpes* and *Thraulus*) represent primitive branches of a major division of the Ephemerida are: (1) the fairly primitive condition of the wings, especially those of *Blasturus;* (2) the simple form of double gill consisting of two foliaceous lamellae without such special modifications as are found in the Heptageniidae and Baetidae branches; and (3) the 3-jointed forceps, lacking any indications of the basal articulation commonly found elsewhere in the order."

Blasturus intermedius Traver

TAXONOMY.—Since Eaton's work, very few ephemerids have been described from the subimago; unless the species is quite distinct from all others, subimaginal descriptions may tend to be misleading and may frequently cause misinterpretations. However, Traver (1932: 136-37) broke with custom and set up a subimago as the holotype of *B. intermedius,* figuring the genitalia and various nymphal structures. By 1935, male imagoes had been reared, and *intermedius* was then found to be distinct on the basis of genitalial differences from all others in the genus, except *B. grandis.* Dr. Traver has designated the genitalia of this complex as the "distinctive 'scarf' type" of penes and in her key to the species separates *intermedius* and *grandis* from *austrinus* and *collinus* (bearing the "hooded" type penes) by the relative length of the reflexed spurs. In the former group, these spurs extend anteriorly from the distal end of the penes to the base of the notch between the paired penes, while in the latter, the spurs are relatively shorter. Absence of a brown cloud in the forewing or absence of a brown stain in the stigmatic area of this wing distinguishes *intermedius* and *grandis* from the remaining species of the genus. Finally, *grandis* is separated from *intermedius* by its larger size and by its short middle caudal filament. The median tail in *grandis* is but one-half the length of the laterals, whereas in *intermedius* the mid-filament is two-thirds the length of the outer ones.

PLATE VII

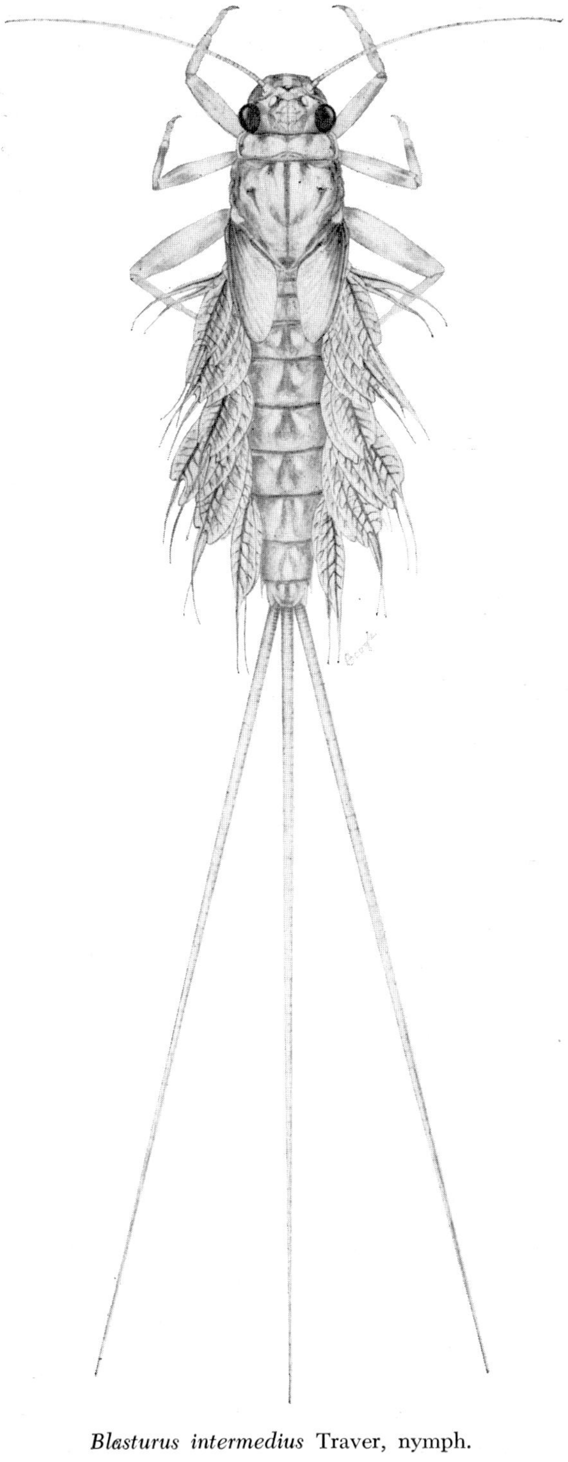

Blasturus intermedius Traver, nymph.

DISTRIBUTION.—On the basis of genitalia, *intermedius* is seen to be most closely related to *grandis*. The distribution of these two species bears out this conclusion as to their relationship, both having been recorded only from North Carolina (prior to my work in Florida); *grandis* from the Piedmont, *intermedius* from the Coastal Plain. Traver (1935: 538) records *intermedius* from "Braden Co.," North Carolina; this is obviously a misprint and should read "Bladen Co."

Continuing southward into Florida, *B. intermedius* is rather widely distributed over the state; in fact, it is found from the eastern portion to the western borders in the continental area. In the peninsular region, the species is much more sporadic in occurrence, but this "spotty" distribution is entirely the result of ecological factors.

The absence of records of *B. intermedius* from South Carolina and Georgia is almost certainly to be explained by lack of collecting. South Carolina offers many opportunities to the Ephemeropterist, and I feel certain that if the Coastal Plain waters were collected, *Blasturus intermedius* would be one of the first species to be found (map 10).

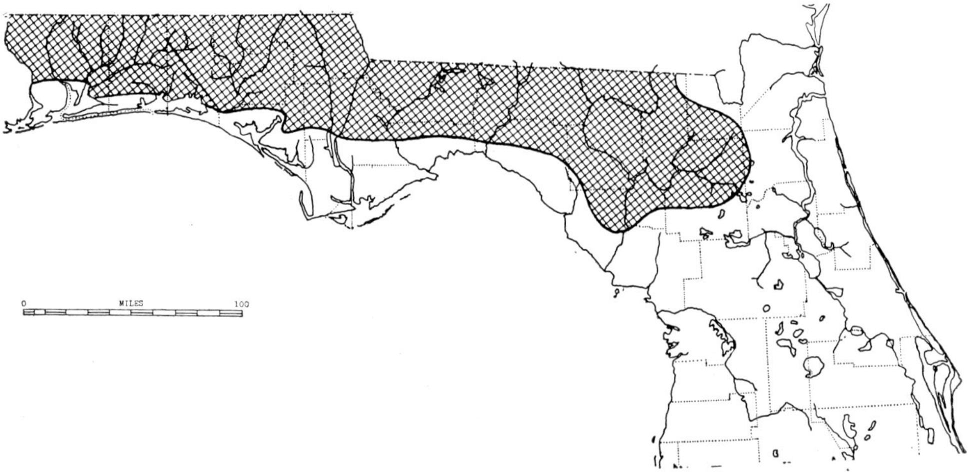

Map 10.—The distribution of *Blasturus intermedius* and *Habrophlebiodes brunneipennis* in Florida.

ECOLOGY.—Ecologically, Florida *Blasturus* nymphs are limited to slow streams, preferably small- to medium-sized creeks. The range of this species in the state is probably limited by physiological as well as physical barriers. Flowing water is relatively scarce in most of the regions from which no specimens have been taken. It is quite likely that with further collecting the genus will be found to extend into the west-central counties where there are streams which appear suitable for the nymphs, but in the short time allotted to collecting in these waters, no immatures were found. From the standpoint of mayflies, the unproductive southeastern regions and the flat pineland and palmetto-scrub area of the east coast prohibit the existence of *Blasturus* in this geologically newly arisen area.

With their enormous expanse of gills, *Blasturus* nymphs are poorly adapted for swift water, and one seldom finds them elsewhere than in the quieter parts of the streams or pools recently cut off from streams. Even though highly tolerant of standing water and conditions where there is little oxygen and much silt, the nymphs occur only in streams or associated waters, and the size of the streams is no criterion for the occupancy of their bed. The presence of the nymphs seems only to demand a permanently flowing stream with quiet areas near the banks, pools where dead leaves may accumulate, or submerged mossy banks where the nymphs crawl amid the bases of the plants. In those parts of the stream where there is submergent vegetation, unlike that described above, the nymphs are usually absent. If both conditions obtain in a stream, that is, if mossy banks and leaf debris are present, the chances are fairly good that *Blasturus* will be there. A small creek near Gainesville yielded over two hundred nymphs in their last instar from an area not over two feet long, three feet wide, and six inches deep. This was a slough of the main stream and connected with it over a shallow sandbank. In this puddle, dead leaves were piled several layers deep, the lower layers partially buried in silt; the first dip with a coffee strainer yielded twenty-three nymphs. Subsequent dips produced an average of fifteen nymphs until most of the leaves had been cleared out, and only then did the catch fall off to four or five nymphs per haul. Downstream a hundred yards, another pool, in the lee of the root of a moderate-sized tree, yielded one hundred more nymphs without clearing out more than half the debris. The latter pool was smaller than the former, yet seemed just as rich in mayflies.

Dr. Ferris Neave has made a study of the migratory habits of *Blasturus cupidus* nymphs in Manitoba. During the spring, this species migrates into and up the temporary streams formed by melting snow, even leaving the water and crawling along the banks in its urgency to move upstream. The nymph may migrate as much as three hundred to four hundred yards per day, which is remarkable in the instance of so small an insect, especially an aquatic and normally slow-moving type. In Florida, I have noted no such migrations, and rather doubt that they occur here, but there may be smaller migrations of the nymphs from the stream bed proper into pools of the type from which so many specimens were collected. The concentration of mature nymphs seems to indicate something of the sort, particularly since younger nymphs are more frequently encountered in the leaf drift along the creek margin than in the pools, while the reverse is true of the mature nymphs.

Connecting with Hatchet Creek, near Gainesville, there is a fairly large, stagnant canal from which water disappears during the dry season. The canal passes through a rich mesophytic hammock, and has its floor paved with leaves partially buried in silt. In February, 1938, several nymphs in their last instar were collected here; I visited the same canal in April and found no trace of *Blasturus.* Previous to their emergence the mature nymphs must certainly have migrated into the canal, since no mature nymphs were found here on other occasions; this migration probably is not comparable to that studied by Neave since his nymphs migrated upstream against swift-flowing water and were not all mature. I believe that quiet water is essential for the emergence of *B. intermedius* and that maturation of the gonads acts as a stimulus, forcing the nymphs to find this ideal situation. The young nymphs, often found in small numbers with the more abundant mature ones in quiet water, have probably wandered there accidentally.

The adults are less tolerant of desiccation than many mayflies, and are confined to mesophytic conditions where the humidity is rather high. If the subimagoes are removed from this moist environment, the final molt is almost impossible.

The oxygen content of the almost stagnant water must certainly be low (no determinations were made), but the large gill expanse with the many tracheal branches provides much surface for gaseous exchange.

Most common associates of *Blasturus* nymphs in the leaf drift are other genera of ephemerids, nymphal Odonata, chironomid larvae, and the small snail *Physa*. Mayflies, however, far outnumber their coinhabitants of these detritus drifts. Besides the *Blasturus* nymphs, there have been found *Habrophlebiodes brunneipennis, Paraleptophlebia volitans, Stenonema smithae, Caenis diminuta, Ephemerella trilineata, Hexagenia munda marilandica,* and *Cloeon* sp. A.

Macroscopic inhabitants of the submerged moss bank are most commonly mayfly nymphs and less frequently damselflies and bloodworms. Nymphs of *Paraleptophlebia volitans* are found more often in this moss than are the immature *Blasturus intermedius*. Some *Stenonema smithae* nymphs also occur here.

SEASONS.—In general, seasons have no great effect on mayflies in Florida, but *Blasturus* seems to be an exception. This genus is the earliest of those mayflies of Florida which are seasonal; its sexual activity gradually increases to a maximum in February and March, and then gradually declines.

There is much overlapping of mature forms with immatures. In February, 1940, I collected numerous nymphs of *B. intermedius* from a small creek near Gainesville; here, mature nymphs outnumbered very young and half-grown nymphs almost ten to one. This would seem to indicate that in Florida this species, unlike the majority of mayflies, has a definite maximum emergence period, with most of the individuals maturing in February to March but with limited emergences throughout spring and early summer.

HABITS.—The nymphs of *Blasturus intermedius* feed upon detritus, diatoms, and algae, scraping the surface of the decaying leaves and indiscriminately eating the animal, vegetable, or mineral materials on the leaves. Only the size of the particles acts as a deterrent and feeding is an almost continuous process.

Nymphs are more or less negatively phototropic while the reverse is true of the subimago and imago. In the laboratory, nymphs seek the dark underside of any materials in their container during the day and whenever artificial light is cast on them. If left in a dark room, they soon find their way to the upper surface of the leaves, but as soon as a light is turned on, they again scurry away into some dark crevice. Nymphs about to emerge undergo a phototactic reversal, and crawl to the upper side of leaves or onto sticks, logs, or any other available support which is near the surface of the water.

LIFE HISTORY.—The length of nymphal life is unknown, but would seem to be one year. This assumption is based on the concentration of mature nymphs during one definite period. Many mature specimens have been reared to adulthood in the laboratory, but I have found it rather difficult to keep younger nymphs alive for more than two months.

Just prior to metamorphosis, the mature nymph moves to the surface, still slowly vibrating its gills. With its mesothorax projecting from the water, it begins to strain and the effects of this straining are soon evident, for a longitudinal split appears down the thorax. Pulling mightily by bracing itself against the nymphal skin, the subimago gradually extricates the wings from their sacs, all the time moving its abdomen from side to side, a movement apparently caused by the tremendous exertion necessary to free the body from the nymphal exuviae. As the wings are pulled free, it can be seen that they are inverted, that is, the costal border faces ventrally. The sudden release of the wings causes them to snap into place and at the same time to twist into their normal position. Immediately after emerging, the subimago is a dull gray with rather translucent wings, but after a short time the animal begins to darken until it is quite blackish. The time consumed between exposure of the split mesothorax and removal of the wings varies from one to three minutes. Once the wings are freed the animal rests, supported partially by the surface film and partially by the floating nymphal skin. During this resting period the wings are frequently vibrated. Having regained its strength, the subimago begins to walk away from its exuviae and slowly pulls its three caudal filaments from their sheaths. Its tails, when freed, are slightly raised above the surface of the water. After resting again for a short time, the winged insect flies or crawls to some nearby support to await the bodily changes which will force it either to undergo a final molt or perish.

To undergo its imaginal molt, the subimago establishes itself firmly on its support and begins to strain at the mesothorax. The straining is probably localized in this part of the body because of the tremendous wing muscles located in the synthorax. The contraction of these muscles forces the wings to spread outwards until they touch the support. As the adult pulls himself from his subimaginal skin, the wings are gradually drawn in towards the body until they lie alongside of the abdomen and are folded fanwise. The withdrawal is slow and when the wings are almost freed, the creature releases its adult legs from the attached skin, fastens its claws to some object, and pulls the wings free. Immediately, the wings are raised, but only to about a sixty-degree angle with the horizontal. The adult rests a moment, flexes its wings, approximates them in the usual manner, and then slowly walks away pulling its tails out as it moves. The subimaginal exuviae is left behind as a gray skin with crumpled sacs, the sole remaining evidence of the subimaginal wings.

During January and February, emergence occurs any time between 8:00 A.M. and 4:00 P.M. with the height of transformation between 1:30 and 4:00 P.M. After a quiescent period varying from eighteen and one-half to twenty-four hours, the last molt occurs and the brilliant adult emerges—its thorax shiny black, its amber wings glistening. In February this final molt takes place in the morning between 8:30 and 10:00 o'clock, reaching its peak about 9:00 to 9:30.

Immediately after emergence, the subimagoes are attracted to the more strongly lighted sides of the cage and here sit waiting for the imaginal molt. After many had died without molting, it was concluded that even though the insects were sitting just over a pan of water, the atmosphere was still too dry. This was remedied by placing the mayflies under a bell jar with a watch glass filled with water and covered with cheese cloth to prevent the mayflies from wetting their wings or drowning in it. Subimagoes introduced into this saturated air were quite successful in molting, but even so, many failed to free their wings completely from the subimaginal skin. It was found advisable not to use forceps to handle the wings in transferring subimagoes to the bell jar as the forewings are very delicate and easily bruised. Once bruised, the wings cannot be freed normally.

Many abnormalities occur in the wings of laboratory-reared adults, particularly in the males. In some there are merely small holes scattered through the middle of the wings, in others there are large breaks, and in still others, the distal half of the wing may be missing.

I have observed no mating flight of this species, but male adults kept in the laboratory became very restless in the afternoon between 2:00 and 4:00. In regard to *B. cupidus* Dr. Traver (1925: 217) states:

> . . . The "stag" flights usually begin at about 3:00 P.M. and last intermittently till nearly 7:00 P.M. At no time were more than a dozen insects seen flying at once, and none of these swarms were directly over the water. One favorite location for this dance was a small grassy plateau about six feet from the water's edge, upon which the last lingering rays of sunlight fell. Here, in the sun's departing beams, the dance went on and on. Often the insects were rising not more than 7 or 8 feet and then with the undulating downward sweep, coming within a few inches of the grass. The wings glistened brightly in the light and were a better guide to the location of the imago than the tails, which could barely be seen a few feet away. The downward movement is apparently a mere falling with the force of gravity, the tails and forelegs serving to increase resistance. Then the insect slowly but surely rises again to repeat the performance. There is a rhythmic swing to the dance, though each dancer keeps his own time.

Two female subimagoes which were reared in the laboratory failed to undergo their final molt; kept under a bell jar, one of these lived as a subadult for approximately fifty hours and thirty minutes, the second for ninety-five hours. Tests for longevity of adults kept under bell jars showed that males remain alive for about thirty-five or forty hours after the subimaginal molt; females live eighty to ninety hours after this molt. Traver's study of *B. cupidus* showed that the length of adult life, when the insects were kept in captivity at room temperature, was twenty-four hours or less.

LOCALITY RECORDS.—Alachua County: 1 mile west of Newnan's Lake, January 8, 1938, nymphs. 3 miles north of Paradise, February 12, 1938, nymphs. Santa Fe River at Worthington Springs, February 12, 1938, nymphs; February 5, 1939, nymphs. Hatchet Creek, February 8, 1938, nymphs and adults; February 26, 1938, adults; March 23, 1938, nymphs and adults; April 2, 1938, adults. 2½ miles west of Gainesville, January 29, 1938, nymphs; February 3, 1938, nymphs and adults; January 7, 1939, nymphs and adults; January 28, 1939, nymphs and adults; February 5, 1940, nymphs and adults; adults reared February 6 - March 10, 1940; January 26, 1947, adults. Blues Creek, February, 1947, nymphs. Columbia County: Falling Creek, February 4, 1938, nymphs. Gadsden County: River Junction, March 17, 1939, nymphs. 4½ miles south of River Junction, March 17, 1939, nymphs. Holmes County: Sandy Creek, December 11, 1937, nymphs. Hamilton County: 0.6 miles north of Live Oak road at U. S. Highway 41, February 4, 1938, nymphs. Jackson County: 3.6 miles north of Altha, December 10, 1937, nymphs; December 1, 1939, nymphs. Jefferson County: Drifton, February 5, 1938, nymphs. Liberty County: Little Sweetwater Creek, December 10, 1937, nymphs. Hosford, March 17, 1939, nymphs. Okaloosa County: Shoal River, December 11, 1937, nymphs. 2 miles east of Crestview, December 12, 1937, nymphs. 1 mile east of Crestview, December 12, 1937, nymphs. Walton County: 2.1 miles west of Walton County line, May 31, 1940, nymphs. Washington County: Holmes Creek, December 11, 1937, nymphs.

Genus PARALEPTOPHLEBIA Lestage

Paraleptophlebia was described in 1917 to receive certain European mayfly species which Lestage felt should not be retained in *Leptophlebia*. His genus was not recognized by American students until 1934, when Traver transferred to it all the North American species that had formerly been placed in *Leptophlebia*, except *gracilis* and *johnsoni*.

At the present time, the status of the genus *Paraleptophlebia* is still in question, at least insofar as the North American species are concerned. Ide (1935, 1937, and 1940) and Ide and Spieth (1939) have disregarded *Paraleptophlebia*. Spieth (1938) apparently accepted this genus but placed all species of *Blasturus* in *Leptophlebia*. This same author (1940), in a paper on the North American species described by Francis Walker, used both of the names which he had formerly discarded. There are certainly at least two distinct groups here, and, if taxonomic confusion is to be avoided, these genera should be redefined. In the papers by Ide and Spieth, neither discussed his reasons for his generic choice.

Paraleptophlebia includes thirty species in North America (north of Mexico), and these are distributed generally throughout the continent.

It is mentioned in the discussion of *Habrophlebiodes* that *Leptophlebia, Paraleptophlebia,* and *Blasturus* form a closely knit group, all derived from a common progenitor. Spieth (1933: 345) considers that the Leptophlebines represent primitive branches of a major division of the mayflies. From a study of the morphological characters, Spieth concluded that *Leptophlebia (Paraleptophlebia)* and *Blasturus* showed a close affinity and that the Leptophlebiinae "stand comparatively low on one of the main branches of the evolutionary tree of mayflies."

Paraleptophlebia volitans (McDunnough)

TAXONOMY.—*Paraleptophlebia volitans* was described by McDunnough in 1924 from Quebec. On the basis of genitalia, it stands alone. There is a pair of U-shaped reflexed spurs on the penes. Several other species also have the reflexed spurs, but in their genitalia the shape of these structures is entirely different. On the basis of presence and shape of the spurs, *volitans* may be closely related to *guttata*.

The nymphs from northwest Florida are fundamentally like those of Alachua County, but differ in having a brownish tinge in the gills; however, adults from the two regions are identical. Upon comparison of Florida nymphs with the Canadian form, Dr. F. P. Ide stated (by letter, 1939) that the Florida species is "very close to our *P. volitans*, which occurs all over the Laurentian Shield region of Ontario. The legs are somewhat hairier and the abdominal markings more prominent than in *P. volitans*. . . ." By rearing and thus correlating nymphs and adults, Dr. Ide's identification has been fully verified.

The nymphs of *Habrophlebiodes* and *P. volitans* are very similar and are easily confused unless carefully examined. The easiest method of distinguishing between them is by means of gill structure. In *volitans*, the gills each bear a bifurcate trachea which has no lateral branches. Although the gills of *Habrophlebiodes brunneipennis* also have bifurcate tracheae, there are prominent lateral branches from the main trunks.

DISTRIBUTION.—There is nothing in the appearance of *volitans* to suggest that it is an unusually vagile form, yet it has been taken over an area stretching from Ontario to Florida. The species has been recorded from the Coastal Plain at Fort Valley, Georgia, close to the Fall Line, but all other published records are more northern.

In Florida, *P. volitans*, *Blasturus intermedius*, and *Habrophlebiodes brunneipennis* have an almost identical distribution over the northern portion of the state. *Volitans* extends southward only to the area around Gainesville; eastward, it has been taken in the western part of Nassau County; westward, it is distributed to the state border and on into Escambia and Baldwin counties in Alabama (map 11).

ECOLOGY.—The ecological distribution of *volitans* is almost identical with that of *Habrophlebiodes brunneipennis*. Ide (1930: 207) and Gordon (1933) allot a few brief sentences to a discussion of the environment of *volitans*. Ide (1935: 64-65) studied the effects of temperature on the distribution of mayflies in a stream and although *volitans* is present in the area in which Ide worked, he did not mention its occurrence in any of the streams that he examined.

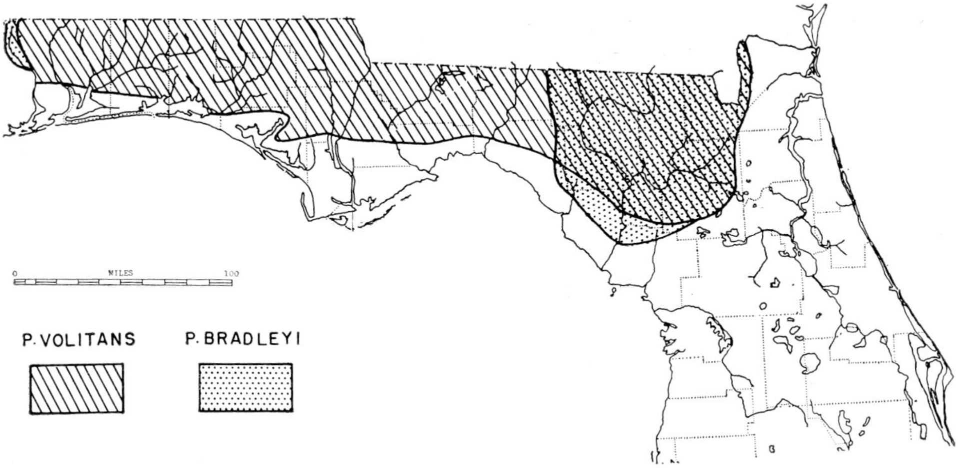

Map 11.—The distribution of *Paraleptophlebia volitans* and *P. bradleyi* in Florida.

However, his conclusions, drawn in part from a study of other species of *Paraleptophlebia,* are very interesting, particularly the following:

> . . . The facts brought out by this study appear to throw some light on the subject of the geographical distribution of mayflies. A spring stream, especially near its source, supplies a remarkably uniform environment for mayfly nymphs, wherever the stream occurs—in warm climes, at high elevations on mountains, or in the north. This uniform environment provides a pathway by which organisms may extend their range wherever this environment is present. As one goes north, however, the warm-season forms will be eliminated and the forms of the source, or near it, will still be accommodated further downstream, and, provided the source becomes colder, there is an opportunity for new forms to appear at the source. The forms near the source are those with a northern distribution in general, and those lower down have a more southerly distribution.
>
> It is probably safe to say that a spring stream in its fauna, taken from the warmer reaches to the source, recapitulates south to north distribution, in the way that a mountain in the torrid zone gives in its fauna and flora, taken from the foot to the top, roughly a vertical section of geographical distribution from torrid to frigid zone.

In Florida the nymphs of *P. volitans* live amid the leaf debris in streams where the current is slow to moderately swift. Though most commonly found in these situations, they may also be taken from riffles, submerged sticks, logs, and vegetation but they seldom venture into the swifter waters. The creeks supporting populations of *volitans* are not more than three to four feet at the deepest point, and the nymphs tend to remain in the shallower areas. The small streams draining the ravines of the Tallahassee Hills region, although heavily populated with *Habrophlebiodes brunneipennis,* are completely lacking in *P. volitans.* Just west of the region, the latter species is found in the larger streams emptying into the Apalachicola. In these larger, but shallow, streams the nymphs can be commonly found near the banks among the roots of terrestrial plants, which project into the water.

The sand-bottomed streams from which nymphs have been taken are mostly circumneutral or acid, and usually drain flatwoods. The water is usually tinted, the tint ranging anywhere from almost colorless to the color of strong tea.

Subimagoes are confined to moist, heavily shaded areas along the creek banks or to any other shaded area above the stream where the humidity is very high. The subimago molts only with great difficulty in the laboratory; in fact, it is rare for a male under such conditions to complete its final ecdysis successfully. Even specimens placed under bell jars failed to shed, and attempts made in a dry atmosphere also failed.

In a letter to me, Dr. B. D. Burks, of the Illinois Natural History Survey, wrote that subimagoes of the Illinois species do not experience any difficulties in molting. He discovered that the subimagoes must be kept in fairly dry jars in cool, dark places, because they need to lose moisture. I tried his method, but on the morning following emergence, as usual, the insects lay dead and withered on the bottom of the container. Of all species of mayflies with which I have dealt, *volitans* has the most difficulty in completing the last molt in the laboratory.

SEASONS.—Available records for Florida adults indicate that in this region the species emerges throughout the year; however, in Canada (Algonquin Park, Ontario), Ide (1940) reported that emergence takes place in the latter half of July and in August. Traver, in recording the species from the South, did not give the date of collection. Among the Florida specimens, there are imagoes taken during the following months: January, February, March, April, May, June, July, October, and November. The peak of emergence is probably during April and May. As might be inferred from this data, nymphs of all ages can be taken during any month of the year. There is some evidence that nymphs in a particular stream may have a tendency toward transformation during a definite period, but there is so much overlapping of instars that this tendency is usually obscured.

HABITS.—Plant matter in one form or another probably constitutes the food of *P. volitans*. When nymphs are placed in a pan with only dead leaves which have been submerged long enough to become softened, they do quite well and have been kept alive in the laboratory for as long as six weeks. Since they are found commonly in leaf drift, it might, therefore, be presumed that they are detritus feeders. They have also been taken from places where algae were quite abundant, and these likewise must play an important part in the nymphal diet.

Discussion of peculiarities of movement among the Leptophlebiinae may be found in the section dealing with *Habrophlebiodes brunneipennis*. The Leptoplebines, when placed in a dish of water, spread their gills and are frequently caught in the surface film where they float helplessly.

As in most other ephemerids, the nymphs are negatively phototropic; the adults are the reverse. This characteristic, however, is not pronounced in adults and subimagoes. I have "lighted" along streams where I knew *volitans* was emerging but not a specimen was attracted to the lighting sheet; on the other hand, when adults are brought into the laboratory, there is a definite orientation to light and subimagoes emerging in cages are immediately attracted to the most strongly illuminated side.

LIFE HISTORY.—Little work has been done on the life history of *Paraleptophlebia,* except for the description of last instar nymphs for taxonomic reasons. A brief discussion of the external features of the eggs of five species of this genus with a figure of the egg has been given by Smith in *The Biology of Mayflies,* and a sentence and figure were devoted to the genus by Morgan (1913: 399).

From Ide's data on the emergence of *P. volitans* in Canada, it is obvious that the life history in that region occupies an entire year. I strongly doubt that the same length of time is required in Florida. Several attempts to hatch eggs have been unsuccessful, perhaps because the females from which the eggs were taken may not have mated. Emergence records and rate of growth of nymphs in the laboratory indicate that six to eight months may be sufficient for nymphal development.

Emergence, as observed outdoors, occurs normally in late afternoon, about two hours before sunset. In the laboratory, most subimagoes appeared at about the same hour, but in some instances individuals have transformed as late as 7:30 P.M. during April, while in October transformation occurred between 6:20 P.M. and 7:00 P.M. Subimaginal life lasts twelve to fourteen hours; the male imago lives a day longer, and the female may sometimes live as long as two or three days after this last molt.

I have never observed mating flights of this species. Another species is thus briefly treated by Morgan (1913: 392): "Mating flights of *Leptophlebia praepedita* have been seen in the middle of a sunny afternoon, and at two, four and five o'clock of bright afternoons in May and June. None of these rose higher than 15 feet and two of the swarms did not fly more than six feet above the ground. One entire swarm which was captured contained forty males and one female."

LOCALITY RECORDS.—Alachua County: 1 mile west of Newnan's Lake, May 11, 1937, nymphs and adults; January 8, 1938, nymphs. Hatchet Creek, March 23, 1938, adults; April 2, 1938, adults; April 18, 1938, adults; May 5, 1938, adults; July 9, 1938, adults; March 22, 1939, adults; April 1, 1939, adults; April 5, 1939, adults; April 13, 1939, adults; May 6, 1939, adults; June 24, 1939, adults. 2½ miles west of Gainesville, January 16, 1938, nymphs; March 5, 1938, nymphs and adults; January 7, 1939, adults; January 14, 1939, adults; January 28, 1939, adults; March 10, 1939, adults; April 21, 1939, adults; October 6, 1939, adults; November 11, 1939, adults; February 5, 1940, adults; March 19, 1940, adults; October 5, 1940, adults. Bay County: 5.6 miles north of Panama City, May 30, 1940, nymphs. 27.4 miles north of St. Andrews, May 30, 1940, nymphs. Pine Log Creek, May 31, 1940, nymphs. Columbia County: Falling Creek, November 13, 1938, nymphs and adults. Hamilton County: 8.3 miles south of Jasper, February 4, 1938, nymphs. Holmes County: Sandy Creek, December 14, 1939, nymphs; May 1, 1946, nymphs. Jackson County: 3.6 miles north of Altha, December 10, 1937, nymphs; June 9, 1938, adults. Jefferson County: April 1, 1938, nymphs. Leon County: 7 miles south of Highway 20, on Sopchoppy road, June 5, 1938, nymphs. Liberty County: Sweetwater Creek, July 1, 1939, nymphs. 10 miles south of River Junction, March 17, 1939, nymphs. Hosford, March 17, 1939, nymphs. Nassau County: 19.1 miles north of Duval County line, August 28, 1938, nymphs. Okaloosa County: 2 miles east of Crestview, December 12, 1937, nymphs. Putnam County: Johnson, April 10, 1948, adults. Santa Rosa County: Pace, June 1, 1940, adults. 2 miles west of Milton, April 4, 1938, adults. Walton County: 5.4 miles east of Freeport, April 2, 1938, adults. 13.8 miles west of Freeport, June 7, 1938, nymphs. 15.8 miles west of Ebro, June 7, 1938, adults. 7.3 miles west of Ebro, June 7, 1938, nymphs. 6.7 miles west of Portland, May 31, 1940, nymphs. 9.5 miles west of Portland, May 31, 1940, adults.

Paraleptophlebia bradleyi (Needham)

TAXONOMY.—Described by Needham in 1932, eighteen years after the specimens were collected in the Okefenokee Swamp by J. C. Bradley, *Paraleptophlebia bradleyi* is unique, especially in the nymphal stage. The female has not yet been found. The male imago can be separated from the other species by its genitalia, lack of pale abdominal segments, and tinted wings. Because of its distinctiveness, it is rather difficult to say in which direction its affinities lie. For the most part this is also true of the other species of *Paraleptophlebia*, since the genitalia, usually one of the best indicators of relationship, differ so greatly among the species of the genus.

The nymphs of *P. bradleyi* cannot be easily confused with those of any other species; in fact, they are so distinctive that Dr. Ide was dubious as to the generic placement. This doubt was resolved by rearing adults. The uniqueness of the nymphs lies in the structure of the gills, which is more like that of *Blasturus* or *Leptophlebia* than that of *Paraleptophlebia*. If, as Ide (1935: 124) and Spieth (1938) seem to think, the species of *Blasturus* should be placed, along with *Paraleptophlebia*, in the genus *Leptophlebia*, then *bradleyi* would be the connecting link between *Blasturus* and *Leptophlebia*. This statement is based on a study of the structure of the gills. In *P. bradleyi* each gill is composed of two distinct plates like those of *Blasturus*, each lamella having distinct and prominent main tracheae provided with numerous lateral branches. As far as I can tell from existing descriptions, no other species of this genus has gills composed of two separate plates. In *Blasturus*, the gill lamellae are produced into one or two blunt lobes at the base of the terminal filament; in *bradleyi*, on the other hand, the gill lamellae are rounded.

P. bradleyi is more primitive than *Blasturus*, but more advanced than *Paraleptophlebia* proper as shown by the following considerations. The first gill of *bradleyi* differs from the other six gills just as does the gill of *Blasturus*. This gill is bifurcate, each branch having a single trachea which lacks lateral branches. The lack of posterior angles on the gills seems to indicate a less highly developed type of gill than that of *Blasturus*. Spieth (1933: 85) believes that "the modifications that the gills have undergone include the flattening of the gill into a foliaceous organ and the changing of the gill from a simple, fingerlike out-pocketing into a double organ consisting of two leaves lying against each other." In all species of *Paraleptophlebia* except *bradleyi*, the seven pairs of gills are similar and without posterior angles. Another major difference between *P. bradleyi* and the remaining species assigned to this genus lies in the structure of gills on segments 2 through 7. In *Paraleptophlebia* s.s. the gills are divided only part way to the base, and overlap, if at all, only to a small extent basally; in *P. bradleyi* and *Blasturus*, on the other hand, the gills are divided to the base and overlap extensively. In all other characters, the nymphs of *P. bradleyi*, the other species of *Paraleptophlebia*, and *Blasturus* are quite similar.

The genitalia of *bradleyi* also seem to indicate that the species occupies a position intermediate between *Blasturus* and *Paraleptophlebia*. The assignment of *bradleyi* to *Paraleptophlebia* has been made on the basis of the equal length of the three caudal filaments of the adult, this character having been used to distinguish *Paraleptophlebia* from *Blasturus* in which the median filament is shorter than the laterals. Ide (1935: 123) has stated with respect to this character:

> . . . Ulmer (1920) remarks: "N. B. Diese Gattung steht *Leptoplebia* so nahe dass sie vielleicht nicht von ihr getrennt zu werden bracht." He is referring to the genus *Blasturus*, erected by Eaton to accommodate species of the *Leptophlebia* type in which the median caudal filament is shortened. Other characters, Eaton found, were not of generic value, the wing venation being similar to *Leptophlebia* and the genitalia of the males very close to the same structures in some of the species of *Leptophlebia*.

Leptophlebia johnsoni, one of the two North American species which Traver refers to this genus, also seems to form a transition between *Leptophlebia* and *Blasturus*.

DISTRIBUTION.—I have found *P. bradleyi* to be scarce and local in distribution in the territory which I have worked. Its range as known at present includes north-central Florida, southeast Georgia, and east Alabama. It seems to be more abundant in the region just south of the Okefenokee Swamp. The species probably spreads throughout south Georgia, but is doubtless confined to the Coastal Plain. The species has hitherto been known only from the type series; all other records are my own (map 11).

ECOLOGY.—The nymphs prefer slow-flowing streams with silty bottoms and dwell amid the leaf drift, on submergent vegetation, sticks, and almost any other available support where they may be found in cracks or crevices, as well as on the protected sides of these objects. As a rule, they shun the center of the stream and live where there is little movement in the water. The nymphs were taken from the middle of only one stream, the Fenholloway River, where the flow was almost negligible. In another locality, the Styx River presented a picture of stagnation with typical vegetation of this type of habitat. Yet *bradleyi* was found here in association with other mayflies, some of which are typical inhabitants of standing bodies of water. These two streams drain flatwoods and swamps which give the water a brownish tinge and also produce a definitely acid condition (pH 5.6 and 5.8, respectively).

The ephemerids which are associated with *P. bradleyi* are *Callibaetis floridanus, Caenis diminuta, Stenonema smithae, Habrophlebiodes brunneipennis, Ephemerella trilineata, Siphloplecton speciosum,* and *Blasturus intermedius.*

SEASONS.—My records, which are inconclusive for other seasons, indicate that emergence occurs throughout the winter. I have records of nymphs in their last instar collected in November, December, and February, and of adults taken in February. Nymphs one-third grown were collected in October; those taken in November were half grown. The February specimens were in either their last instar or penultimate instar. The type specimens taken by Bradley were captured on December 21, 27, 28, and 29. The single adult from Hatchet Creek collected in February was first seen as it was being delicately manipulated in the jaws of a spider, but the mayfly was soon taken away from him to become a valuable addition to my collection. After discovering this mayfly to be *bradleyi,* I searched the stream repeatedly for nymphs but never found them.

HABITS.—Nymphs have not been observed, except superficially, in the field. As my collecting trips to distant localities have been merely short excursions to a stream rather than a thorough working of the situation, very little has been learned about *P. bradleyi.*

LIFE HISTORY.—Though nothing definite is known of the life history, the nymphal data available provide a reasonable basis for estimating the length of the life cycle as approximately twelve months. Nymphal history, emergence, and adulthood of related species have been discussed, and *bradleyi* does not differ much from them in these respects.

LOCALITY RECORDS.—Alachua County: Hatchet Creek, February 8, 1938, adults. Columbia County: Falling Creek, February 4, 1938, nymphs; November 13, 1938, nymphs. 11.5 miles north of Lake City, October 27, 1938, nymphs. Hamilton County: 0.6 miles north of Live Oak road at U. S. Highway 41, February 4, 1938, nymphs and adults. Taylor County: Fenholloway River, November 30, 1939, nymphs.

Genus HABROPHLEBIODES Ulmer

Ulmer established *Habrophlebiodes* in 1919—including *betteni* and *americana*—chiefly on genitalia characters and peculiarities in the shape and venation of the metathoracic wings. At present, the genus is known to include four species which can be placed in two groups: *annulata;* and *betteni, americana,* and *brunneipennis. Betteni* and *americana* are very similar and have even been regarded as synonymous. *H. brunneipennis,* although resembling them in genitalia structure, differs in other characteristics, principally wing color.

The species of *Habrophlebiodes* are widely distributed over eastern North America and in the Southwest. *H. betteni* is known from Quebec and North Carolina; *americana* from Ontario to South Carolina; *annulata* has been collected in Oklahoma; and my work has revealed the presence of *brunneipennis* in southern Alabama and in Florida. This discovery is not surprising, since the Coastal Plain has been consistently neglected by the Ephemeropterists. Mountainous and hilly regions are the ideal places to secure abundant and varied collections, but it now begins to be evident that the Coastal Plain, likewise, has its interesting forms. The Appalachians at present constitute the region of greatest apparent abundance for the species of the genus, but future collecting may well show that species of *Habrophlebiodes* are common throughout the mid-portion of the continent. There is as yet no reason to suppose that they have spread westward into the Rocky Mountains.

Habrophlebiodes, Habrophlebia, Choroterpes, Thraulus, and *Thraulodes* among the Leptophlebiinae are all closely related, and are most likely derived from a common basic stock which split off from the ancestors of *Blasturus, Leptophlebia,* and *Paraleptophlebia.*

Habrophlebiodes brunneipennis Berner

TAXONOMY.—For some time the Florida species of *Habrophlebiodes* was thought to be *H. betteni;* however, examination of specimens of both *betteni* and *americana,* identified by McDunnough, immediately proved that the Florida form was new. The most obvious distinction between *brunneipennis* and the two more Northern species lies in the wings. Those of *brunneipennis* are deep amber in color and have strong, dark venation, whereas the wings of *americana* and *betteni* are colorless, or almost so, and the venation is weaker and paler. The genitalia of the three species are similar.

Nymphs of *H. brunneipennis* differ from those of the previously known species of the genus in having spinules on tergite 6, as well as on 7 through 10. It has hitherto been thought that the absence of spinules from tergites 1 through 6 and their presence on 7 through 10 constituted one of the generic characters of *Habrophlebiodes.*

DISTRIBUTION.—In Florida, *H. brunneipennis* is confined to the northern part of the state, the southeasternmost limit of the known range being Alachua County. As is usual with stream forms, even very tolerant ones, *brunneipennis* is absent from the more or less stagnant waters in the lowlands of southern Florida (map 10).

ECOLOGY.—*H. brunneipennis* is found commonly in slow- to moderately swift-flowing streams, where it dwells in the leaf debris. These streams are usually of the sand-bottomed type, but occasionally the nymphs can be taken from slow-flowing, silt-bottomed streams. If leaf debris is scarce, the nymphs may tend to gather on vegetation near shore where they are fairly well protected. *Brunneipennis* inhabits any portion of the stream where the current is moderate to slow and in which there is either vegetation or leaf drift. Nymphs have

PLATE VIII

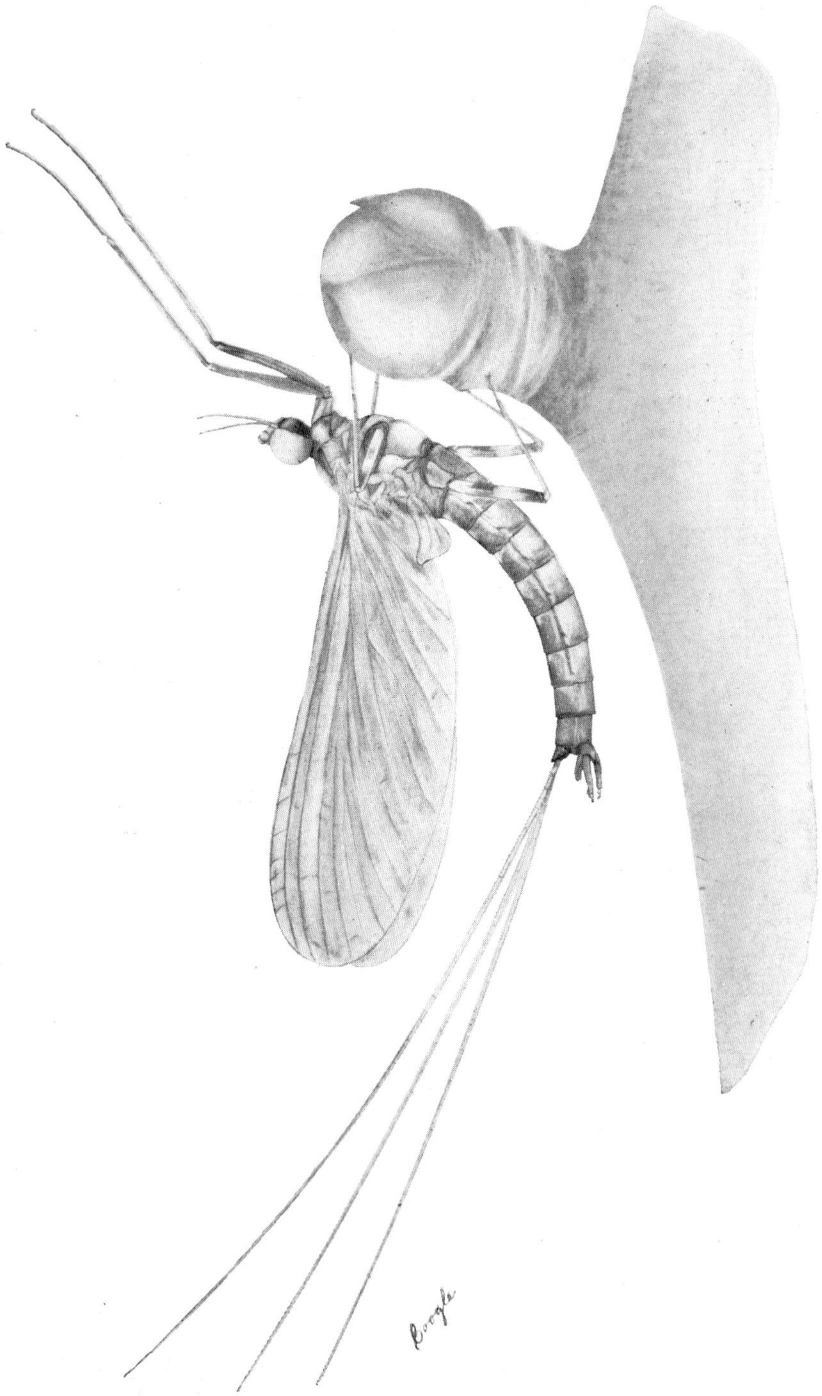

Habrophlebiodes brunneipennis Berner, male imago.

never been taken from rivers nor large streams, but occasionally they occur in medium-sized creeks and are most common in small brooks. The small brooks flowing through the ravines of the Tallahassee Hills have a mayfly fauna consisting almost entirely of *brunneipennis,* but the species is much less frequently encountered in the larger streams of this area.

The well-tracheated gills allow these nymphs a great deal of freedom in moving from one part of the stream to another. I have studied this species most thoroughly in a small sand-bottomed creek about two and one-half miles west of Gainesville. The nymphs are exceedingly common here, particularly in the leaf drift along the margin of the stream and in quiet pools. Roots of terrestrial plants are exposed over large areas in the bottom of this brook, forming a dense mat where the nymphs are easily located. Among the roots, the immatures become clingers rather than sprawlers, because the current is much swifter over these root beds than in the leaf drift. *Brunneipennis* nymphs, in the earlier instars, may often be found among the pebbles in riffles, but they are less numerous in this situation than in quieter, more silty zones.

The streams inhabited by this species drain heavily wooded areas, and in consequence are usually somewhat acid and tinged with brown. Temperatures in the streams rarely fall to freezing, and then only in the almost stagnant areas. The temperature of the water seldom varies more than 15° C.

Common associates of *brunneipennis* are other mayflies and Odonata. The mayflies are *Paraleptophlebia volitans, Blasturus intermedius, Stenonema smithae, Baetis spinosus, B. spiethi, Acentrella ephippiatus, Pseudocloeon alachua,* and *Hexagenia munda marilandica.* Odonata include: *Dromogomphus spinosus, Libellula* spp., *Macromia* sp., *Boyeria vinosa, Ischnura* spp., *Enallgma* spp., *Haeterina* sp., and *Agrion maculatum.* In the leaf drift there are present large numbers of chironomids, caddisfly larvae (*Hydropsyche, Molanna,* and other genera), occasional stoneflies (particularly *Taeniopteryx nivalis*), larvae of the cranefly *Tipula abdominalis,* various other dipterous larvae, and, very commonly, the snail *Physa.* These associates make up the principal macroscopic inhabitants of the leaf drift and riffles of small Florida streams.

SEASONS.—*H. brunneipennis* is not limited by seasons, nymphs of all ages being present in Florida streams throughout the year. Adults have been taken in seven months of the year, leaving gaps only during May, August, September, October, and December. There seems to be no peak of emergence, for examination of the nymphal fauna indicates that throughout the year the distribution of immatures of all ages is approximately the same.

HABITS.—The feeding habits of this species are not definitely known, but I believe that they do not differ radically from those of other mayflies. The well-developed molar area of the mandible proves that grinding of food materials must be habitual. When a nymph is placed in an aquarium with only plant materials (and of course numerous protozoans and microcrustacea), it flourishes and does not seem to lack sufficient food for normal growth. Examination of the nymphal gut of this species indicates that the food materials are probably

PLATE IX

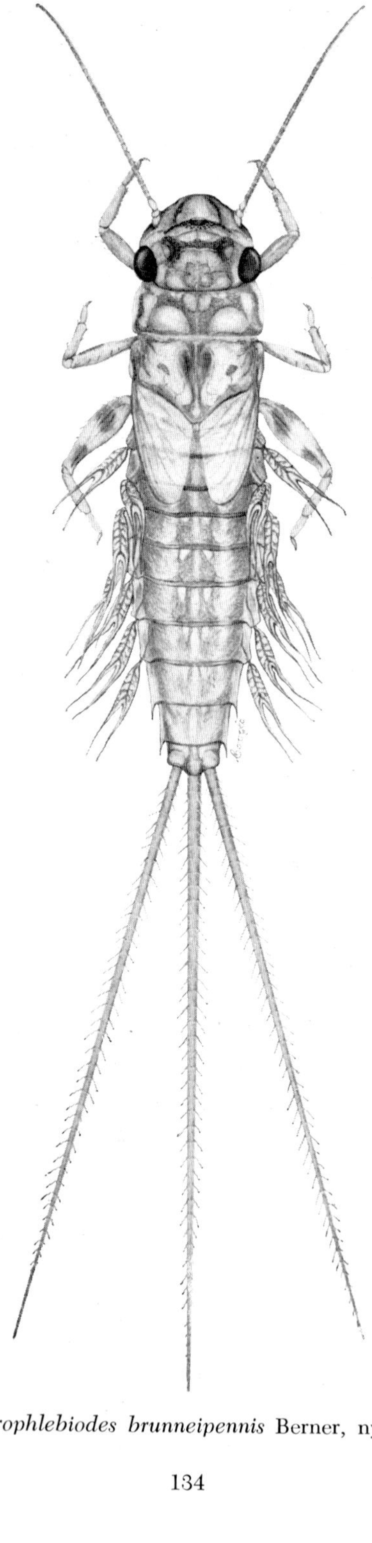

Habrophlebiodes brunneipennis Berner, nymph.

obtained by scraping the surface of dead leaves. Organic substances as well as small grains of quartz sand were found in the enteron.

Although the gills are well tracheated, single, and exposed, the nymphs keep them in constant motion. When the nymphs are at rest, the gills are held stiffly out from the body at a 45° angle to the plane of the dorsum, and slowly waved forward and backward. In swimming, the nymphs avoid the interference of the gills by appressing them against the abdomen.

Like all Leptophlebines, these nymphs are rather awkward swimmers that usually move most efficiently by crawling. Swimming requires a great amount of exertion and is not at all natural in this genus; the swimming movements can in no way compare in gracefulness with the darting of the Baetinae. The nymphs swim by a series of undulatory motions, the wave beginning cephalically and traveling to the tails which are depressed or raised as a unit. The caudal filaments aid very little in propelling the nymph; almost naked, they are little more than long hairs trailing behind.

Leptophlebine nymphs can be recognized immediately by their movements on the surface of a white-enameled pan. When the water is drained away from them, the insects crawl in the wake of the liquid, vigorously wriggling the abdomen and tails from side to side in a sort of scrambling effort to hide themselves beneath any convenient object.

As in many of the Ephemeroptera, *brunneipennis* exhibits a definite negative phototropism in the nymphal stage and a reversal of this condition in the subimaginal and imaginal stages. Rheotropism is also exhibited by the nymph when it is placed in any sort of current.

The orientation of *Hexagenia variabilis* adults to air currents was discussed by Krecker in 1915. *H. brunneipennis* exhibits this reaction extremely well. By blowing gently on the adult, the insect can be made to turn in almost any direction, and it is only when the strength of the air current becomes excessive that the adult is forced to fly. Krecker pointed out (1915: 384) that orientation normally occurs with the head turned toward the air current. From his experiments, he concluded that the Ephemeroptera do not orient positively to a breeze because of sensations derived from the breeze itself, but rather because of tension exerted on the leg and wing muscles.

LIFE HISTORY.—Development of nymphs from egg to adult probably takes six to eight months, but the length of this developmental period is not definitely known. Numerous attempts to hatch eggs have failed. I have repeatedly tried to bring about artificial fertilization and have attempted to induce parthenogenetic development, but without success. In addition, eggs which were stripped from apparently impregnated females were placed in small dishes, some of which were unaerated, others well aerated. At the end of three weeks no development could be discerned in any of the eggs. Clemens (1915) found that the eggs of *Ameletus ludens,* a parthenogenetic mayfly, took five months to incubate. He correlated this long incubation period with the fact that the nymphs of this species live in intermittent brooks, and he explained this relationship by the assumption that drying is necessary for development. This is certainly not the case with *H. brunneipennis,* for the streams which it occupies are permanent. The eggs are elongated, with the sides parallel, and the ends are rounded. Upon deposition they immediately settle to the bottom, where they adhere to any solid object. These eggs do not send out tenacious threads as do the eggs of many other species.

The oviposition of *betteni* was briefly mentioned by Morrison (1919: 144) in a discussion of the mayfly ovipositor. The ovipositor is a structure peculiar to the Leptophlebine mayflies, and is best developed in *betteni* and *brunneipennis* among the North American species.

The oviposition of *H. brunneipennis* was observed under artificial conditions in the laboratory. Believing that I had a female which had mated, I held her firmly by the wings and raised and lowered the body, touching it to the surface of the water rhythmically. Soon the female began to release her eggs upon each contact. They came out usually two at a time, but occasionally singly, passing between the egg valve and the ovipositor (fig. 86) and slipping out as though in a greased trough. As oviposition occurred, the forward end of the abdomen was depressed, the posterior portion (that is, segments 8 through 10 and the tails) was elevated. At first the eggs were ejected in a continuous stream, but as the female became spent, the flow of eggs gradually grew slower until only an occasional egg left the body. In another female, as the eggs were released the abdomen became somewhat transparent and the release of the eggs into the posterior part of the oviducts could be easily observed. They literally popped into place with machine-like regularity and precision. As the eggs were freed from the body, the female would twitch her abdomen, apparently to disseminate them. Oviposition continued until not an ovum remained in the body. Once started, the process continued uninterruptedly, eggs passing from the body even when the female was not in contact with the water. They accumulated around the ovipositor, but as soon as the female touched the surface, the eggs were dispersed.

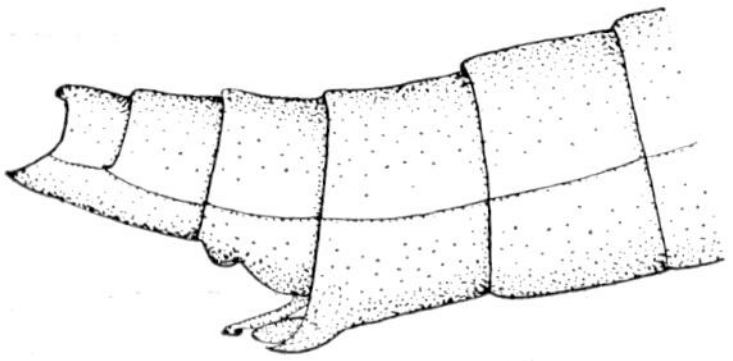

Fig. 86.—The terminal abdominal segments of a female *Habrophlebiodes brunneipennis* showing the ovipositor and the passage of an egg from the body.

Transformation usually occurs in the early afternoon from 12:00 noon to 3:00 P.M., depending on the season, but it has been known to take place as early as 10:30 A.M. The process has been observed only in the laboratory. When ready to transform, the mature nymph slowly crawls out of the water, sometimes as far as an inch above the surface. After a few moments, the thorax begins to pulsate, the nymphal skin splits, and the adult thorax and abdomen are gradually drawn forth, followed by the wings, legs, and tails. The subimago slowly crawls away from the skin for a short distance, then flies to a nearby support. Along the banks of streams, subimagoes can be found on the undersides of leaves about four or five feet above the water. The shaded sides of bridges also are favored resting places. Subimaginal life lasts only twelve to fourteen hours. In captivity, this species has little difficulty in undergoing its ultimate molt so long as the room is not too dry.

Life as an adult is little more than a day. Two males were collected as subimagoes at 3:00 P.M. on November 11; by 9:20 P.M. on November 12 one male was dead and the other showed only slight movements. Other records on length of life of adults have confirmed the conclusion that *H. brunneipennis* is typically a very short-lived species.

The mating flight of *Habrophlebiodes* has never been described. I was, however, fortunate enough to observe it on a number of occasions. Mating occurs in early afternoon, and as with emergence, the time depends on the season.

My first observation of a mating flight happened on March 17, 1940, as I was wading in a stream at 2:15 P.M. I sighted the small insects flying about six inches above the water, and captured five of them, all of which were males of *H. brunneipennis*. They were flying in an area cleared of vegetation and bathed in sunlight. The mayflies were in two separate swarms, both small, consisting of only five or six members. After sweeping my net through one of the swarms, it was reduced to two individuals which flew continuously for ten minutes, when two other males joined them. Two females then flew into the small swarm and began rising and falling with the males. The males then approached the females from below, and, as soon as coupled, the two pairs flew to shore and were lost in the shadows. Twenty minutes later the third male paired. Within a few minutes another male joined the remaining one. These flew in unison and then the first male approached the new arrival to assume the copulatory position, but seemed to be repulsed as the two immediately separated and began flying in unison again. Once more copulation was attempted, the attempt lasting not more than two or three seconds. The new arrival finally withdrew and was lost to sight. After fifty minutes, a female approached the remaining male and coupled with him. The two flew slowly toward shore and parted after about ten seconds, the male returning to midstream, the female becoming lost in the shadows. Exactly one hour after I began my observations, the last male ceased its flight.

The insects did not at any time rise more than twelve inches above the surface of the stream and horizontal distance covered did not exceed three feet. The flight was fairly rapid. As I swept my net, many of the mayflies were able to escape the gentle movements that it was necessary to use in order to capture the males without injury.

The entire flight took place in bright sunlight except toward the last. As the sun shifted, trees began to cast shadows over the area in which the remaining single male was dancing, but the shadows had little effect on his position. The flight was usually in a downstream direction against the wind, but occasionally it became transverse and, once or twice, the insect was headed upstream. The male seemed to fly backward, forward, sideways, or obliquely with equal ease. When directed downstream, the flight of the insect was forward and backward, but when crosswise, the flight was from side to side. During the flight, the abdomen drooped considerably, being held at an angle of about sixty degrees or more with the horizontal, and sometimes becoming almost perpendicular. The tails could sometimes be seen if the light happened to strike them at the correct angle.

The insect markedly resembled a mechanical toy attached to an invisible string. Running the length of the string, the toy is suddenly jerked backward, then immediately runs out again. The movement during the mating flight is diagramed in figure 87.

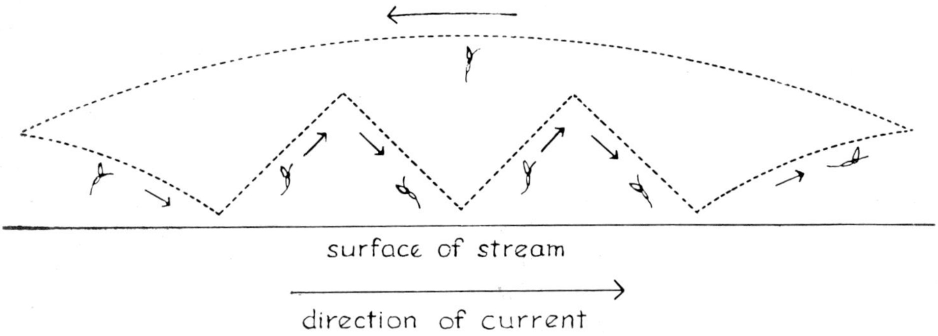

Fig. 87.—A diagrammatic representation of the characteristic movements of the males of *Habrophlebiodes brunneipennis* during the mating flight.

Almost never during the period of swarming did the insect do an about face to move upstream. Now and then, the mayfly seemed to touch the surface of the water and to rise immediately.

Spider webs (of *Tetragnatha* sp.) are common along the creek where flights have been observed, and in nearly every case the webs were spun approximately six to eight inches above the water. Numerous *H. brunneipennis* adults have been taken from the webs. Every specimen taken here on various occasions in March and April, 1940, between the hours of 2:00 P.M. and 3:00 P.M., was still alive; apparently the spiders were not yet aware of their catches. These adults had undoubtedly come from mating swarms.

On May 6, 1940, I again visited the creek where my first observations of mating behavior were made. This time I observed numerous flights of *H. brunneipennis*. The flight resembled that of March 17; however, the swarms were much larger and were composed of seven to ten individuals, all of which were males. The body of the flying insects was usually perpendicular and now and then even seemed to be leaning over backward. I saw no copulating pairs on this occasion and only a single female was spotted as it was flying alone. After flying for a considerable time, the males dispersed, marking the end of the flight. Seemingly, there was no reason for this breaking up of the swarm as no mating took place. The males flew in unison but frequently they would "get out of step." The insects were so insensitive to outside influences that once I managed to get within twelve inches of a small swarm without disturbing it. Males were exceedingly common. Sweeping the bushes after the flights had culminated, I secured a large number of them, but not a single female was taken. Perhaps it is because they fly away from the stream after mating. The flights of the males took place most often in the shade, but frequently they moved into the sunlight. Even though subimagoes emerge early in the afternoon in the laboratory, I did not take one all afternoon, in spite of the fact that adults were so common. Spider webs were not so numerous as on previous dates.

The female which I noted flying alone was near a spider web. After a few backward and forward movements, she became entangled in the web. This occurred as she was backing up. It is likely that it is on the backward flight that the adults are caught by the spiders.

Again on January 30, 1941, I was fortunate enough to observe mating flights, but this time two distinct types of behavior were exhibited. As I walked downstream for a distance of about 250 yards, I saw twenty-one separate mating flights, with a swarm in every sunlit spot. The swarms varied in size from those with two males to those with nearly one hundred. I also observed two pairs in copulation, but only for a moment before they were lost in shadows. There seemed to be a definite correlation between size of swarm and height of flight. The very small groups flew six to twelve inches above the water whereas the large swarms of twenty to one hundred males flew from two to five feet above the water. The manner of flight in large swarms did not differ from that of small except in height above the stream.

LOCALITY RECORDS.—Alachua County: Devil's Mill Hopper, April 18, 1933, nymphs; October 25, 1937, nymphs; March 5, 1938, nymphs and adults. Near Worthington Springs, February 5, 1939, nymphs. 2½ miles west of Gainesville, January 16, 1938, nymphs and adults; January 29, 1938, nymphs and adults; February 3, 1938, adults; February 6, 1938, adults; March 5, 1938, nymphs and adults; June 18, 1938, nymphs and adults; January 7, 1939, nymphs and adults; January 28, 1939, nymphs and adults; March 10, 1939, nymphs and adults; November 11, 1939, adults; February 5, 1940, nymphs and adults; March 18, 1940, adults; March 19, 1940, adults; April 17, 1940, adults; January 30, 1941, nymphs and adults; January 9, 1949, nymphs. Experiment Station, University of Florida, January 15, 1940, nymphs and adults. Bay County: 5.6 miles north of Panama City, November 5, 1938, nymphs; May 30, 1940, nymphs. Columbia County: Falling Creek, November 13, 1938, nymphs. Hamilton County: 8.3 miles south of Jasper, February 4, 1938, nymphs. Jackson County: 2.9 miles north of Altha, July 1, 1939, nymphs and adults. 3.6 miles north of Altha, July 1, 1939, nymphs and adults. Jefferson County: April 1, 1938, nymphs. Drifton, February 5, 1938, nymphs. Leon County: 11.2 miles west of Tallahassee, March 16, 1939, nymphs and adults. 16.9 miles west of Tallahassee, March 17, 1939, nymphs. 7 miles north of Tallahassee, March 18, 1939, nymphs. 13 miles west of Tallahassee, November 30, 1939, nymphs. Liberty County: Sweetwater Creek, June 10, 1938, nymphs; May 2, 1941, nymphs; April 29, 1946, nymphs. Santa Rosa County: 7.1 miles west of Milton, April 4, 1938, nymphs. Wakulla County: Smith Creek, June 5, 1938, nymphs. Walton County: 7.3 miles west of Ebro, June 7, 1938, nymphs. 5.4 miles east of Freeport, April 2, 1938, adults; May 2, 1946, nymphs. 13.8 miles west of Freeport, June 7, 1938, nymphs. 10.6 miles west of Washington County line, May 31, 1940, nymphs. 2.1 miles west of Washington County line, May 31, 1940, nymphs.

Genus CHOROTERPES Eaton

In the nymphal stage, *Choroterpes* is quite distinctive by virtue of the morphology of its gills; the adults are easily separated from other mayflies by the structure of the male genitalia and the venation of the metathoracic wings. Eaton erected this genus in 1881 to include the one species which he knew, *C. picteti* of Europe.*

Even though the genus is easily distinguished, the species within the genus are difficult to separate. There are nine species known at present. For the most part the species are separated by color pattern and by the shape of the basal segment of the forceps of the males.

Choroterpes is a Neotropical, Holarctic, and Indo-Australian genus. In the Nearctic it is widely distributed; published records are from Alberta to California, and from Ontario to Texas. The Florida species is the first record of the occurrence of the genus in the Southeast.

Leptophlebia, Blasturus, Thraulus, and *Choroterpes* form a closely knit group, but Spieth has concluded in his phylogenetic study that *Leptophlebia* and *Blasturus* present a closer affinity to each other than to *Thraulus* and *Choroterpes.*

Choroterpes hubbelli Berner

TAXONOMY.—*Choroterpes hubbelli* can be distinguished from all other species in the genus by its dark abdomen and pale, unbanded caudal filaments. The species was described in 1946. Since the nymph of *C. basalis* is the only described immature of the genus, it would be fruitless to discuss the taxonomic relations of *C. hubbelli* nymphs.

DISTRIBUTION.—In Florida the species is widely distributed, its range extending from the western part of Duval County to the northeastern part of Hillsborough County, with a greater concentration in the north-central portion of the state; *Choroterpes* next is found in Bay and Walton counties. No other Florida records are known, but nymphs have been taken in Thomas County, Georgia, just above Jefferson and Leon counties in Florida. Other nymphs in my collection (species uncertain) are from Toombs County, Georgia, and Nashville, Tennessee. It is very likely that *C. hubbelli* occurs in the streams between Alachua and Bay counties, even though it has not appeared in the many collections made between the two regions.

*Eaton also had an undescribed species from Arizona before him at the time.

It is probable that *C. hubbelli* is a Coastal Plain species occurring throughout that province in Florida, Georgia, and Alabama wherever ecological conditions permit (map 12).

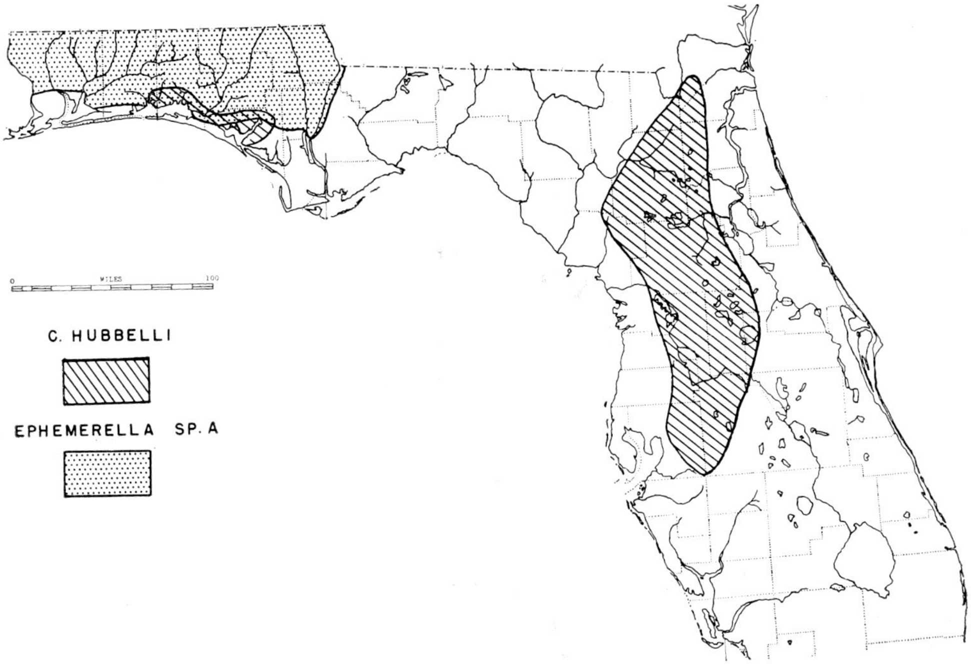

Map 12.—The distribution of *Choroterpes hubbelli* and *Ephemerella* sp. A in Florida.

ECOLOGY.—Among the unexpected places in which *hubbelli* has appeared, the most spectacular is Jerome Sink. This sink is located approximately two miles north of Newberry in a perennially arid region. Drainage in this part of Florida is subterranean, the nearest surface stream being the Santa Fe River, which is about twelve miles distant. Underlying the Newberry region, there are many subsurface streams, but very few are exposed at the surface. Of those which are exposed, only a small portion of the water is in evidence. The opening to these subterranean waters is usually a steep-sided sinkhole, sometimes even well-like, the sides of which may be covered with various types of ferns. Flow is not evident in any of those which I have observed, though the various underground channels do seem to connect as indicated by the distribution of the white cave crayfish, *Procambarus lucifugus alachua* (see Hobbs, 1942: 138 and 1944: 6-8). The walls of Jerome Sink extend thirty feet downward to the surface of the water, and thence an unknown distance to the bottom of the sink. That part of the open east wall lying below the surface of the water probably connects the sink water with the subterranean drainage. Along the west shore of the sink, the bottom drops off rapidly, but there is sufficient shore to allow an accumulation of dead leaves and other debris. It is here that *Choroterpes* nymphs occur along with *Epiaeschina heros*, a dragonfly which

PLATE X

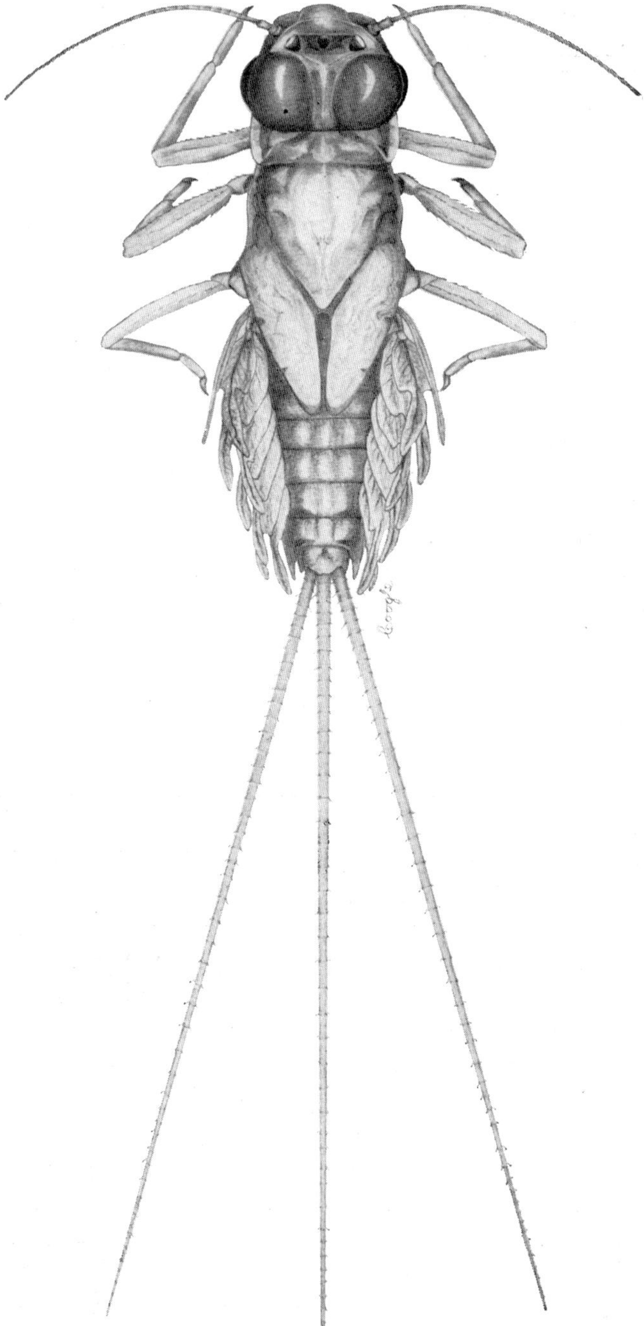

Choroterpes hubbelli Berner, nymph.

Byers (1930: 256) says "lives clinging to debris caught in moving water." Other insects are rare, but *Erimystax* sp., one of the small, stream-inhabiting minnows, is rather common.

Very few species of mayflies inhabit the shores of sand-bottomed lakes in Florida, but *hubbelli* is one of the exceptions. Partially buried sticks, logs, and boards usually appear on the bare sand in shallow water, and their underside crevices sometimes provide habitats for three species of ephemerids—*Choroterpes hubbelli, Ephemerella trilineata,* and *Stenonema proximum*—of which *Choroterpes* is the scarcest.

C. hubbelli apparently prefers streams where it may inhabit the leaf drift in the slow-flowing reaches. One of the richest *Choroterpes* streams which I have examined is located in Walton County. At that time it coursed over a bed of silt varying from one to three or more feet; to reach the main channel, two smaller branches, each with its own bed of soft silt, had to be traversed. In several places it was only by holding to the pilings of the bridge over the stream that I was able to manipulate myself into a position suitable for collecting nymphs. The floor of the creek was paved with many layers of leaves intermingled with silt, and the margins of the bed were bordered with a dense growth of *Isnardia* and *Pontederia,* which were absent from the main flow. Among the silt-imbedded leaves there were numerous mayfly nymphs, the majority being *hubbelli.*

The nymphs frequently show up in unusual places. While collecting in a small, sand-bottomed creek emptying into the Choctawhatchee Bay, Dr. F. N. Young and I overturned a large log lying not more than twelve feet from the salt water. The underside of this old stump was literally covered with mayfly nymphs, chiefly *Choroterpes hubbelli,* which were scrambling in all directions. The water around the log from which the mayflies were collected was perfectly fresh, there being a continuous flow unaffected by the tides. I have been informed by the Coast and Geodetic Survey that tidal fluctuation in the Choctawhatchee Bay amounts to only a few inches, hardly sufficient either to overflow the sand bar separating the fresh from the salt water or to cause the salt water to back up into the creek.

The streams in which nymphs of *C. hubbelli* live are either slightly acid or circumneutral. Red Water Lake, one of the two lakes from which specimens have been taken, derives its name from the color of its water which is definitely acidic. As can be seen from the distribution of the mayfly, the nymphs shun the nonflowing waters of the eastern and southern part of Florida.

In sand-bottomed streams, *C. hubbelli* may be associated with numerous other mayfly species—*Acentrella ephippiatus, Baetis spiethi, B. spinosus, B. australis, Centroptilum viridocularis, Habrophlebiodes brunneipennis, Paraleptophlebia volitans, Tricorythodes albilineatus, Ephemerella trilineata, Stenonema proximum, Stenonema smithae,* and *S. exiguum.* In the silt-bottomed creeks, its associates are *Blasturus intermedius, Habrophlebiodes brunneipennis, Stenonema smithae, Cloeon rubropictum, Baetis spinosus,* and *Hexagenia munda marilandica.* The lake associates have been mentioned previously, and the only mayfly inhabiting Jerome Sink is *C. hubbelli.*

SEASONS.—I have taken or reared adults during the early part of March, in April, May, June, July, August, and October. Though there are no records for nymphs or adults during any other month, I believe that the emergence, which occurs throughout the year, reaches its peak during the spring and early summer. Emergence records for other species show that the Northern forms emerge from July to the early part of September. *C. nanita* from Texas was described from specimens taken in June; the types of *C. oklahoma* were collected on March 20. Eaton, in describing *C. inornata* from Mexico and Arizona and *C. nervosa* from Guatemala, failed to give the date of collection. It seems likely, however, that these species follow the same departure from the Northern species as does *hubbelli.*

HABITS.—As in other species of the subfamily, *C. hubbelli* is negatively phototropic and strongly thigmotactic in the nymphal stage. The adult, particularly during the subimaginal stage, shows a strong positive reaction to light and this phototactic response has frequently been demonstrated in the laboratory.

Food materials are much the same as those utilized by *Habrophlebiodes brunneipennis,* that is, plant materials and probably some protozoans. When nymphs are kept in an aquarium with only dead leaves, they seem to grow perfectly well on a diet of scrapings from the surface of these leaves, and have been kept alive almost two months on such food.

Though body movements of nymphs are similar to those of *Habrophlebiodes,* the nymphs are easily distinguished from other Leptophlebiinae by the fact that the gills, which are not spread when the insect is submerged, are held above and close to the abdomen where they are frequently vibrated.

LIFE HISTORY.—Nymphal data indicate that underwater life may occupy a period of from six to eight months. So few adults were reared that the length of life of the imago was not determined, but it appears likely that death will take place within fifteen to twenty-four hours after the final molt. Transformation in the laboratory, which occurs after dark, has taken place as late as 9:30 P.M.

When the subimago is ready to emerge, the nymph, like *Blasturus,* swims vigorously; however, it does not crawl from the water, but bursts free at the surface, floats for a moment on the exuviae as a subimago, and then flies to a nearby support. Transformation is over in a few seconds. The whole process, after the swimming movements have been completed, takes no more than thirty seconds. Once emergence has taken place, the subimago quietly awaits the final molt—an event which almost never happens in the laboratory under normal room conditions. I obtained one male and one female imago by placing subimagoes in paper sacks with wet blotters in the bottom. Two other adults molted when I used the bell-jar method discussed under *Blasturus intermedius.* Subimaginal life lasts eight to ten hours, and in some cases may be even shorter.

The female of this species, as of most other mayfly species, molts with greater ease than does the male.

Choroterpes hubbelli mating flights have never been observed. A search of the literature reveals that the only published observations of this phenomenon in other species are those of Needham (1905: 38-39) and Morgan (1913: 392). Needham has described the flight of *C. basalis* as occurring in small swarms in early afternoon. The insects flew high above the water, seldom descending to lower levels. Morgan, who has observed the mating flight of the same species, noted a swarm of three or four hundred individuals of *Choroterpes basalis* flying over Fall Creek at about 4:00 P.M. The swarm included both males and females.

LOCALITY RECORDS.—Alachua County: 2½ miles north of Newberry, Jerome Sink, April 21, 1939, nymphs and adults; April 22, 1939, nymphs; adults reared April 22 - May 17, 1939; March 1, 1940, nymphs; adults reared March 4 - 13, 1940; March 8, 1946, adult. Hatchet Creek, May 6, 1939, adults; April 24, 1948, adults. Bay County: 5.6 miles north of Panama City, May 30, 1940, nymphs. Duval County: 11 miles north of Jacksonville, August 28, 1938, nymphs. Marion County: Rainbow Springs, March 9, 1940, nymphs; adults reared March 13, 1940. Putnam County: Red Water Lake, March 26, 1939, nymphs; adults reared March 31 - April 3, 1939; October 20, 1940, nymphs. Walton County: 2.1 miles west of Walton County line, May 31, 1940, nymphs. 9.5 miles west of Portland, June 7, 1938, nymphs; May 31, 1940, nymphs and adults.

Genus HABROPHLEBIA Eaton

The genus *Habrophlebia* is a somewhat more diversified genus in Europe than in North America, several species having been recorded from the former, whereas only two have been described from the latter.

The nymphs are most easily distinguished from other Leptophlebiinae by the structure of the abdominal gills, which consist of two clusters of slender filaments. Likewise, the adults are easily identified by the shape of the male genital organs, and by the venation of the metathoracic wings.

Habrophlebia is known in North America only from the East and Southeast, but in the Southeast, published records of its distribution include only mountainous regions.

In this genus, gills and genitalia, two excellent phylogenetic indicators, differ markedly from those of any other in the subfamily. The hind wings offer still a better indication of relationships, and on this basis, *Habrophlebia* would seem to be derived from the same stock which gave rise to *Choroterpes, Habrophlebiodes, Thraulodes,* and *Thraulus,* though its closest affinities are with *Thraulodes.*

Habrophlebia vibrans Needham

TAXONOMY.—*Habrophlebia vibrans* was described from New York by Needham in 1907 and recorded from the Black Mountains of North Carolina by Banks in 1914 as *H. jacosa.** These are the only references to this species until 1935 when Traver redescribed the form in *The Biology of Mayflies.* During Dr. Traver's study of the mayfly fauna of North Carolina, she did not take a single specimen of *vibrans,* but did take a new species which she designated *H. pusilla.*

H. vibrans is either very elusive or very rare in Florida, for during the entire period of my research, only three nymphs have been collected. My identification of the nymphs is based on a comparative study made with a nymph taken from a stream in the mountains of north Georgia where, at the same time, two male imagoes were also captured. Color pattern, gill structure, mouth parts, and other structures correspond very closely in the two groups of nymphs (map 8).

ECOLOGY.—So far as I can determine, there has been nothing written concerning the ecology of the North American species of *Habrophlebia,* and very little can be added about the Florida nymphs. I took the first immature in December from the leaf debris of a small, sand-bottomed stream in northwest Florida. The bottom of the creek, which was heavily covered in places with leaf drift, was quite silty. As I did not collect the second nymph, I have no information about its habitat, except that it was a swift-flowing, sand-bottomed stream. The third nymph came from the same habitat as the first and was collected in February, 1949.

Associates of the nymph from the leaf drift were *Blasturus intermedius, Habrophlebiodes brunneipennis, Paraleptophlebia volitans, Stenonema smithae,* and *Hexagenia munda marilandica.* Damselflies, dragonflies, stoneflies, and the usual associates of the sand-silt stream bed were also present. This stream was acidic, having a pH of 6.2.

SEASONS.—The December and February nymphs are somewhat better than half grown; the April specimen is in its last instar. No conclusions can be drawn from such scanty data, but it is evident that emergence must take place during the spring. Of *vibrans* in Quebec, McDunnough says that adults are very common in late June; my Georgia specimens were taken in late April.

HABITS.—There is nothing known concerning the habits of the North American nymphs, but I assume that they are much the same as those of *Habrophlebiodes brunneipennis.* Dr. Needham's comment (1935: 102) on their mating flight is the only reference to the habits of adults. He describes *H. vibrans* as flying in small, compact swarms in forest openings beside brooks.

LIFE HISTORY.—Nothing is known of the life history, but considering that of related genera, I would say that to pass from egg to adult requires eight to ten months.

**H. jacosa* was synonymized with *vibrans* by McDunnough in 1925.

LOCALITY RECORDS.—Jackson County: 3.6 miles north of Altha, December 10, 1937, nymphs; February 3, 1949, nymphs. Okaloosa County: 2.3 miles east of Niceville, April 3, 1938, nymphs.

Genus BAETISCA Walsh

The genus *Baetisca* is not a particularly large one, being composed of only eight species, three of which were described since 1934. These insects are so distinctive that Ulmer (1933: 209) placed them in a separate family (Baetiscidae), Spieth (1933: 357) in a separate superfamily (Baetiscoidea), and Traver (1935: 555) in a separate subfamily (Baetiscinae).

Baetisca nymphs are the most unusual looking mayfly nymphs in North America. With their stout, spinous, humped-up bodies, they might well rival some of the most bizarre types of dinosaurs of the Mesozoic. In general, the adults of the genus are not as readily identified as are the nymphs. As Needham (1935: 235) observes, "In such genera *(Baetisca* and *Ephemerella)* differential characters are often better developed in the nymphs than in the adult. In general we believe that species are best described in the state that is most differentiated, and therefore, most easily recognized." Frequently, nymphs of one group of a genus may be readily identified, whereas in another group of the same genus, there are no really good separating characters.

Baetisca can be divided into two groups on the basis of the presence or absence of an orange or reddish coloration in the basal part of the wings. The *obesa* group, in which there is no coloration, would include *B. obesa, B. lacustris, B. laurentina,* and *B. bajkovi* of which the nymph is the only form known. Into the *rubescens* group would fall *B. rubescens, B. carolina, B. thomsenae,* and *B. rogersi,* all of which have some coloration in the basal portion of the wings. To some extent, the geographic distribution of these species bears out this division—the former group being predominantly Northern and the latter seeming to show tendencies toward a Southern distribution.

Though the genus *Baetisca* is Nearctic, occurring in eastern North America, except for *B. bajkovi* from eastern Manitoba, it has never before been recorded from the Coastal Plain.*

According to Spieth (1933: 359), *Baetisca* seems to be a distinct entity in the phylogenetic arrangement of mayflies with an ancestral stock that presumably separated early from the remainder of the Ephemeroptera. *Baetisca,* although a highly specialized form, exhibits such ancestral characters as a relatively primitive venation, a large hind wing, a highly developed cubito-anal area in the forewing, the forceps structure, the similar laciniae mobiles, and the highly modified, but single, gills. Needham considers *Baetisca,* along with *Oreianthus,* to be one of the isolated remnants in the mayfly fauna of North America.

* *B. obesa* recorded from Sandy Creek, Georgia, by Traver (1935: 561) may occur in the Coastal Plain.

Baetisca rogersi Berner

TAXONOMY.—*Baetisca rogersi,* the most recently described species of the genus, is so distinctive in the nymphal stage that it is not likely to be confused with any other known species. Its outstanding characteristics are the presence of four prominent lateral thoracic projections, small dorsal thoracic prominences, serrate margins on the mesonotum, and prominent elevations on abdominal segments 6 through 9. The adults may be distinguished chiefly by the reddish-brown coloration in the basal third of the forewings and the basal three-fourths of the hind wings. They are very bulky creatures with a robust thorax, short abdomen, long forewings, and large hind wings.

DISTRIBUTION.—*B. rogersi* is known to occur in Florida only in the northwest portion. The species has been taken from one locality east of the Apalachicola River drainage, which is not more than fifteen miles distant from the river. *Rogersi* is recorded as far westward as Santa Rosa County in Florida and Escambia County in Alabama.

The species most closely related to *B. rogersi* are *carolina* and *thomsenae,* both of which are North Carolina forms. *B. carolina* is known from the Piedmont and Appalachian provinces; *B. thomsenae* has been found only in the Appalachian (map 8).

ECOLOGY.—With so few rocks in the stream beds of the sand-bottomed streams of Florida, the peculiar *B. rogersi* nymphs must adapt themselves to the next best situations. The pebbly riffles in shallow water seem to meet their requirements better than any others, as is indicated by the number of nymphs in them. Dr. Traver's description of the habits of *B. thomsenae* (1937) indicates that the immatures of the two species occupy similar habitats. The greater development of the lateral spines in the short, rotund nymphs of these two species is probably correlated with the ability to maintain themselves in moderate- to swift-flowing water. Nearly fifty nymphs were collected from two small riffles in a sand-bottomed creek in Gadsden County, where they lay among the small pebbles, partly covered by sand and with their heads upstream.

A few nymphs have been taken from streams emptying into the Choctawhatchee Bay. One specimen was collected by sweeping a dip net through the dense mass of *Vallisneria, Potamogeton,* and *Sagittaria* which choked the stream. In the same stream, another nymph was seen clinging to the underside of a submerged log caught in the vegetation. A second creek near Niceville yielded three nymphs from the undersides of boards which were partially imbedded in the stream. The boards were partly covered with sand, and many sand particles adhered to the undersurfaces, making it difficult to discern the nymphs. One of the unusual aspects of these habitats was their proximity to salt water. The former stream is completely fresh beyond the margin of the bay, even though the flow is not extraordinarily rapid. Though *Baetisca* was taken not more than one hundred yards from the salt water, there is no doubt that the nymphs in this stream could have lived much closer, for nymphs of *Choroterpes hubbelli* and *Ephemerella trilineata* were taken within twenty-five feet of the bay.

PLATE XI

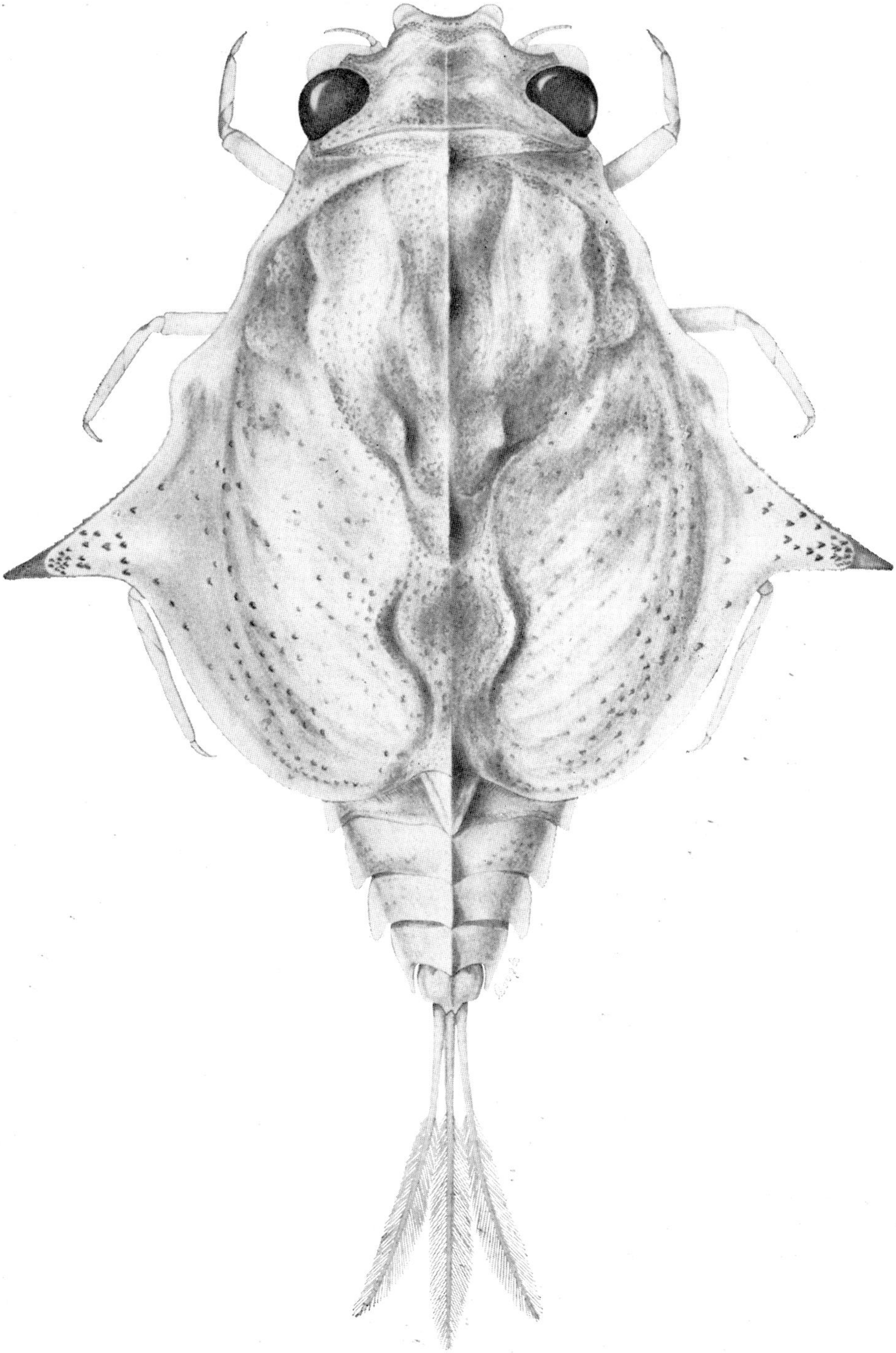

Baetisca rogersi Berner, nymph.

At the Niceville stream, nymphs were collected from within five feet of the salt water, where the creek, though very shallow and wide, had a moderately swift flow.

All streams in which I have found *Baetisca* nymphs have rather clear water, and are sand-bottomed as well as slightly acid or circumneutral. Because these streams drain upland regions where there is little swampland, the water picks up only a small amount of humic acid. During the winter, I did not take temperature readings in these creeks. The usually heavy frosts in the western part of the state, however, would seem to indicate that the water temperatures were low. In some of the more slow-flowing streams west of the Apalachicola River, I have seen ice near the banks in December. Growth of nymphs is slowed considerably by the cold, but as soon as warm weather returns, more rapid growth is resumed. Ide has shown that emergence is correlated with temperature. Even though many Florida species are little affected by winter temperatures, several of those in the northwestern part of the state most certainly are.

The mayfly nymphs associated with *Baetisca rogersi* nymphs in the riffles are *Stenonema smithae,* of which younger forms predominate, *Baetis spinosus, B. australis,* and *Pseudocloeon punctiventris.* Now and then, a specimen of *S. exiguum,* one of the Isonychias, or a young Leptophlebine may be taken from the riffles. On the underside of the boards on which *Baetisca* nymphs were found were also *S. smithae* and *S. exiguum.* The eelgrass association is the same as that of *Pseudocloeon bimaculatus.*

The adults, like other mayflies, are normally confined to the mesophytic hammocks lining the stream margins. Here the humidity is fairly high and the adults experience little difficulty in undergoing their imaginal molt. The subimagoes, upon emerging, fly to the lower branches of the marginal trees and bushes where they await the final molt.

SEASONS.—The emergence of *B. rogersi* corresponds with that of Northern relatives in that it is strictly seasonal; the season, however, is somewhat extended in Florida. Adults begin to emerge early in March and continue until late summer. My earliest record for mature nymphs is March 17; my latest for two-thirds grown nymphs, the early part of June. I have collected in many Florida streams for the nymphs of *B. rogersi,* but for some unknown reason, they have appeared only occasionally in collections. A stream from which many nymphs were taken in March, 1938, did not yield a single specimen in June, 1940. I have found no habitat where nymphs can be obtained regularly. Repeated collecting from other streams where I had previously recorded them failed to reveal a single immature. In March I collected nymphs one-half and two-thirds grown, as well as those in the last instar—the younger nymphs were most likely destined to emerge sometime in middle summer. One of the most interesting records is that of a very young nymph taken in early June. Very likely this specimen, which was one of a brood hatched from eggs laid during the early spring, would have emerged the following spring. A nymph that I collected at Niceville in November was two-thirds grown, and undoubtedly was one of the spring hatch. Adults have been reared in April and others captured in June.

HABITS.—Examination of food contents of the alimentary tract showed that the immatures are vegetarians. The material that could be identified consisted mostly of diatoms, algae, and plant epidermis, the last probably obtained from leaf drift which accumulated in the lee of pebbles in the riffles. The predominant food seemed to be diatoms, of which *Navicula* was very common; the other substances were not identifiable.

When nymphs of *B. rogersi* are quiet, they hold their caudal filaments over the abdomen. They swim by pulling the legs under the body, drawing together and depressing their tails, and rapidly vibrating the posterior portion of the body. The nymph swims in short spurts, and as it comes to rest, spreads the legs and seizes some supporting object.

On March 21, 1939, I noted a peculiar behavior in one of the nymphs which I brought back from northwest Florida. At 9:45 A.M., I examined the rearing cage and found that one of the *Baetisca* nymphs had crawled out of the water, along the air tube, on to the floor, and then up the cloth side of the cage. The nymph had progressed fourteen inches above the floor and, still alive, was hanging on the cloth. Further examination showed that the animal was quite immature—not over two-thirds grown—and yet had acted like a mature nymph which was ready to transform. On March 29 I found another individual from the same group lying on its back on the floor of the cage. When I righted the nymph, it immediately crawled toward the cloth and started climbing. It then fell to the floor and soon began moving. Even though the nymph was out of water for two hours, it showed signs of life, but was unable to emerge.

The nymphal habits of this unique genus have been little studied, and have been discussed only by Walsh and Traver. One of the interesting statements of Walsh (1864: 206), which later proved to be erroneous, concerns the respiration of the nymphs: "When they were taken out of the water the tip of the notal shield generally after a few seconds gaped apart from the fifth dorsal joint of the abdomen, and the palpitation and structure of the branchiae became plainly visible under the shield. The reason is obvious. They were then compelled to breathe through their spiracles, instead of through their branchiae." It is now known that the nymphs of mayflies do not have open spiracles, and that the only method of respiration is by diffusion of gaseous matter through the thin membrane of the gills. In this connection, it might be mentioned that all the gills of *Baetisca* nymphs are completely covered by the mesonotal shield. I have observed that the shield is raised almost rhythmically and that the sixth pair of gills is projected from below the median posterior tubercle to form a tube. At the same time, a current of water is led under the huge cover to aerate the more anterior branchiae. The last pair of gills is quite different from the other four pairs in that each gill consists merely of a curled plate fitted into the median prominence of the sixth tergite in such a way that the two gills produce a neatly formed tube which opens anteriorly a short distance under the shield. Once under the mesonotum, the oxygen-bearing water spreads out over the other, multibranched, more anteriorly situated gills.

LIFE HISTORY.—When ready to transform, the nymphs crawl onto a projecting surface until they are entirely clear of the water. In the laboratory I kept a rock partly in and partly out of the water, and this made an admirable holdfast for young nymphs, as well as a place for those preparing for emergence to crawl from the water. In the field, I have seen only a single cast skin. This was found, with the claws firmly attached, on a bridge piling in a sand-bottomed stream. The exuviae were about two or three inches above the surface of the water and were accompanied by numerous stonefly skins. On the piling, as well as on the rock in the laboratory, the nymphs assumed a nearly vertical position with the head upward; on the rock, they were on the upper side at its highest point. The nymphs seem instinctively to crawl to the highest place possible before emerging.

It takes *B. rogersi* approximately one year to grow from egg to adult. This conclusion is based on preserved material rather than on rearing. The section on seasonal distribution includes a discussion of the sizes of nymphs at various times of the year. Growth, as would be expected, is greatest during the warm season, and those nymphs hatched from eggs laid early in the emergence period are approximately two-thirds grown by late fall. Those hatched late in the season reach their maximum growth the following spring and during the early summer, so that by late summer all the eggs of the previous year are accounted for in one way or another.

Emergence data for only two specimens are at hand. A male emerged on March 29 about 3:00 P.M., after remaining in the aquarium from March 17; the imaginal molt occurred March 30 between 9:30 A.M. and 10:00 A.M. The second specimen, a female, emerged at 2:45 P.M., March 31, and underwent its final molt April 1 at 8:45 A.M. The male took slightly more than eighteen hours, and the female, eighteen hours. A female subimago taken at light on June 3, 1940, had molted by the following morning. Traver (1931a: 50) found that the subimaginal stage of *B. carolina* lasted from twenty-one to fifty-two hours; she attributed the great discrepancies in this period to temperature, for during a cold period, a female remained in the submature stage for at least fifty hours. During 1930, there was a spell of extreme heat and the period was shortened to as little as twenty-four hours. Dr. Traver has likewise found that *B. thomsenae* emerged and transformed in the forenoon. I was unable to secure any longevity data for *B. rogersi,* and there are none available for other species of *Baetisca.*

LOCALITY RECORDS.—Escambia County: Perdido Creek, June 3, 1940, nymphs and adults. Gadsden County: 4½ miles south of River Junction, March 17, 1939, nymphs; adults reared March 29, 31, and April 3, 1939. Liberty County: Hosford, March 17, 1939, nymphs. Sweetwater Creek, May 2, 1941, nymphs; April 29, 1946, nymphs; February 3, 1949, nymphs. Okaloosa County: 5 miles west of Walton County line, June 7, 1938, nymphs. Niceville, June 7, 1938, nymphs; November 6, 1938, nymphs. Santa Rosa County: 2 miles west of Milton, April 4, 1938, nymphs. Walton County: Freeport, April 2, 1938, nymphs. 5.4 miles west of Washington County line, May 31, 1940, nymphs.

Genus EPHEMERELLA Walsh

Needham, in *The Biology of Mayflies* (1935: 209) concluded that, although mayflies are only remnants of a disappearing order, new species are still being formed in it because of the large number of rather similar species in the genus *Ephemerella*. By 1939, sixty-nine species of *Ephemerella* had been listed for North America alone, and with the discovery of three additional species from Florida the number now totals seventy-two.

McDunnough (1931b: 188) stated that *Chitonophora* could not be separated from the genus *Ephemerella*, and he divided the genus into four sections. Traver (1934: 206) used McDunnough's divisions of the genus with only slight modifications, but in 1935 she divided *Ephemerella* into seven groups, which were retained in her 1937 treatment of the genus. McDunnough (1939: 50) apparently did not agree with Traver's conception of the groups, for in describing *E. jacobi* he asserts that it obviously belongs in a group with *hystrix* and *heterocaudata*. According to Traver's classification, *heterocaudata* belongs in the *serrata* group and *hystrix* in the *needhami* group.

From the nymphal standpoint, Traver's *needhami* and *invaria* groups are rather poorly defined. McDunnough obviates this difficulty by placing three of the species of the *needhami* group in his section 1, which includes the species of the *invaria* group. This was done in 1931 before *angustata, catawba, concinnata, euterpe, hystrix*, and *maculata* were described; all of these have been included in Traver's *needhami* group. Traver (1937) described another nymph, tentatively placed in the *needhami* group, and intermediate in its characters between *E. catawba* and *E. rotunda*—*rotunda* being a member of the *invaria* group.

The *simplex* group is one of the few taxonomic groupings in *Ephemerella* about which McDunnough and Traver seem to agree. Both authors include in this division three species—*simplex, attenuata,* and *margarita*. The group is quite distinct in three respects: (1) all the nymphs possess overlapping gills on abdominal segments 4 through 7 only; (2) all have well-developed maxillary palpi; and (3) the caudal setae show at least sparse long hairs in their apical portion. In 1937, Traver described another nymph of this group, giving it no name, but stating that it appeared to be a distinct species. *E. hirsuta,* which I described in 1946, also falls into the *simplex* group.

The *bicolor* group is one of the more highly evolved groups of the genus when considered from the standpoint of nymphs. This is borne out by morphological evidence: the absence of gills from segments 2 and 3; the modification of the fourth pair to form elytroid gill covers; and the vestigial condition of the first pair. On the basis of the nymphal structure, the *bicolor* group may be subdivided by use of the condition of the occipital tubercles—*doris, temporalis,* and *trilineata* all have strongly developed tubercles. In the other species, these structures are either moderately developed or reduced.

Ephemerella is a Holarctic genus and is widely distributed throughout the Nearctic. Very few species of *Ephemerella* have been recorded from the

Coastal Plain, but the presence of four species in Florida certainly indicates that when this province is more thoroughly investigated, the genus will be found to be abundant in the region.

Spieth (1933: 351-53) concluded that *Ephemerella,* along with *Tricorythodes,* occupies a distinct and separate place in the phylogenetic history of mayflies. Needham, Traver, and Hsu (1935: 564) did not accept his placement for reasons which they did not indicate. My reasons for accepting Spieth's conclusions on the relationships of *Tricorythodes* and *Ephemerella* have been given in the generic treatment of the former.

Ephemerella trilineata Berner

TAXONOMY.—For some time, I considered *Ephemerella trilineata* to be a variant of *E. doris.* Further study convinced me that I was dealing with a new species that can be distinguished from *doris* by the coloration of the tenth tergite, by the presence of a median dorsal abdominal stripe and of ruddy bands on the distal ends of the femora, as well as by reddish markings on the femora.

According to the classifications of *Ephemerella* as set up by McDunnough and Traver, *E. trilineata* belongs in the *bicolor* group, one of the larger ones of the genus *Ephemerella.* The Florida species is very close to *E. doris* and is also closely related to *E. temporalis,* but differs from the latter by its much paler coloration.

DISTRIBUTION.—In discussing the range of this new species, it will be best, at the outset, to treat that of *E. doris,* for *trilineata* may prove to be a subspecies of the former. *E. doris* was described from the Piedmont of North Carolina and paratypes were recorded from the Withlacoochee River near Macon, Georgia—obviously an error, as the Withlachoochee does not flow within a hundred miles of Macon. Traver has, in addition, recorded the species from the Coastal Plain of North Carolina at Lake Waccamaw, from Valle Crucis, North Carolina, and from Cooley Creek in the northern part of Alabama.

West of the Apalachicola River there occurs a group of nymphs very similar in appearance to those of north-central Florida, but with the spines of the dorsum more erect and thinner and the body somewhat shorter; however, I also have a typical nymph of *trilineata* from this side of the river. The former individuals may be merely variants of *trilineata,* or they may represent *E. doris* in Florida. I hesitate to say that *trilineata* occurs on the east side of the Apalachicola and *doris* on the west, for, with their powers of flight, the adults should certainly be able to span the river. Records of *doris* from Georgia also preclude the possibility of such an abrupt break in distribution of the two species.

E. trilineata occurs throughout the west-central, north-central, and northwest portions of Florida as far as the Apalachicola River, and if the specimens taken west of the river are *trilineata,* then the range of the species extends to the western border of Mobile County, Alabama, and northward into Escambia County, Alabama (map 13).

PLATE XII

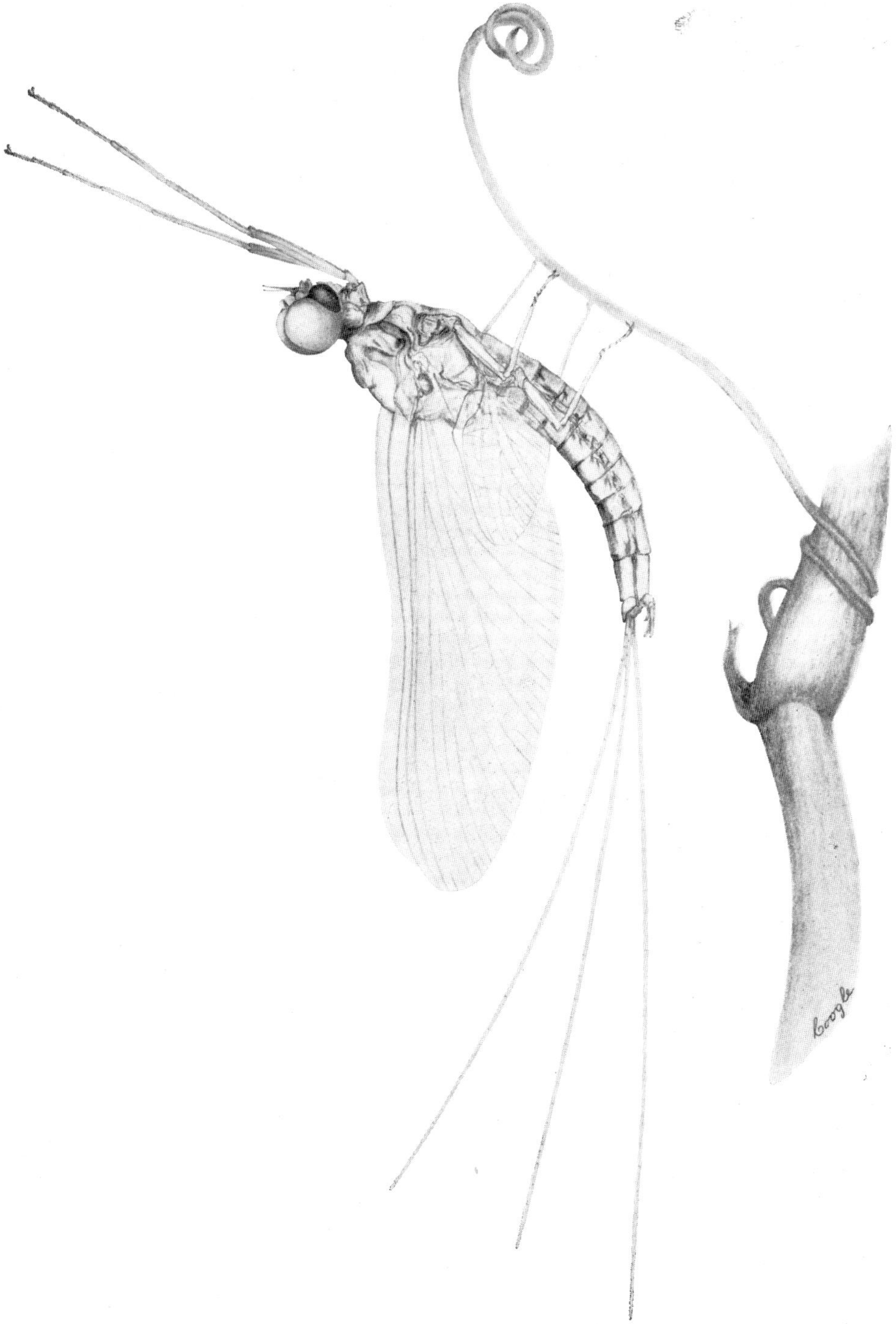

Ephemerella trilineata Berner, male imago.

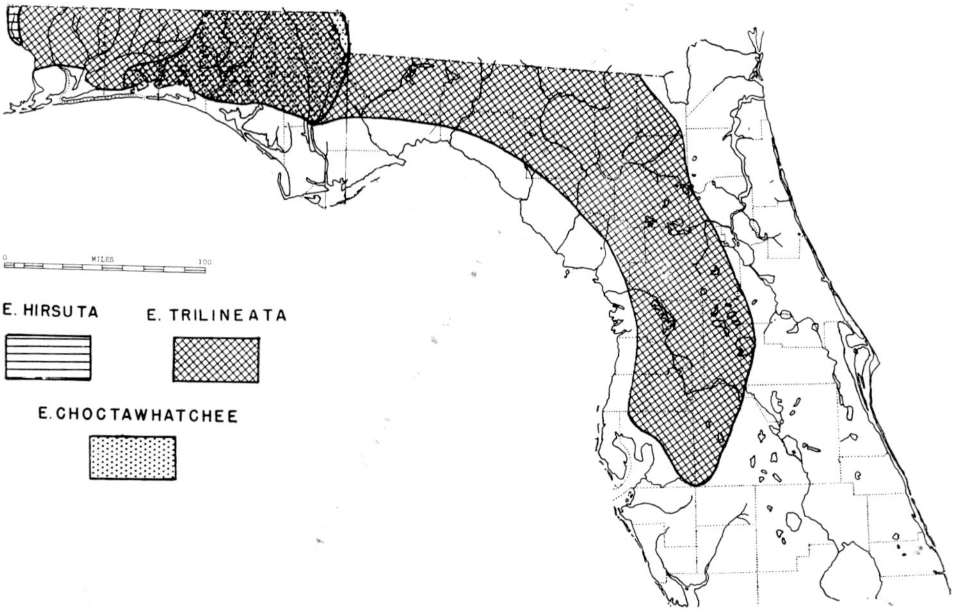

Map 13.—The distribution of *Ephemerella hirsuta*, *E. trilineata*, and *E. choctawhatchee* in Florida.

ECOLOGY.—Ecologically, the nymphs are confined either to flowing water or lake margins where there is sufficient stirring of the water to supply conditions suitable for development. Both silt-bottomed and sand-bottomed creeks support populations of nymphs equally well, and in them the immatures favor vegetation where the flow is not too rapid and where there may be thick algal mats. I have found many specimens entangled in masses of *Spirogira,* where they seemed to be thriving. Another favored habitat for more mature nymphs is among large masses of leaf drift in slow water. Submerged logs, boards, sticks, and other rather solid materials where the nymphs hide in cracks and crevices, beneath the bark, or in any other protected place, provide a third important habitat. Nymphs have been found most abundantly at Hatchet Creek near Gainesville. This stream is a moderately swift-flowing, sand-bottomed, acid creek that drains flatwoods. That part of the stream most thoroughly examined was dredged some years ago, and is now rather deep and slow flowing. Near the shore of the dredged area in shallow water, there are dense growths of parrot's-feather, *Myriophyllum,* where the flow of water is negligible. *Ephemerella* nymphs are quite common during the fall and spring on the submerged stems and leaves of *Myriophyllum,* but they are also frequently found on other vegetation near shore where the water is not too swift.

Small populations of *Ephemerella trilineata* nymphs inhabit those lakes with sandy shores where they seldom venture beyond depths greater than three feet. They are not found on the vegetation or among the smaller debris, as in streams, but are confined to submerged sticks, boards, logs, and other large objects

PLATE XIII

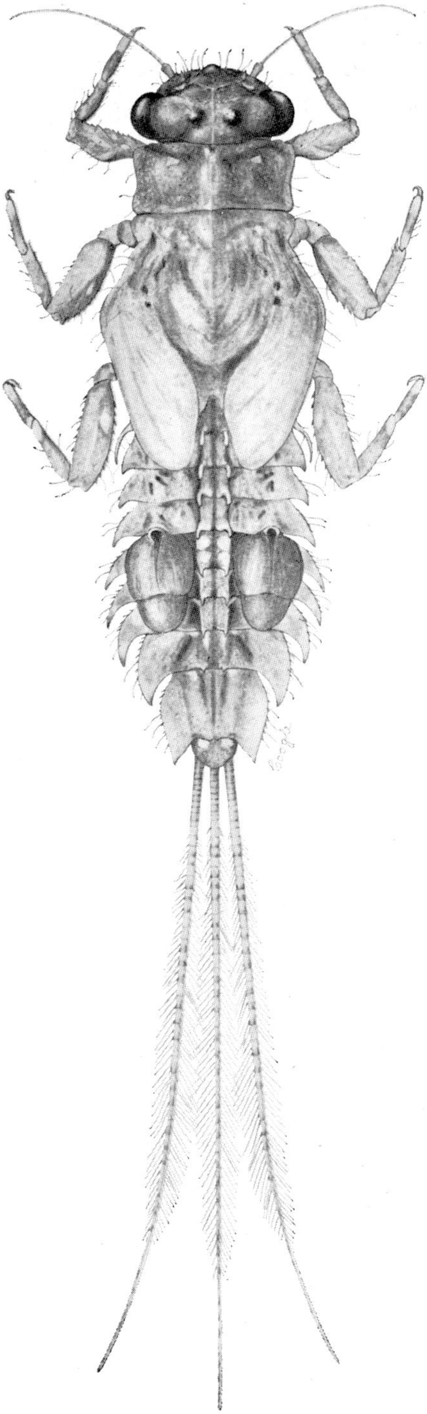

Ephemerella trilineata Berner, nymph.

partially buried in the sand. The immatures live only on the undersurfaces in company with *Stenonema proximum* and *Choroterpes hubbelli.*

The streams in which the nymphs of this species live drain flatwoods and are usually acid or circumneutral, and the water is nearly always tinted. Nymphs have not been taken from spring runs or any other basic streams. The mayfly associates are the same as those listed for *Choroterpes hubbelli* in the sand-bottomed streams; its lake associates have been listed above.

SEASONS.—*E. trilineata* is a late winter, spring, and early summer species. The period of principal emergence takes place during the spring in April, but with scattered emergences from early February through late June. Traver has recorded the emergence of *E. doris* on March 21 and 22, May 6 and 30, and July 4; the latter collection dates are from mountainous regions where transformation would be expected to occur later than in the lowlands.

HABITS.—*E. trilineata* nymphs blend perfectly with their background when lifted from the water while they are attached to sticks, boards, or other dark objects. It is only when they begin moving about that they become discernible. Their movements are slow and deliberate, each step seeming to be measured. When out of water, the nymphs flick the tails forward over the abdomen and then straighten them, repeating the act continually. The movement is very reminiscent of that of the Caeninae and of *Oreianthus.* When placed in water, the nymphs, like these two groups of mayflies, swim awkwardly with undulatory movements of the abdomen. They then head for some submerged object to which they can cling and, once attached, become difficult to see. The nymphs are easily identified by their actions on the surface of a white-enameled pan. Characteristically, they move forward slowly, not wriggling as do the Leptophlebines, and occasionally flick the tails as described above. The movement may be confused with that of the Caeninae, but the nymphs can be separated from this group by the easily seen, sawlike margins of the abdomen and the widely separated gill elytra.

I have been able to keep nymphs alive in the laboratory for as long as two months, and during this period have fed them only dead leaves. Examination of the enteron of wild nymphs indicates that they, too, eat the epidermis of decaying vegetation, as well as that of living plants, for the alimentary canal was packed with plant fibers and other cellulose materials.

Nymphs of the *bicolor* group of the genus *Ephemerella* are characterized by the absence of gills from segments 2 and 3 and by the modification of the fourth gills into gill covers. When they respire, water is circulated about the gills by the lifting of the operculate covers, and the rapid vibration of the fifth, sixth, and seventh pairs. Normally, both gill covers are raised at the same time, but I have observed on numerous occasions that one gill cover may remain tightly closed while the other is raised and the gills under it rapidly vibrated.

LIFE HISTORY.—A study of the material in my collection indicates clearly that *trilineata* requires approximately one year to pass through its life history. Last instar nymphs have been taken from January through the middle and latter part of June, but during July, August, September, and October there are no records, although I have collected from streams in which nymphs are plentiful during the remaining months of the year. Since I collected very young nymphs in November, it seems reasonable to suppose that the eggs were developing during the middle summer and early fall seasons. Nymphs one-third to two-thirds grown were found in December and some were even in the penultimate instar. In March, rather immature specimens, which were obviously the summer brood, were taken in Hillsborough County.

I have not observed transformation in the field, but in the laboratory the nymphs, just before emergence, swim vigorously and then float freely at the water surface for a few moments. Soon the thorax splits and the subimago appears on the floating skin where it rests quietly. After recovering from the exertions necessary for its emergence, the subimago flies away to the nearest shaded support where it remains until the imaginal molt. I have noted that when partially submerged sticks were placed in the aquarium, the nymphs tended to crawl onto them until a portion of the thorax was above the water; in this position they transformed.

On one occasion in February, I noted an *Ephemerella trilineata* female emerge at 6:35 P.M., just at dark. The female nymph came to the surface of the water about 6:25 P.M. and remained in the quieter portion of the aquarium away from air bubbles which were rising from the air hose. Now and then it swam vigorously for a short distance and again came to rest. I stimulated the nymph to move; it then submerged, swam into the region where the water was agitated, and immediately returned to the quiet water at the surface. There it strained so intensely that its head and mesothorax broke through the surface film. It repeated the straining after each spurt of swimming, but for the most part, it lay quietly at the surface. At 6:35 P.M. it reared its thorax almost completely out of the water and, by the count of ten, the wings were free. The subimago did not move after freeing its abdomen, but sat quite still, partially on the surface film and partially on the exuviae; after two minutes, it became active again, crawled up the side of the aquarium, rested briefly and then flew to the wall of the cage.

Emergence which normally takes place in late afternoon, sometimes occurs after dark in the laboratory. Life as a subimago extends from late in the afternoon of the emergence date until late afternoon of the following day, the number of hours varying from twenty-two to twenty-five for both sexes.

The mating flight of *E. trilineata* has not been observed, and descriptions of the phenomenon in other species of the genus *Ephemerella* are almost nonexistent.

LOCALITY RECORDS.—Alachua County: Santa Fe Lake, April 2, 1935, nymphs; April 7, 1937, nymphs and adults; January 30, 1940, nymphs; December 15, 1947, nymphs. 1 mile west of Newnan's Lake, May 11, 1937, nymphs; January 8, 1938, nymphs; January 25, 1938, nymphs; 3 miles north of Paradise, February 12, 1938, nymphs. Near Worthington Springs, February 5, 1939, nymphs. Hatchet Creek, March 22, 1937, nymphs; February 8, 1938, nymphs; February 26, 1938, adults; March 23, 1938, adults; April 2, 1938, adults; April 18, 1938, adults; May 5, 1938, adults; November 13, 1938, nymphs; March 5, 1939, nymphs; adults reared March 15-26; April 1, 1939, nymphs and adults; April 5-13, 1939, nymphs and adults; May 6, 1939, adults; June 24, 1939, adults; February 16, 1940, nymphs and adults; March 2, 1941, adults; January 15, 1949, nymphs. 2½ miles west of Gainesville, February 5, 1940, nymphs; adults reared March 1-9, 1940. Columbia County: Falling Creek, November 13, 1938, nymphs. Gadsden County: River Junction, March 17, 1939, nymphs. 4½ miles south of River Junction, March 17, 1939, nymphs. Hamilton County: 0.6 miles north of Live Oak road at U. S. Highway 41, February 4, 1938, nymphs and adults. Hillsborough County: Hurrah Creek at Picnic, March 26, 1938, nymphs. Holmes County: Sandy Creek, December 11, 1937, nymphs. Jefferson County: Drifton, February 5, 1938, nymphs. Liberty County: Hosford, March 17, 1939, nymphs. Little Sweetwater Creek, December 10, 1937, nymphs. Sweetwater Creek, November 4, 1938, nymphs; December 1, 1939, nymphs. Marion County: Oklawaha River at Eureka, February 12, 1938, nymphs. Oklawaha River, April 10, 1948, nymphs. Putnam County: Red Water Lake, March 26, 1939, nymphs and adults. Orange Creek, April 10, 1948, nymphs. Washington County: Choctawhatchee River, April 2, 1938, adults. Holmes Creek, December 11, 1937, nymphs. 5.8 miles south of Vernon, April 2, 1938, nymphs.

Ephemerella sp. A

TAXONOMY.—*Ephemerella* sp. A is closely related to *E. deficiens.* In the nymphal stage, however, it differs from *deficiens* in that it has yellow-banded femora, longer and more attenuated tarsal claws, and a narrow brown band on each of the caudal filaments. Moreover, the nymphs of species A are slightly smaller than *deficiens.* According to Traver's division of the genus *Ephemerella,* species A would fall into the *serrata* group along with *deficiens.* Characters shared in common with *deficiens* and different in other members of the group include the absence of maxillary palps and of dorsal abdominal spines.

The claws of the nymphs of *deficiens,* as drawn by Morgan (1911: 111), are short and thick, and adapted for maintenance of the little nymphs in swift water. The claws of *Ephemerella* sp. A are proportionately longer and thinner than those of the former species, and this size relationship seems to be correlated with the slower waters it inhabits and the consequently lessened tension on the claws. The character may, however, be only an expression of habitat difference and of no value taxonomically.

From the nymphal standpoint, *deficiens* and species A would seem to be two of the more primitive species of *Ephemerella* were it not for the loss of the maxillary palpi. The lack of any spining on the body of the nymphs, the presence of gills on segments 3 through 7, and the lack of elytroid gill covers indicate that the *serrata* group, to which *deficiens* and species A belong, is most likely primitive.

DISTRIBUTION.—The geographic distribution of *Ephemerella* sp. A is very interesting because in only two localities is the species known to occur east of the Apalachicola River drainage. The first of these, about ten miles north of Atlanta, Georgia, is a small stream which drains into the Chattahoochee

PLATE XIV

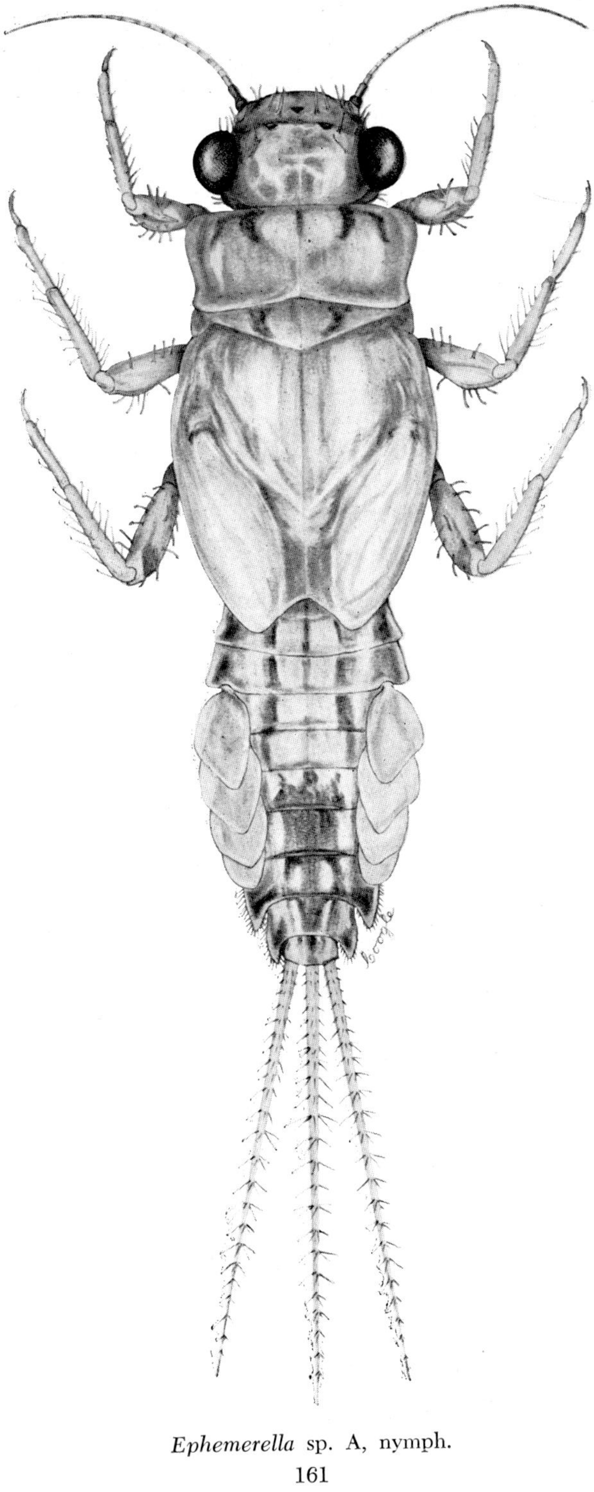

Ephemerella sp. A, nymph.

River; the second, Sweetwater Creek, empties into the Apalachicola River a few miles north of Bristol, Florida. I have nymphal records extending westward from here as far west as Kushla, Alabama, about ten miles north of Mobile; these nymphs were collected from Seabury Creek, a stream which drains eastward into the Mobile River. It would seem, then, that species A stops just east of the Apalachicola River, but westward is not so closely confined. Even the change from one physiographic province to another does not appear significant, for the Georgia specimen was recorded from the Piedmont and all others came from the Coastal Plain. If my nymphs are merely aberrant forms of *deficiens*, the distribution would still be quite logical, for Traver records *deficiens* from both the Piedmont and Appalachian regions of North Carolina. The headwaters of the Apalachicola River drain the foothills of the Appalachians, and in the tributaries of this large stream, *deficiens* could find a ready entrance into the Coastal Plain of Alabama and Florida. The true *deficiens* is known to occur from Ottawa to Nova Scotia and from Nova Scotia south to the highlands of North Carolina. If species A proves to be synonymous with *deficiens* when adults are known, the species will range from southern Canada to northern Florida, an exceptionally large area for a mayfly (map 12).

ECOLOGY.—I have collected nymphs of *Ephemerella* sp. A from streams of both moderate and fairly swift rate of flow. In Sweetwater Creek, one of the westward-flowing tributaries of the Apalachicola, the nymphs are not especially abundant. In this stream, there are no rocks and very few logs or other footholds for clinging forms like *Ephemerella*. Nymphs which inhabit the creek usually live on the roots of terrestrial plants which project into the water. The roots are numerous, for the banks of the stream are being continually undercut. These root masses form silty mats where frequently dead leaves, Spanish moss, and other detritus are caught. Many mayfly nymphs of several species live in this mass of debris.

Farther west, in Walton and Okaloosa counties, the nymphs have been taken in swifter water on the masses of vegetation which choke the streams. In none of these creeks, however, have I found species A nymphs in large numbers. I collected from a stream in Escambia County in which root masses like those of Sweetwater Creek and dense growths of aquatic vegetation similar to those of the Walton and Okaloosa county streams were present. There the nymphs were found to be a more conspicuous element of the ephemerid fauna than in any other Florida stream which I have examined. The nymphs were commonest in the dense mats of *Potamogeton* in the more gently flowing water, but were also present on the roots projecting from the undercut banks. The species was also found along the undercut banks of a small, moderately swift-flowing stream.

The streams, in which this species of *Ephemerella* lives, vary from strongly to moderately acid and drain the red-clay hills of the Citronelle Formation or the flatwoods and swamps of the lowlands. All the streams are permanent and have a moderate to rather swift flow; none is more than three feet nor less than six inches in depth.

In the root masses of Sweetwater Creek, the mayfly associates were *Isonychia pictipes*, *Oreianthus* sp. No. 1, *Caenis hilaris*, *Baetis spiethi*, *B. intercalaris*, *B. australis*, *B. spinosus*, *Paraleptophlebia volitans*, *Habrophlebiodes brunneipennis*, *Ephemerella trilineata*, *Stenonema smithae*, *Tricorythodes albilineatus*, and occasionally a nymph of the burrower *Hexagenia munda marilandica*. In other streams, ephemerid relatives of *Ephemerella* sp. A are not so varied, and in addition may include *Choroterpes hubbelli*, *Pseudocloeon bimaculatus*, and *Acentrella ephippiatus*.

SEASONS.—My only records of species A are for the months of April, May, June, and July. The April and May nymphs were all in their last instar, and the June records include nymphs of all ages, ranging from those less than one-third the length of mature individuals to those in the last instar. Though this range of size seems to indicate a year-round emergence, I do not believe that it actually occurs. Rather, the data point more clearly to an emergence beginning in spring and extending through early fall, with slow growth of the nymphs during the winter months, or perhaps a long egg period. The gradation in size of nymphs may be explained by a very rapid growth period allowing several broods per year. The life span would then be approximately six to eight months. From the small amount of material at hand, it is impossible to draw conclusions regarding the peak of emergence. No nymphs were found in any of the streams of northwest Florida during November of 1938 and December of 1937 and 1939.

HABITS.—The difficulties of transporting live nymphs of stream mayflies from west Florida to Gainesville are very great, and while two species have been transported successfully, no nymphs of *Ephemerella* sp. A survived the trip.

The nymphs, which live in well-protected places where there is usually some accumulation of silt, are very difficult to see when mingled with debris. In a small stream in Alabama, one specimen was collected from the moss covering the upper side of a large rock that was near the middle of the rapids below a dam. Traver (1937: 177) described the habits of the related *deficiens* in North Carolina. She found the nymphs among the moss and other plant growth that grew thickly on the upper side of all the larger rocks, where they could be seen only after the rock had been left out of the water for some time.

In the field, young *Ephemerella* sp. A nymphs may easily be confused with those of *Oreianthus* or *Tricorythodes*. When taken from the water, members of all three of these genera walk slowly about flicking the caudal filaments over the dorsum. *Ephemerella* sp. A nymphs are easily distinguished from both *Oreianthus* and *Tricorythodes* by their lack of gill covers.

LOCALITY RECORDS.—Bay County: 27.4 miles north of St. Andrews, May 30, 1940, nymphs. Escambia County: Bayou Marquis, June 1, 1940, nymphs. Jackson County: 12.2 miles southeast of Marianna, June 9, 1938, nymphs. Liberty County: Sweetwater Creek, July 1, 1939, nymphs; May 2, 1941, nymphs. Okaloosa County: 5.1 miles west of Walton County line, June 7, 1938, nymphs; May 31, 1940, nymphs. Walton County: 7.3 miles west of Ebro, June 7, 1938, nymphs.

Ephemerella choctawhatchee Berner

TAXONOMY.—This rather rare species is represented in collections by only nine nymphs. These differ from all described *Ephemerella* nymphs, but clearly fall into the *needhami* group as defined by Traver. They are relatively small (five to six millimeters in length), the tibiae have a pale band at the apex, the tarsi are banded with brown in the proximal third, the lateral extensions of the abdomen are moderately developed, and small dorsal abdominal spines are present.

The *needhami* group is one of the more primitive groups of the genus, as is indicated by the presence of a maxillary palp and gills on segments 3 through 7, and, in some species, absence of spining.

DISTRIBUTION.—*E. choctawhatchee* is the first species of the *needhami* group to be recorded from the Coastal Plain; it is, in fact, the first to be recorded south of the mountains of North Carolina. As there are few records for the species in Florida, it is hazardous to draw more than very tentative conclusions as to its distribution. *E. choctawhatchee* is found as far eastward as the tributaries of the Apalachicola River, where it has been taken from a small stream in Gadsden County about one or two miles from the river, and from Sweetwater Creek, Liberty County, at a point not over two miles distant from the parent stream. Other records are from small streams in Okaloosa County that empty into Choctawhatchee Bay. The distribution of this species seems to parallel that of *Ephemerella* sp. A, although it has been taken on fewer occasions and in fewer places. There is no apparent reason for the eccentricity of distribution shown by these two species. Other species which are relatively new to the region have managed to escape the hold of the Apalachicola drainage (map 13).

ECOLOGY.—Clear, sand-bottomed streams with a rather strong flow provide an ideal habitat for the majority of the stream forms of Florida. *E. choctawhatchee* is no exception. In Sweetwater Creek, the nymphs inhabit the same root beds from which *Ephemerella* sp. A was taken, and in the other streams they live either in the vegetation in more rapidly flowing water, or near the shore amid the projecting roots.* The ephemerid companions of *E. choctawhatchee* are the same as those listed for *Ephemerella* sp. A.

SEASONS.—With so few nymphs and no adults available, it is impossible to be definite as to the seasonal relations of the species, though some tentative conclusions can be drawn. The earliest records for nymphs in their last instar are in March, and records for this stage extend to May 31. In March, I collected nymphs in the tertiultimate instar, which probably would have emerged in April. On December 1, I took another group of half-grown nymphs which would have emerged during the spring. It seems, then, that emergence probably starts in late winter and proceeds through the spring and even well into

* For a description of the streams in Okaloosa County from which nymphs were taken, refer to *Pseudocloeon bimaculatus*, p. 252.

the summer, but that there is no transformation during fall or early and middle winter.

LIFE HISTORY.—The small amount of data concerning life history has been summarized above under seasonal range.

LOCALITY RECORDS.—Gadsden County: 4.5 miles south of River Junction, March 17, 1939, nymphs. Liberty County: 10 miles south of River Junction, March 17, 1939, nymphs. Sweetwater Creek, December 1, 1939, nymphs; April 30, 1946, nymphs. Okaloosa County: 5.1 miles west of Walton County line, May 31, 1940, nymphs. 3.6 miles north of Niceville, April 3, 1938, nymphs.

Ephemerella hirsuta Berner

TAXONOMY.—In 1946 I described *Ephemerella hirsuta* from two nymphs taken by Drs. H. H. Hobbs and L. J. Marchand in April, 1938, from Perdido Creek less than one mile north of the Florida state line in Alabama. The specimens belonging to Traver's *simplex* group are very distinctive and cannot be confused with any other Florida species. Subsequently, I collected from Sandy Creek two adult *Ephemerella* which also fall into the *simplex* group. Since the adult differs from other described species of the group and came from an area not far distant from the type locality, I am associating these adults with the nymphs of *hirsuta*.

DISTRIBUTION.—*E. hirsuta* is known only from Perdido Creek, just north of the Florida-Alabama state line and from Sandy Creek in west Florida. The distribution of *E. hirsuta* is of particular interest when compared with that of the previously described species of its group. *E. attenuata,* the closest relative, has been recorded from Ottawa, Ontario, southern Quebec, and the Potomac River; *simplex,* from southern Quebec, Ontario, New Brunswick, western New York, and the mountains of North Carolina; *margarita,* from Utah, Wyoming, Montana, Alberta, and New Hampshire; and *Ephemerella* sp. No. 1 (Traver, 1937), from Blowing Rock, North Carolina. No member of the *simplex* group has been previously found in the Coastal Plain, or even in the Piedmont (map 13).

ECOLOGY.—In June of 1940 I collected specimens from Perdido Creek, but found no trace of *E. hirsuta.* The creek was swift-flowing and fairly deep. The deeper portion, which was three to four feet deep, had almost no plant growth, but the shallower and quieter margins were choked with vegetation, principally *Vallisneria, Potamogeton,* and a sedge, among the higher plants, together with the alga *Batracosperum* sp. The latter plant was growing very thickly in the stream, and formed dense masses covering the other vegetation. In this alga were found numerous mayfly nymphs, and they were also very common on the blades of the *Vallisneria* and on the leaves and stems of the *Potamogeton;* but no specimens of *E. hirsuta* were taken in either of these places. Along the shore where the flow was almost negligible, *Persicaria* and leaf drift were the dominant mayfly refuges.

Although I have not been able to find the nymphs of *E. hirsuta,* a study of the external anatomy of the specimens taken by Hobbs and Marchand leads me to believe that the nymphs must live either in the algal mats or among the leaf drift and silty matter deposited about the bases of the *Persicaria* stems. The nymphs, as the name of the species indicates, are covered on the dorsum of the thorax and head with long, somewhat matted hairs. The hairs also cover the femora and the anterior part of the abdomen, acting as a protection from the debris which might collect on the insects. Another strong point supporting my contention as to their habitat preference is based on a comparison of the body shape of *E. hirsuta* with that of *E. trilineata.* The latter species, discussed elsewhere in this paper, is constructed very much like *hirsuta:* a somewhat flattened body, elongated legs and claws, operculate coverings for the gills, hairy body, and occipital tubercles. It is probable that these characters may be associated with moss-covered rocks, but in Perdido Creek there is only a sandy bottom with some dead wood but no rocks.

An examination of the literature shows that the only reference to ecological habits of any species of the *simplex* group is that of McDunnough (1931: 209). He noted that *E. simplex* nymphs live in very swift water, where they cling to the underside of limestone slabs and are so covered with silt as to be almost unrecognizable.

Perdido Creek, which separates Florida and Alabama, is unusual among Florida streams in that it is comparatively rich in the variety of mayflies present. The associates which live in the stream, though not perhaps in the same minor habitat, include *Acentrella ephippiatus, Stenonema exiguum, S. smithae, S. proximum, Tricorythodes albilineatus, Baetisca rogersi, Caenis hilaris, Baetis spinosus, Pseudocloeon bimaculatus, Baetis spiethi, Isonychia pictipes,* and *Isonychia* sp.

SEASONS.—The two nymphs in my collection were taken on April 5, and the adults on May 1. As one of the nymphs was in the last instar and the other in the tertiultimate, emergence presumably occurs during April and May. It is likely that the species emerges only during the summer, because in this colder region of Florida and southern Alabama, emergence of other mayflies, particularly in the genus *Ephemerella,* appears to take place chiefly in spring and summer.

LIFE HISTORY.—Nymphal life probably lasts from nine months to one year.

LOCALITY RECORDS.—Escambia County: Perdido Creek, April 5, 1938, nymphs. Walton County: Sandy Creek, May 1, 1946, adults.

Genus TRICORYTHODES Ulmer

Until Ulmer established the genus *Tricorythodes,* the North American forms were included under *Tricorythus.* The males of species of these two genera may be separated by the length of the foreleg which, in the African *Tricorythus,* is half the length of the body; in the American *Tricorythodes,* it is almost, if not fully, as long as the body.

McDunnough in 1931 summarized the knowledge of the North American species of *Tricorythodes,* and described four—one of which was new. In 1935 Traver limited *T. explicatus* (Eaton) to a Texas form and described four additional species. McDunnough (1939: 52-54) described another new species, and also agreed with Traver's limitation of *explicatus.* A tenth species of *Tricorythodes* was described in 1946.

Concerning the distribution of the group of mayflies, Lestage (1924: 262) says that *Tricorythus* is present in Africa and Java and does not inhabit Europe. The North American species belong to Ulmer's genus, *Tricorythodes.* According to Needham and Murphy (1924: 7), one species of *Tricorythodes* is known from South America, its range extending from Guatemala to Argentina. In North America, this is a widely distributed genus, occurring from North Sonora, Mexico * to California to British Columbia to eastern Canada to Florida. *Tricorythodes,* which has never previously been recorded south of West Virginia, nor from anywhere else east of the Mississippi River in the Coastal Plain, is common in Florida.

Only rarely has a species of *Tricorythodes* been treated in other than a purely taxonomic sense. Needham and Christenson (1927: 12-13) discussed the tiny "snowflake mayfly," *Tricorythodes minutus* (as *T. explicatus*) in Utah and described flights which were miles long. In 1905 and 1908, Needham briefly mentioned the habits of nymphs of *allectus;* and in 1927, of *explicatus.* Morgan (1911: 115) also briefly discussed *T. allectus.*

Spieth (1933: 353-56) placed this genus in the same family with *Ephemerella,* removing it from its classical position in the Caenidae. Traver (1935: 631) disagreed with this view, considering *Tricorythodes* to be an aberrant member of the Caeninae, but presented no evidence, merely stating that the venation of the imago is much nearer to the other members of the Caeninae than to *Ephemerella.* I believe that Spieth's conclusions indicate the true relations of the genus. As he pointed out, there is a difference in venation between *Tricorythodes* and *Caenis:* the loss of hind wings with a corresponding increase in the anal area which is merely a parallel development in the two stocks is an example of convergent evolution. The structure of the genitalia of the males seems to be one of the strong points in Spieth's argument. In both *Tricorythodes* and *Ephemerella,* the forceps are three-segmented, whereas in *Caenis* and *Brachycercus* there is but a single segment. Likewise, the penes of *Tricorythodes* are similar to those of *Ephemerella* in that they are fused for about three-fourths their length and prolonged into a prominent, somewhat conical structure; in the other two genera the penes, which are united, broad, and platelike, are considerably different from those of *Tricorythodes.* Substantiating the evidence from the adults, Spieth also pointed out that such nymphal characters as mouth parts and gills show closer affinities to *Ephemerella* than to *Caenis* and *Brachycercus.* In a verbal communication, Dr. Spieth also stated that the eggs of *Tricorythodes* and *Ephemerella* are fairly similar.

* I have taken *Tricorythodes* in eastern Mexico.

Tricorythodes albilineatus Berner

TAXONOMY.—*Tricorythodes albilineatus,* described in 1946, can be distinguished from other species of the genus by differences in color pattern. The sprinkling of numerous black dots on the femora, a characteristic given by Traver as diagnostic, would place the Florida specimens as *atratus.* Judging from McDunnough's discription, my insects have more limited dark markings on the abdomen than does *atratus* and they are characterized by a pale median line on the dorsum of the abdomen. There is also a slight difference in the leg measurements of the males of the two species. The colorational characters are quite constant in *albilineatus,* particularly the pale line, even in series from widely separated localities.

DISTRIBUTION.—*T. albilineatus* is generally distributed over Florida wherever there is permanently flowing water. Nymphs or adults are known from as far south as Hillsborough County, and as far north and west as the southern part of Escambia County, Alabama. The species occurs in an almost continuous belt, starting in Hillsborough County and passing westward to the Alabama locality. The geographic distribution is probably limited by the ecologic factors of flowing water and by the acidity of the water. Although I sampled numerous streams in southern Alabama, I found no nymphs. But it does not seem probable that the genus is absent from the region between Florida and Texas, where *Tricorythodes* is known to occur. Very few localities in Georgia have been

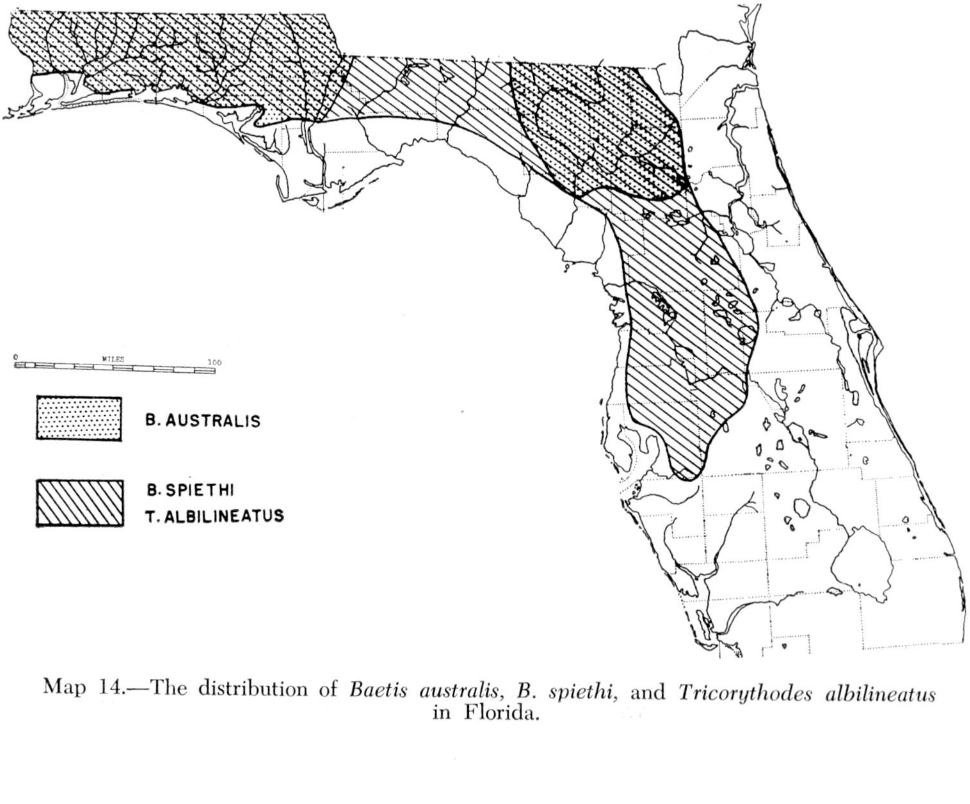

Map 14.—The distribution of *Baetis australis, B. spiethi,* and *Tricorythodes albilineatus* in Florida.

PLATE XV

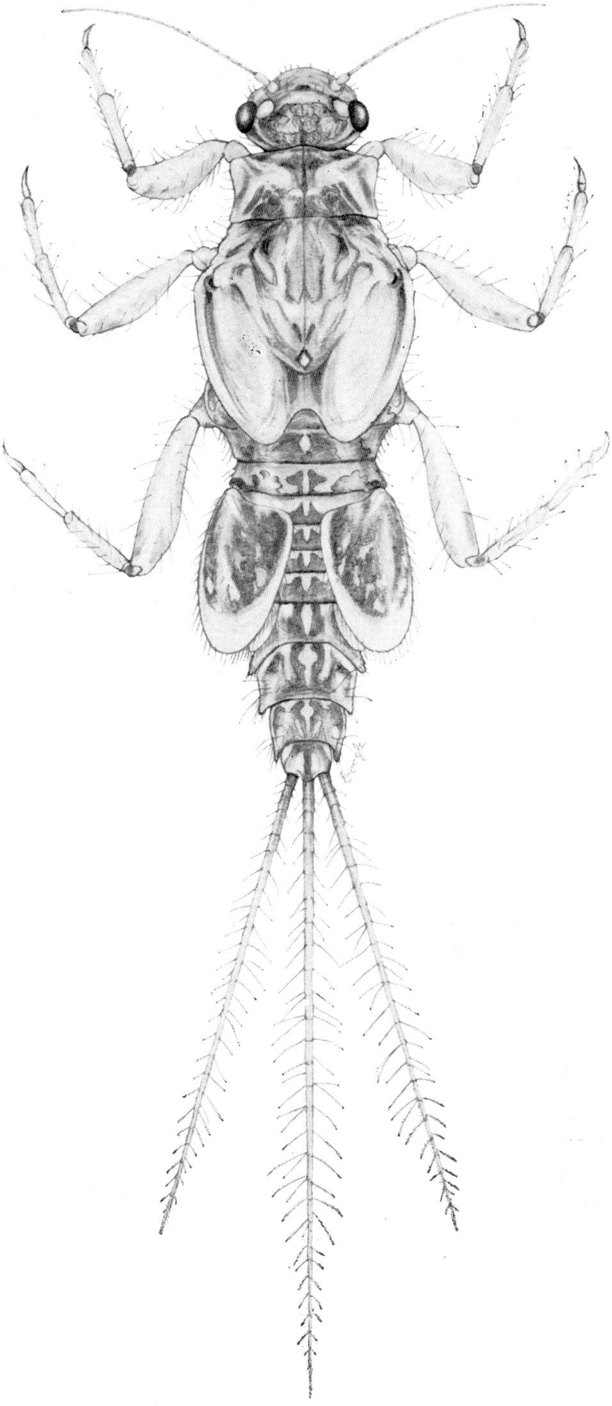

Tricorythodes albilineatus Berner, nymph.

examined, and no specimens of *Tricorythodes* have been taken from that state, but I believe that *T. albilineatus* is more generally distributed throughout the Coastal Plain than my collections would indicate (map 14).

ECOLOGY.—Ecologically, the nymphs are confined to streams with at least a perceptible current. They are moss and silt dwellers, occurring occasionally on vegetation in the more swifty flowing water. The Santa Fe River is an ideal stream for the rheotropic *T. albilineatus.* There, below a dam where the river spreads out and becomes shallow, the bed of the stream is strewn with pebbles and rocks. The upper surfaces of the larger stones are covered with dense mats of *Fontinalis,* one of the mosses which grows profusely when completely submerged. The flow of water over these rocks is quite rapid; nevertheless, silt becomes lodged between the closely packed moss "leaves." When one of the rocks is removed from the water and allowed to dry for a few minutes *Tricorythodes albilineatus* nymphs begin to crawl slowly about, meanwhile flicking their tails over the abdomen just as do most of the other nymphs which possess operculate gills. If the moss is examined while it is wet, the nymphs are very difficult to see, their movements being slight until the drying action begins to take effect.

Below the dam, and to some extent above it, the Santa Fe River is normally choked with *Vallisneria, Sagittaria,* and *Nais.* Frequently, when a dip net is run through these tangles of submerged vegetation in swift-moving water, *Tricorythodes* nymphs will be found along with the usual inhabitants of this biotope.

In smaller streams, predominantly of the sand-bottomed type, nymphs can be found clinging to vegetation, occasionally on detritus, and most frequently among roots of terrestrial vegetation which project into the stream. Silt and various other detrital elements accumulate among these roots. The nymphs of *Tricorythodes* tend to congregate in this mass of roots, but are much less numerous than they are in the Santa Fe River.

Many sand-bottomed streams in Florida have dense growths of *Vallisneria* and *Sagittaria* like those in the Santa Fe River, but without the exposed rock surfaces. In these streams, which include the runs of most of the springs in Florida, the nymphs are commonly found clinging to the vegetation, sometimes very close to the base of the plants, but they are not limited to this situation. Frequently they occur near shore among leaf drift and on other debris.

The nymphs, which are found more often and in greater numbers in alkaline waters, are also fairly common in circumneutral or slightly acid streams. *T. albilineatus* is one of the representative species of mayflies of the spring runs and has been collected from many of them. The nymphs become noticeable about one-fourth of a mile below the head of the springs where the water has been exposed for a time to surface conditions and has become aerated.

T. albilineatus is normally the only mayfly living on the upper side of the moss-covered rocks; on the underside occur the three Florida species of *Stenonema.* Also associated with the nymphs in the moss, but most commonly found on the underside of the rocks, are *Corydalis cornutus* larvae, caddisfly larvae, sometimes cranefly and various other dipterous larvae, as well as snails. In the eelgrass are *Baetis spinosus, B. intercalaris, Pseudocloeon parvulum, P. punctiventris, Centroptilum viridocularis, C. hobbsi, Stenonema smithae, S. exiguum,* and *S. proximum,* together with the snail *Goniobasis,* one of the most numerous and characteristic animals of the basic streams in Florida. Many other aquatic animals are also found in the eelgrass association.

The usual associates in the roots of terrestrial plants projecting into the water are listed under the discussion of *Ephemerella* sp. A. The coinhabitants of the underside of boards are *Stenonema proximum, S. exiguum,* and *S. smithae,* as well as many *Goniobasis.*

I have never found nymphs in the very small permanent streams, but I have taken them from the slightly larger streams which are one to two feet deep and at least ten feet in width. The limiting factors involved in stream size are unknown, but they may be linked with the relative rate of flow and the type of vegetation present. *Vallisneria* is seldom present in very shallow streams where the detritus and silt may be so markedly subject to drying at low water that the nymphs cannot withstand such great changes.

The adults are sensitive to drying and must remain in a very moist situation at all times. A large number of subimagoes were attracted to my light one evening. I treated these as I would most other species of mayflies, placing them in a paper bag to await their ultimate molt. After a few hours, however, every individual was dead and was beginning to shrivel—not one had shed. Many adults have emerged in the laboratory but, here too, none have undergone the final molt.

HABITS.—Examination of the alimentary canal of nymphs showed that those specimens which are root inhabitants feed on epidermis of dead leaves, other debris, and occasionally on diatoms; the nymphs living among water grasses feed on the epidermis of these plants, as well as on the diatoms covering them, although the latter are not eaten in great numbers. The moss-dwelling individuals also feed to a large extent on plant epidermis, but the diatoms form a more important element in their diet than in that of the two other groups.

The nymphs are awkward swimmers, using an undulating action of the abdomen for propulsion. The nymphs seldom swim, and move about almost entirely by crawling. The operculate gills are more or less rhythmically raised to allow aeration of the lower or posterior pairs. In the lowered position, the operculate gills cover rather completely all others behind them, though not so well as do those of *Ephemerella trilineata.*

SEASONS.—From a study of the material in my collection, it would appear that *Tricorythodes albilineatus* emerges throughout the year. For that portion of the state east and south of Leon County, I have records of adults taken during early February, March, April, June, August, and October; my records of adults from west Florida do not indicate such a disregard for season. In that region I took adults in April, May, and June; and mature nymphs, in July. Returning to a consideration of nymphs, west Florida records of the immatures may belie adult collection, for very young nymphs have been taken in May, June, July, and November. If the life cycle of the species is comparable to that of *Caenis,* then adults would emerge each month of the year. If, on the other hand, nymphal existence is at least one year, then the occurrence of these immature forms during summer and early fall may be explained by assuming that the young spring and summer nymphs would emerge during the corresponding seasons of the following year and that the young fall nymphs would emerge during the late summer of the next year.

It appears that in the west-central and northern part of the peninsula, excluding the panhandle, *T. albilineatus* has a year-round emergence, but that in the colder western area, emergence is limited to the spring, summer, and early fall.

LIFE HISTORY.—Just as in more northern species, large numbers of adults emerge at the same time, and I have seen literally thousands of adults floating on the surface of the Santa Fe River. They were present in such numbers that they might be scooped up in handfuls, but most of them were in a somewhat advanced state of decomposition. Emergence probably occurs after dark. I could not find a single adult before dusk, but when my lantern was turned on just after dark, the lighting sheet was very soon covered with subimagoes which were predominantly males.

Many of the subimagoes seem to experience difficulties in molting even under natural conditions. In the Santa Fe River, there are many large boulders projecting one to two feet above the surface of the stream. Upon emergence large numbers of *Tricorythodes* subimagoes alight on them and remain to shed; however, sprinkled among the many shed skins, I found a considerable number of exuviae still retaining the partially freed imagoes—males and females in approximately equal numbers. These were dead but not yet dried when discovered. All of them seemed to have experienced great difficulty in freeing the wings, for it was by these structures that they were held.

I have never witnessed the mating flight of this species, but Needham (1908: 192) reported that *T. allectus* swarmed in great numbers at midday over an open area above a bridge. The same author (1927: 13) described the flight of *T. minutus* (as *T. explicatus*): "They fly by day, the swarms being thickest in the forenoon. The males fly highest, up to 50 feet or more, in vertical lines, up and down incessantly. The females, after mating, fly low over the rippled surface of the stream, each carrying her little bunch of eggs extended at the tip of the abdomen and ready to let fall into the water. On sweeping the wet hand-screen thru the swarm, it was covered at one stroke with hundreds of females and their loosened egg masses."

My first attempts to rear the nymphs in the laboratory in still water resulted in the death of the insects within one or two days after collecting. After installation of an aerator, the immatures were kept for as long as a month after collection and many of these emerged.

Emergence was not seen in the field; in the laboratory, there seemed to be no set time for the event to occur. All my observations were made in February, during which the hour of emergence ranged from 9:15 A.M. to 7:30 P.M. with a tendency toward a morning predominance. The subimaginal and imaginal life spans were not determined, because, as mentioned above, the subimago in every case died before molting.

LOCALITY RECORDS.—Alachua County: Santa Fe River at Poe Springs, May 14, 1934, nymphs; May 21, 1934, nymphs; March 19, 1935, nymphs; March 24, 1937, nymphs; March 12, 1938, nymphs and adults; March 18, 1938, adults; May 14, 1938, nymphs; February 11, 1939, nymphs and adults; February 18, 1939, adults; March 4, 1939, adults; March 24, 1939, nymphs and adults; October 25, 1939, nymphs and adults; April 6, 1940, nymphs and adults. Bay County: 27.4 miles north of St. Andrews, May 30, 1940, nymphs. Pine Log Creek, May 31, 1940, nymphs. Citrus County: Withlacoochee River, April 2, 1937, nymphs. Chassahowitzka Springs, May 10, 1941, adults. Gadsden County: 10 miles south of River Junction, July 1, 1939, nymphs. Gilchrist County: Suwannee River at Fannin Springs, April 5, 1938, nymphs. Hernando County: Southern county line, March 27, 1938, nymphs; Weekiwatchee Springs, August 20, 1938, adults. Hillsborough County: Six-mile Creek, March 26, 1938, nymphs. Hillsborough River, February 11, 1939, nymphs; June 18, 1939, adults. Jackson County: Blue Springs Creek near Marianna, June 9, 1938, nymphs. Leon County: 7 miles south of Highway 20 on Sopchoppy road, June 5, 1938, nymphs. Liberty County: Sweetwater Creek, June 10, 1938, nymphs; November 4, 1938, nymphs; July 1, 1939, nymphs. Madison County: At Jefferson County line, Aucilla River, June 4, 1938, nymphs and adults. Marion County: Silver Springs, May 7, 1934, nymphs. Rainbow Springs run, February 26, 1939, nymphs; March 9, 1940, nymphs. Oklawaha River at Highway 42, March 19, 1938, nymphs and adults. Okaloosa County: 5.1 miles west of Walton County line, June 7, 1938, nymphs; May 31, 1940, nymphs. Wakulla County: Wakulla Springs run, May 29, 1940, nymphs and adults. Walton County: 13.8 miles west of Freeport, June 7, 1938, nymphs and adults. 9.5 miles west of Portland, May 31, 1940, nymphs. 10.6 miles west of Walton County line, May 31, 1940, nymphs. Washington County: Holmes Creek at Holmes County line, July 2, 1939, nymphs.

Genus CAENIS Stephens

The genus *Caenis* was described in 1835 by Stephens, who made the European *Ephemera ahlterata* the genotype. At the time of publication of Eaton's monograph, only two species of *Caenis* were known from North America, seven from Europe (he included *Brachycercus harrisella*), one from Asia, and two from Africa. Hagen (1861: 54-55) listed three species in his *Neuroptera of North America,* but Eaton synonymized *diminuta* and *amica.* The next general review of the genus in North America was that of McDunnough (1931: 254-64) in his paper on the Caeninae. In this, he treated ten species, of which five were described as new. Traver (1935: 643-54) in *The Biology of Mayflies* again treated the North American species of *Caenis,* describing two new species, and redescribing the ten that had been discussed and described by McDunnough. Since this publication, there have been no new species of *Caenis* recorded for North America proper, but Traver (1938: 22-24) described three species from

Puerto Rico without naming them. As these are known chiefly from the nymphs, their relationships with the Florida species cannot be determined at this time.

Campion (1923) discussed in full the synonymy of the generic name *Caenis* and erected the genus *Ordella* to include part of *Caenis;* the remainder was placed under the name *Brachycercus.* Lestage (1924d) followed Campion and suggested that the family name be changed from Caenidae to Brachycercidae; however, McDunnough, without stating a reason, restored the American species to *Caenis* and used Caenidae to include *Brachycercus* (= *Eurycaenis*), *Tricorythodes* (= *Tricorythus* as used for North American species), and *Caenis.* My treatment of the genus will follow that of McDunnough.

No studies have been made on the ecology of North American species of *Caenis;* American authors, in fact, have treated the species chiefly from a taxonomic point of view. Needham (1935: 180) has had more to say about the habits than anyone else, but his statements have been little more than short comments on the habits of the nymphs. He characterized the nymphs as "the sprawlers amid the silt in still water."

The species of this genus are generally distributed throughout North America, although there are relatively few records from the West. However, *C. simulans* is known from Wyoming and British Columbia, and I have collected an unidentified species from eastern Mexico.

Phylogenetically, *Brachycercus* and *Caenis* form a closely knit unit. The relations of these two genera to *Tricorythodes* have been discussed under the latter genus. I agree with Spieth (1933: 357) in his conclusion that "apparently the ancestral stock from which *Caenis* arose differentiated long ago, and it has since then become highly specialized. In doing so it has reached, both in the nymphal and adult stage, a condition superficially—but only superficially—like that in *Tricorythus [Tricorythodes].*"

The high state of development of *Caenis* is indicated by the complete absence of hind wings. Diminution in size leading to loss of the hind wings began in the Jurassic. Reduction of the number of segments in the forceps of the male to a single one further bears out the idea that the genus is an advanced one; however, the eyes of the male have not become enlarged as in many other members of the order, but have remained rather close to the primitive ancestral type.

Caenis diminuta Walker

TAXONOMY.—Francis Walker described this minute form in 1853 from specimens collected at the St. Johns Bluff, Florida, a point on the south side of the St. Johns River about six miles from the ocean. Since then, the species has been redescribed by Hagen (1861: 55), who relates it to *C. lactea* of Europe; by Eaton (1871: 95 and 1884: 147); by McDunnough (1931: 257); and by Traver (1935: 648-49). Spieth (1940: 337) supplemented the description. All these authors agree that the species is quite distinct, but McDunnough and Traver have shown that in addition to the true *diminuta,* the material of earlier writers also included five other species. The adults of *C. diminuta* may be distinguished

PLATE XVI

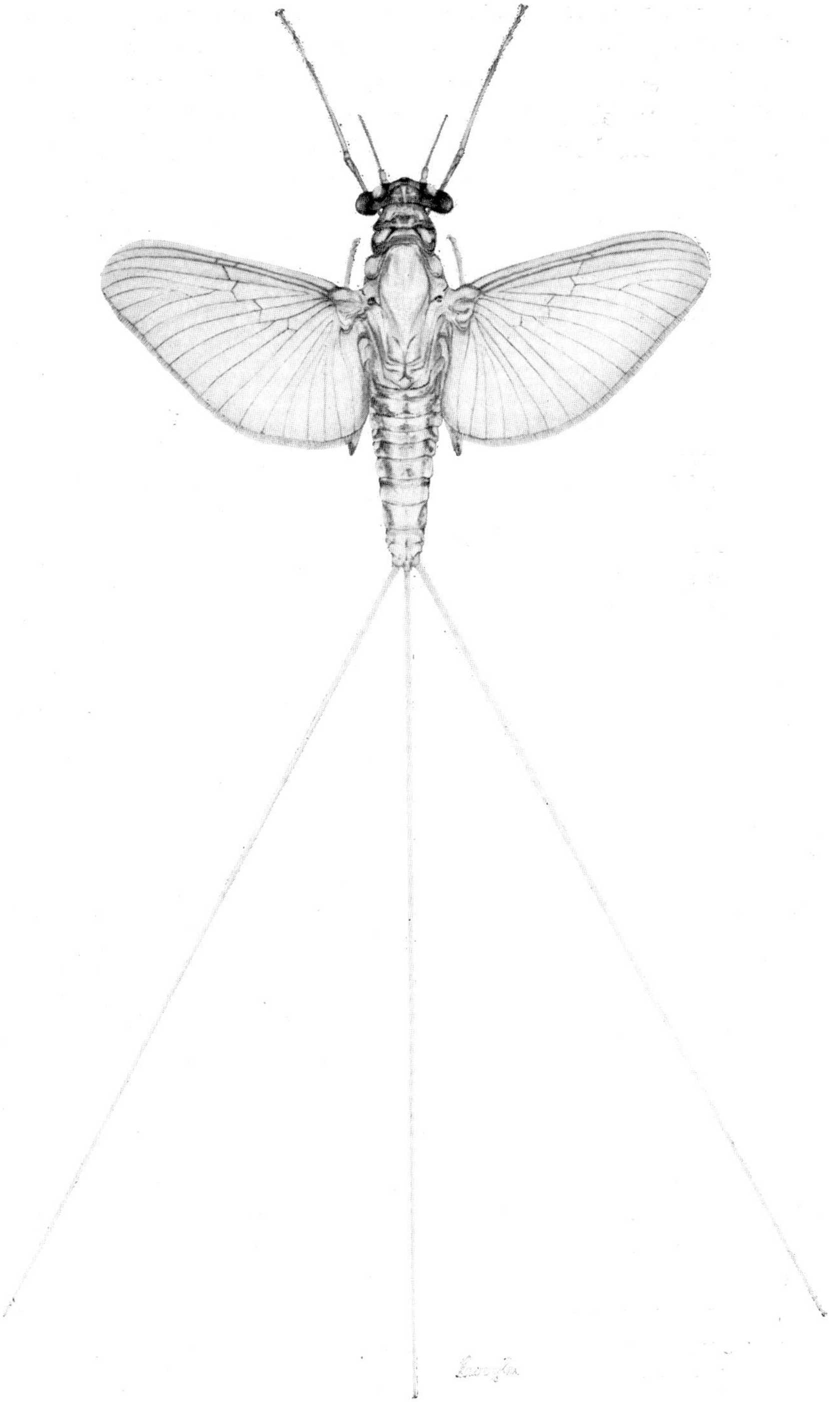

Caenis diminuta Walker, male imago.

from all other North American species of *Caenis* by the dark stigmatic streaks present on the anterior abdominal tergites, by the presence of a dark apical band on the hind femora, and by the absence of the black dots from the femora. Nymphs of *Caenis* are almost impossible to separate, but if only the two Florida species are considered, the problem is somewhat simplified. In general, *diminuta* nymphs in their last stage range from 3.9 to 6.7 millimeters, and usually have numerous yellow spots scattered over the pronotum and mesonotum; in *hilaris,* the size range is from 2.2 to 3.8 millimeters. Yellow spots are usually absent from the pronotum of *hilaris;* the mesonotum has one pair of submedian spots and yellow spots at the bases of the wing pads. The difference in size between the two species is the significant feature, the smaller measurements given above being those of the male nymphs, the larger those of females.

DISTRIBUTION.—This species is the most widely distributed of the mayflies occurring in Florida. McDunnough has recorded specimens taken near Gravenhurst, Ontario, and from Stuart, Florida, and the species is now known from many localities in Florida. There are no published records of *C. diminuta* between these two widely separated regions, except at Spring Creek, Georgia. In Florida, the species is known to occur from the southernmost county as far north and west as Bay and Washington counties in the panhandle. It may be found in almost any Florida locality where there is standing, fresh water.

ECOLOGY.—Even very small bodies of fresh water serve as habitats for *Caenis* nymphs. The smallest biotope in which I have found nymphs was a puddle about a foot wide, which was produced by the recession of a creek that had overflowed. The species is most successful in standing bodies of water of pond size or smaller. In lakes of the type represented by Lake Wauberg, the species is very common, but in such sand-bottomed lakes as Lake Geneva, they are much scarcer and may even be absent. The nymphs are sprawlers in the mud and silt of the pond bed. Here they crawl slowly over the plant stems, where they are more or less protected from predators. Occasionally, they leave the bottom to climb a short distance up the plants, but they are more frequently found clinging to the bottom of the stems. Leaf debris and other trash accumulating in shallow water also harbor *Caenis* nymphs, but they do not seem to be as numerous here as in the plant association. The margin of the vegetation also appears to limit the distribution of the nymphs, and, even though there may be debris beyond this shore zone, the insects seldom venture into the deeper water.

Caenis nymphs rarely inhabit ponds and the edges of lakes that are covered with water hyacinths, because there is usually an inadequate supply of oxygen and not enough bottom vegetation to support even these rather tolerant nymphs. Likewise, sinkhole ponds covered with duckweed and *Brunneria punctata* seldom have populations of this species, as in most of these waters attached vegetation is lacking and there is much decaying material. The combination of these factors may be sufficient to render the bottom of these ponds unsuitable for habitation. Two sinkhole ponds, approximately one-fourth of a mile apart

PLATE XVII

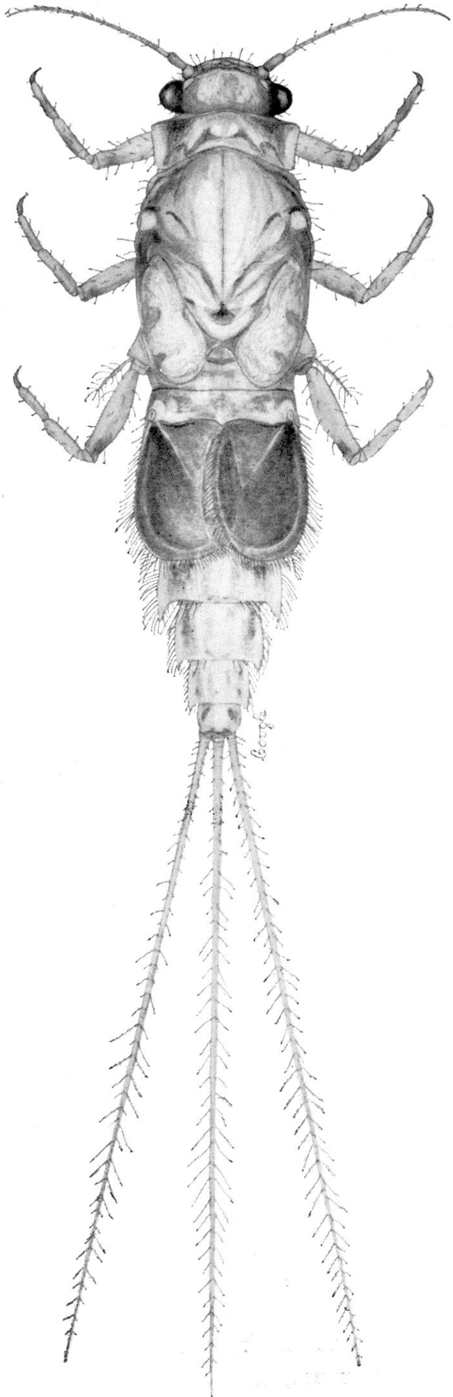

Caenis diminuta Walker, nymph.

on the grounds of the experiment station of the University of Florida, exemplify this phenomenon. One of these ponds is completely covered with duckweed; the other is entirely free of the plant but has a luxuriant growth of *Riccia fluitans*. In the former there are no mayfly nymphs; in the latter there is an abundance of *C. diminuta* immatures. Mayflies are not the only animals affected by the paucity of oxygen in the duckweed-covered pond, for the usual pond associates of *C. diminuta* are also either entirely absent or extremely rare.

Roadside ditches are among the most fruitful places for collecting *C. diminuta* nymphs, because comparatively few predators inhabit the ditches, submergent vegetation is plentful, and oxygen and food are abundant. Cypress swamps and marshes in which there is vegetation or leaf debris also maintain populations of nymphs. Even streams may be inhabited by the nymphs of this species. In flowing water, they live in shallow regions among the rooted aquatic plants, on the roots of terrestrial plants projecting into the water, on submerged logs, sticks, or other trash, and even on rocks in moderately swift water. When in the flowing parts of the stream, they must cling more closely than in the quiet areas, and the current limits their ability to move about easily. In still or slow-flowing water they move more freely.

The nymphs have surprisingly wide limits of toleration: they withstand not only marked variations in acidity and alkalinity, but also marked fluctuations in temperature and in the amount of decomposition taking place in the water. The lowest pH recorded in water from which nymphs were taken was somewhat less than 4.0, the highest well above 8.0. The nymphs must also be able to withstand great temperature variations as is indicated by the fact that they thrive in shallow pools continually exposed to south Florida sunlight. Yet *C. diminuta* apparently flourishes in the very cold waters of Canada. Nymphs in all stages of development have been collected from odorous, stagnating water. Possibly decomposition had recently begun; even so, the nymphs showed remarkable powers of resistance to the putrefaction.

Associated in the ponds and ditches with *C. diminuta* are only two other species of mayflies, *Callibaetis pretiosus* and *C. floridanus*. There are large numbers of insects, many of them predators, living in the ponds along with small fish and the other usual organisms of the pond biocoenosis. In the slow-flowing parts of streams, the other mayflies found include *Stenonema smithae*, *Baetis spinosus*, *B. spiethi*, *Callibaetis floridanus*, *C. pretiosus*, *Centroptilum viridocularis*, and, more rarely, *Paraleptophlebia volitans*, *Habrophlebiodes brunneipennis*, *Blasturus intermedius*, and *Ephemerella trilineata*.

SEASONS.—There is no definite season for emergence of *C. diminuta* in Florida. I have records of adults for all months of the year and there are no clear-cut broods. McDunnough recorded the species from adults taken on August 20 at Gravenhurst, Ontario. This is apparently the sole reference in the literature to an emergence date for the species. In the North, *C. diminuta* must be seasonal; it could not, in fact, be otherwise, for the ponds in which it lives are frozen over during the winter months. Though in Florida emergence occurs throughout the year, the species is limited to some extent by the temperature. I have observed on a number of occasions that during a cold spell emergence comes to a standstill, yet as soon as there is a change in weather and the shore zone warms sufficiently, subimagoes begin rising.

Collections at a light on warm nights during the winter indicate that the rate of emergence is somewhat less than during the summer. Nevertheless, the number of emerging adults is large. Lighting in a particular area on consecutive nights frequently does not yield the same results, although lighting conditions may be approximately the same. This may be due to emergence of different broods from the same pond, but as there is obviously so much overlapping in generations, the existence of distinct broods is somewhat doubtful.

HABITS.—The food of the nymphs consists of plant debris, algae, and any other organic materials which chance may throw their way; even inorganic substances may happen to be encompassed by their raking feet. One nymph was observed feeding on a leg of a dead immature. When feeding, the nymphs cling with the two posterior pairs of legs and one of the forelegs and at the same time extend the other foreleg; this leg, which is then drawn in toward the mouth, pulls along with it the debris lying within reach of the claw. When the food reaches the mouth, the maxillary and labial palpi begin an active movement, sorting and shoving the food into the mouth. As soon as one clawful of material is disposed of, the other leg may reach out, or the first may again be used if there is still material within reach; if not, the nymph moves slowly forward until it encounters more food. Pieces of the epidermal covering of plants are detached by the sharp-pointed claws and pulled into the voracious jaws. They seem to feed continually, and the anterior part of the alimentary tract is always well packed as in all other mayfly nymphs.

The adults of *C. diminuta* are among the most pronouncedly phototropic of mayflies. A light set up near any body of water where they are known to be present will attract males and females equally. On some nights, either males or females may be dominant at the lighting sheet; on other occasions the two sexes may be attracted in approximately equal numbers.

When most mayflies alight, the wings are held vertically over the back. *Caenis*, however, alights with its wings spread wide, leaps up again, and repeats the performance several times, each time moving more closely to the lantern. Finally flying stops and the adult begins to crawl toward the light—its wings spread in a horizontal plane, or even depressed until they almost touch the object on which the insect is walking. Once attracted to a light, the adult seems almost helpless and is an easy prey for many insects, particularly ants.

LIFE HISTORY.—I have been fortunate in being able to rear this species from egg to adult. Many nymphs have hatched, but only two have been successfully reared, the others having died during some preadult stage. The entire period of development was approximately three hours less than 124 days. I was unable to determine the number of instars, but some data have been obtained on growth rate, hatching, and other features of life history. These are presented in Table 4.

TABLE 4

RATE OF GROWTH OF *Caenis diminuta* NYMPHS
(Average measurements covering about one-fourth of early life)

Date Specimen Killed	Body Length	Antennal Length	Number of Antennal Segments	Length of Lateral Cerci	Number of Segments in Lateral Cerci	Length of Median Filament	Number of Segments in Median Filament	Length of Fore Tibio-tarsus	Length of Foreclaw	Gill Present
Oct. 30	0.42 mm.	0.2 mm.	5	0.35 mm.	3	0.36 mm.	5	0.1 mm.	0.045 mm.	No
Nov. 6	0.49 mm.	0.32 mm.	6 (?)	0.55 mm.	6 (?)	0.57 mm.	6 (?)	0.11 mm.	0.055 mm.	No
Nov. 13	0.50 mm.	0.38 mm.	6-7	0.68 mm.	6-7	0.68 mm.	6-7	0.11 mm.	0.055 mm.	No (except in one specimen)
Nov. 22	0.83 mm.	0.73 mm.	9-10	1.0 mm.	9-11	1.1 mm.	9-10	0.17 mm.	0.07 mm.	In some
Nov. 27	0.93 mm.	0.80 mm.	11-12	1.2 mm.	14-15	1.2 mm.	14-15	0.21 mm.	0.075 mm.	Yes
Dec. 4	1.16 mm.	0.85 mm.	14-16	1.55 mm.	25 (?)	1.6 mm.	?	0.3 mm.	0.075 mm.	Yes

On October 23, 1939, at 6:40 P.M., I visited a light near the University of Florida campus and collected fifteen female *C. diminuta.* I took these within ten minutes, even though it was not yet dark and there was a bright moon. I separated the specimens and placed each one on the surface of a bowl of water. The adults began to oviposit immediately and completed the process by 7:20 P.M. Most specimens voluntarily completed oviposition, but others which did not were aided by squeezing the abdomen with forceps. The females, once placed on the water, showed no inclination to leave, even after emptying the abdomen. I numbered each dish and female correspondingly and preserved the specimens in alcohol.

In some of the bowls all the eggs had hatched by 7:15 P.M. on October 30; and in the remainder, hatching of the other eggs had begun. By December 4, all the nymphs were healthy and apparently growing rapidly. Two of the nymphs, though, were very immature, hardly larger than a normal week-old individual. This immaturity may have been due to delayed hatching.

While I was preparing the bowls for the *C. diminuta* eggs on October 23, a few of the ova adhered to the vial in which the females were brought to the laboratory. I had filled the bottle with water, added a pinch of dried organic matter and then completely forgotten about it. On February 7, 1940, while cleaning the laboratory, I found the vial. The water in it was almost gone, but there were seven *Caenis* nymphs still alive and active. These were all decidedly dwarfed. Two of the nymphs had not even developed gills, but only a pair of tubular outpocketings on segment 1. The stunting was evidently the result of an insufficiency of food. The small mass of material which I had placed in the bottle in October was still hard packed and apparently had only been nibbled at by the nymphs.

By February 19, only four nymphs still remained alive of those which were hatched on October 30. Three of these were in the last instar by this time, and one of them measured 4.5 millimeters in length. Finally, on February 24, the wing pads of one of these last-instar nymphs began to blacken, and on March 3, the subimago was ready to emerge.

The first of the laboratory adults emerged sometime during the late afternoon or evening of March 3. On the morning of March 4 it was found floating on the surface of the water in the bowl in which it had been reared. The shed nymphal skin of another specimen was also found in the rearing dish, but the imago escaped.

Table 4, together with the observations, gives a partial picture of the development of *C. diminuta* from egg to adult. Many difficulties beset the rearing of the nymphs. The eggs are easily hatched, but when food material is introduced, rotifers, flatworms, and small oligochaets may be accidentally transferred with it. In most cases, these other forms furnish too much competition for the delicate nymphs, many of which perish. The four specimens mentioned above must have been exceptionally sturdy, for several times the water almost disappeared from the dish. There were many Protozoa in the bowl, and most of the time very little light reached the insects.

Table 5, gives the time required from oviposition to hatching for many of the nymphs hatched in the laboratory.

The figures in Table 5 indicate that the fall and winter broods hatch much more rapidly than do those of the spring. No eggs were hatched during the summer; the time required for hatching in this season would probably be between that of the spring and fall (five to eleven days). Then again, the

TABLE 5

HATCHING TIME FOR EGGS OF *Caenis diminuta*

Date of Oviposition	Date of Hatching	Approximate Time Required for Hatching
September 4, 7:45 p.m.	September 11	7 days
September 13, 8:10 p.m.	September 19, 9:30 a.m.-12:40 p.m.	135 hours; 6 days
September 28, 7:30 p.m.	October 3, 11:10 a.m.-7:15 p.m.	115 hours; 5 days
February 24, 6:15 p.m.	March 3, 2:00 p.m.	164 hours; 7 days
March 18, 8:00 p.m.	March 29, 9:20 p.m.	265 hours; 11 days
April 26, 8:20 p.m.	May 6-7, 12:30 p.m.	256 hours; 10½ days
September 14, 7:45 p.m.	September 21, a.m.	156 hours; 6½ days

figures may be of no significance, for all specimens were hatched indoors where the fluctuation of temperature was slight as compared to that outdoors. The time necessary for hatching is one of the shortest of any of the species for which I have been able to secure data. Listed below are hatching times for a number of other species of mayflies hatched under laboratory conditions:

Baetis vagans—11 days (Murphy, 1911: 41)
Ameletus ludens—5 months (Clemens, 1922: 78)
Heptagenia hebe—12 days (Needham, Traver, Hsu, 1935: 90)
Hexagenia recurvata—15 days (Needham, Traver, Hsu, 1935: 90)
Ephemera varia—15 days (Needham, Traver, Hsu, 1935: 90)
Stenonema tripunctatum—11-23 days (Needham, Traver, Hsu, 1935: 90)
Stenonema interpunctatum—13-15 days (Needham, Traver, Hsu, 1935: 90)
Ephemera vulgata—10-11 days (Needham, Traver, Hsu, 1935: 90)
Hexagenia bilineata—9 days (Needham, Traver, Hsu, 1935: 90)
Stenonema smithae—13 days (Original observation)
Caenis diminuta—5-11 days (Original observation)

When the nymph of *C. diminuta* is ready to hatch, the chorion splits cleanly along the longitudinal axis. One nymph had freed its head when first observed. The antennae were still twisted, but free; also two of the left legs had been released. By working the body from side to side, the other legs finally were pulled loose and with these, the nymph then seized the bottom trash, and clinging, even though his legs had not yet straightened, pulled the remainder of his body from the egg. The tails, however, adhered to the egg membranes so that I finally had to detach the insect. The entire process took about two minutes. Many nymphs when first hatched have the antennae and tails crinkled, the legs bowed, and the abdomen twisted to one side, but this condition generally corrects itself and within a few hours these nymphs are normal. At times the condition persists for several days, probably until the first molt.

The eggs of *C. diminuta* are among the most interesting produced by the mayflies. Smith (*in* Needham, Traver, Hsu, 1935: 82) described the surface of the ovum but mentioned little else concerning the eggs. The eggs are released by the female in a somewhat spherical mass, and as she touches the water, this mass immediately breaks up into the individual eggs which scatter and sink. As they fall, they release a tangle of threadlike, sticky fibers that cling to the first solid object with which they come in contact. Once an egg is attached, there is some difficulty in tearing it loose without injuring the filaments; if the egg is freed, the remaining threads may again become entangled. There is no swelling of the chorion nor any formation of a protective jelly-like coat as in the eggs of many other species of mayflies.

Upon hatching, the nymphs begin to search for concealment and food. Although in many cases the spreading caudal filaments are evident, the young insects are rather transparent, almost any background being sufficient for their concealment. Their swimming actions are very awkward. The young nymph lashes out with its caudal filaments, then bends them completely back over the abdomen and lashes out again, producing a series of jerky, somewhat undulatory, motions. During this process, the legs are held widely spread and ready to grasp any support. Until the operculate gills are formed, the nymphs retain their whitish, opaque appearance. Shortly after the formation of these gills, brown pigment begins to color the immatures, which are further concealed by the detritus that adheres to the long hairs covering the body.

For some time I entertained the idea that this species was parthenogenetic, but numerous attempts to hatch eggs from last-stage female nymphs, subimagoes, and virgin imagoes were unsuccessful. For controls in these experiments I used females which were taken at the campus light mentioned above. These had almost certainly mated, and their eggs invariably hatched.

Emergence of *C. diminuta* adults occurs late in the afternoon or at night. The nymphs, when ready to transform, float freely at the surface in shallow water from three to four feet deep and just inside the limits of attached vegetation. The subimago soon bursts from the nymphal skin, but seems to experience great difficulty in taking flight. In many instances the subimago is unable to rise from the water at all, and during its struggles the wings become wet, thus spelling the doom of the insect. Frequently the subimago will take off from the nymphal exuviae, but the flight results in only a short hop. After resting briefly on the surface film, the insect makes another attempt to fly. Once free and safely in the air, the subimago flies to the nearest support and very shortly molts. In the instance of specimens reared in the laboratory, the molt occurred approximately five to six minutes after emergence from the nymphal skin.

This species is one of the shortest-lived of all mayflies. Laboratory-reared adults remained active about four hours, although a slight reflex movement of the legs could still be observed thirty minutes later.

I have never observed the mating of this species, and there seem to be no written accounts of the mating flights of any species of *Caenis*. I did, however, note a flight which may have been the mating flight of *C. diminuta*. About 3:00 P.M., January 25, many subimagoes were seen rising from the surface of Orange Lake. They flew to the nearest available support where they molted and, as imagoes, became very active. At 3:45 P.M. a single male imago was noted rising and falling in a typical mating-flight behavior about three hundred yards from shore, but still within the bonnet zone. Flying about five feet above the water, it rose and dropped vertically about two feet, but seldom moved horizontally. The male was soon lost to sight.

While collecting at Orange Lake a few days later, Dr. H. H. Hobbs saw the coupling of a pair of *C. diminuta* adults that were flying with little rise and fall about five feet above the water. After about thirty seconds the two individuals separated and soon disappeared.

LOCALITY RECORDS.—Alachua County: 3 miles southwest of Gainesville, April 9, 1934, nymphs. Hogtown Creek, February 15, 1937, nymphs; May 30, 1937, nymphs; September 25, 1937, nymphs. 5 miles northwest of Gainesville, November 8, 1937, nymphs. 15 miles north of Gainesville, May 12, 1937, nymphs. 7 miles northwest of Gainesville, May 4, 1937, adults. 7 miles northeast of Gainesville, May 11, 1937, nymphs. Newnan's Lake, March 16, 1932, nymphs; September 20, 1938, adults; September 27, 1939, nymphs. Santa Fe River at Worthington Springs, September 2, 1939, nymphs and adults. Freezers Pond, September 25, 1937, adults; September 29, 1937, adults; October 5, 1939, adults. Lake Alice, March 17, 1937, nymphs; November 15, 1937, adults; February 24, 1940, adults. Santa Fe River at Poe Springs, March 21, 1933, nymphs. Gainesville, March 1, nymphs, April 5, nymphs, May 10, nymphs, October 4, nymphs, October 8, nymphs, October 24, nymphs, 1937; April 1, adults, October 21, adults, 1938; March 23, adults, March 24, adults, September 4, adults, September 6, adults, September 13, adults, September 28, adults, October 6, adults, October 14, adults, October 23, adults, October 24, adults, 1939; March 18, nymphs and adults, March 19, adults, March 29, adults, April 26, adults, September 14, adults, October 2, adults, October 25, adults, 1940. Lake Wauberg, 1932-33, nymphs. 5 miles west of Gainesville, March 6, 1937, nymphs. Goat Sink about 12 miles southwest of Gainesville, April 23, 1939, nymphs. Montenocha Creek, October 28, 1939, nymphs. Hatchet Creek, March 22, 1937, nymphs; March 23, 1938, adults; April 24, 1948, adults. Biven's Arm, September 22, 1946, adults. Bay County: 3 miles west of county line at U. S. Highway 98, May 30, 1940, nymphs. Citrus County: 1½ miles south of Withlacoochee River, March 25, 1938, nymphs. Floral City, April 2, 1939, nymphs. Clay County: Green Cove Springs, October 15, 1939, nymphs. Columbia County: 6 miles east of Lake City, May 12, 1937, nymphs. 11.6 miles west of Lake City, October 27, 1938, nymphs. Dade County: Miami, April 15, 1937, nymphs; August 20, 1937, nymphs; August 26, 1937, adults; September 4, 1937, adults; September 7, 1937, adults. 5 miles west of Miami, September 2, 1937, nymphs.

Flagler County: Andalusia, April 4, 1939, nymphs. Gadsden County: River Junction, June 30, 1939, adults. Glades County: Moorehaven, May 20, 1940, nymphs. Hamilton County: 6 miles north of Live Oak road at U. S. Highway 41, February 4, 1938, nymphs. Hernando County: Bayport, November 1, 1947, adults. Hillsborough County: Tampa, March 21, 1937, nymphs. Mullis City, December 31, 1937, nymphs. Hurrah at Picnic, March 26, 1938, nymphs. Six-mile Creek, March 26, 1938, nymphs. 2 miles east of Tampa, March 26, 1938, nymphs. Jackson County: 14.2 miles north of Bay County line, June 8, 1938, nymphs. 3.6 miles north of Altha, July 1, 1939, nymphs. Blue Springs Creek near Marianna, April 13, 1935, nymphs; June 9, 1938, nymphs; July 1, 1939, adults; December 1, 1939, adults; June 5, 1940, adults; May 3, 1941, adults. Lake County: Tavares, March 23, 1936, adults. Crows Bluff on St. Johns River, September 12, 1938, nymphs. Umatilla, October 2, 1938, nymphs. Leon County: 13.8 miles west of Tallahassee, June 5, 1938, nymphs and adults. Liberty County: Hosford, March 17, 1939, nymphs. Near Ochlocknee River, March 16, 1939, nymphs. Madison County: 5.9 miles west of Madison, February 5, 1938, nymphs. 11.3 miles west of Madison, February 5, 1938, nymphs. Marion County: Ocala National Forest, February 12, 1938, nymphs and adults; March 19, 1938, nymphs. Eureka, February 12, 1938, nymphs. Lake Bryant, April 15, 1938, nymphs. Juniper Springs, September 1, 1938, adults. Salt Springs, March 23, 1941, adults. Oklawaha River, April 10, 1948, adults. Nassau County: Gross, December 19, 1939, nymphs. Orange County: Lake Ola, April 4, 1939, nymphs. Orlando, November 10, 1938, adults. Pasco County: 20 miles north of Tampa, March 25, 1939, nymphs. Elfers, March 21, 1947, adults. Putnam County: 30 miles east of Gainesville, October 2, 1937, nymphs. 1.2 miles north of Palatka, October 2, 1937, nymphs. Palatka at bridge over St. Johns River, October 15, 1940, adults. Welaka, July 5, 1939, adults; July 16, 1939, adults; August 18, 1939, October 12, 1939, adults; October 19, 1940, adults. St. Johns County: At Highway 16, April 23, 1938, nymphs. Taylor County: 6.6 miles east of Perry, February 5, 1938, nymphs and adults. Near northern county line, February 5, 1938, nymphs. Volusia County: De Leon Springs, August 5, 1939, adults. Washington County: Holmes Creek, July 2, 1939, adults; April 30, 1946, adults.

Caenis hilaris (Say)

TAXONOMY.—*Caenis hilaris,* one of the first recorded North American species of mayflies, is the smallest ephemerid found in Florida as well as one of the smallest known, being surpassed in this respect by only one other member of the same genus. The body and wing measurements are slightly over two millimeters. The species was described as *Ephemera hilaris* by Say in 1839; later it was redescribed from Illinois material by Walsh (1862: 381), after first having been placed in the genus *Caenis* by Francis Walker (1853: 583). Hagen (1861: 54-55) briefly described the adults, and Walsh, in his description, pointed out differences between Hagen's specimens and his own. Eaton (1871: 96 and 1884: 147) redescribed *C. hilaris.* After an interval of nearly fifty years, McDunnough (1931: 255-56) revised the descriptions of this species. Its most recent treatment is by Traver (1935: 650) in the taxonomic section of *The Biology of Mayflies.* My specimens fit the descriptions by McDunnough and Traver fairly well, and the few minor differences are probably individual variations.

DISTRIBUTION.—In Florida, *C. hilaris* is confined to an area in the northwestern part of the state extending from the western boundaries to somewhat east of the Apalachicola River. The species is rather widely distributed over the eastern United States, but is unknown west of Oklahoma and in Canada. It inhabits a wide variety of physiographic provinces, ranging from the Coastal Plain into the Appalachian Province. The Florida records locate *hilaris* much farther south than it was previously known (map 15).

ECOLOGY.—Nymphs have been taken only from gently flowing streams. These streams are predominantly of the sand-bottomed type, with debris accumulated near the shore. The habitats of the nymphs are identical with those of *Ephemerella* sp. A, as described for Sweetwater Creek. In other streams, nymphs have been found in the vegetation near shore, among leaf drift, and on submerged logs, and sticks. I have never collected any specimens from the deeper, swifter parts of the streams, even where debris was present. Adults have been taken at Blue Springs Creek near Marianna. Nymphs have not been found at this locality, and it is uncertain whether they inhabit the stream itself or the lake formed by a dam. This stream, which is decidedly basic, would be a rather exceptional habitat for the species, as it has a very soft bottom, an abundance of snails, and no vegetation.

C. hilaris nymphs are commonly found with the same mayfly species as those associated with *Ephemerella* sp. A.

SEASONS.—Because of the distance, I have been unable to make the trips to northwest Florida which would have permitted obtaining fuller data on the seasonal distribution of *C. hilaris*. However, some evidence is at hand which leads me to believe that, like *diminuta, C. hilaris* emerges throughout the year. My collection includes adults taken in June and July, as well as last-instar nymphs recorded in November and December. Although my records have many gaps, it would seem logical from a study of younger specimens to assume that emergence must occur during the warm periods of other months.

HABITS.—The nymphs are extremely minute, some last-stage specimens being less than two millimeters in length and, consequently, difficult to see. I have found that the most efficient method of taking nymphs is to put debris collected by means of a fine-screened coffee strainer into a white-enameled pan. When the pan is tilted and the water drains away from the nymphs, a slight movement is discernible as the insects attempt to follow in the wake of the liquid. As with *C. diminuta* nymphs, the movement is a slow crawl combined with a somewhat wriggling motion. The nymphs of this species do not tend to flick their tails over their backs, and both the young *Tricorythodes* and the young damselfly nymphs may easily be confused with *C. hilaris* unless separated with a hand lens.

The adults, which are strongly phototropic, are attracted to lights along with *C. diminuta*. At Blue Springs Creek a large electric plant, the lights of which are kept burning all night, faces the lake formed by the damming of the stream. The walls and windows of the illuminated porch are covered with spider webs, the spiders hiding in every crevice to catch the myriad of insects which are attracted to the lights. I have examined the webs on several occasions and have been well rewarded for my efforts; the predominant insects present were midges, but *Caenis* adults were second in abundance, other insects occurring in much smaller numbers.

LIFE HISTORY.—In spite of the large amount of taxonomic literature concerning *C. hilaris,* its life history remains almost completely unknown.

LOCALITY RECORDS.—Bay County: 14.1 miles north of Panama City, June 8, 1938, nymphs. 24.7 miles north of Panama City, June 8, 1948, nymphs. 28.7 miles north of Panama City, June 8, 1938, nymphs. Gadsden County: River Junction, July 1, 1939, nymphs. Jackson County: 3.6 miles north of Altha, December 1, 1939, nymphs. Blue Springs Creek near Marianna, July 1, 1939, adults; June 5, 1940, adults. Leon County: 7 miles south of Highway 20 on Sopchoppy road, June 5, 1938, nymphs. Liberty County: Sweetwater Creek, June 10, 1938, nymphs; November 4, 1938, nymphs; July 1, 1939, nymphs; December 1, 1939, nymphs; May 2, 1941, nymphs; April 30, 1946, nymphs. Eastern county line at Highway 20, November 30, 1939, nymphs. Okaloosa County: 5.1 miles west of Walton County line, May 31, 1940, nymphs. Wakulla County: Smith Creek, June 5, 1938, nymphs. Washington County: Holmes Creek, July 2, 1939, nymphs.

Genus BRACHYCERCUS Curtis

As there has been considerable confusion regarding the proper application and status of the generic names *Brachycercus, Caenis, Eurycaenis,* and *Ordella,* it may be well to review briefly the justification for the use of the first two names by recent American authors. Curtis (1834), in describing *Brachycercus,* established the genus with *B. harrisellus* Curtis as the only included species. Hence it became the type of the genus by monotypy. In 1835 J. F. Stephens proposed the genus *Caenis,* dividing it into two sections. The first (*Caenis* s.s.) included two "species" with long caudal filaments [= males] (*macrura* Steph. and *dimidiata* Steph.); the second (*Brachycercus* Curtis) comprised five "species" with short caudal filaments [= females] (*harrisellus* and four others). In 1837 Curtis fixed the type of *Caenis* s.s. as *macrura* Steph. and that of *Brachycercus* as "*B. harrisii*" [= *harrisella*]. In 1839 Burmeister erected the synonymous *Oxycypha,* citing under it, in synonymy, Curtis' *Brachycercus* and including in it *harrisella,* the genotype of *Brachycercus.*

Because the length of the caudal filaments is a purely sexual character, the long-tailed "species" placed by Stephens in *Caenis* being males and the short-tailed ones assigned to *Brachycercus* being females, subsequent authors, including Pictet, Hagen, and Eaton, used *Caenis* in preference to *Brachycercus,* despite priority, on the ground that the latter name was unsuitable. In 1917 Bengtsson described the new genus *Eurycaenis,* again with *harrisella* as the type; this genus is evidently a strict synonym of *Brachycercus* Curtis. Campion in 1923 reviewed the history of the genus *Brachycercus* to that date; he stated that Bengtsson's action had restricted the generic name *Brachycercus* to the single species *harrisella,* and that this left the species which had been associated with *harrisella* under the name *Caenis* (properly *Brachycercus*) without a generic name, as *Caenis* had been "invalid from the beginning." This conclusion, however, is incorrect because it assumes that Curtis' inclusion of *harrisella* in *Caenis (Brachycercus)* invalidated *Caenis* s.l., whereas in reality the type of *Caenis* s.l. must be chosen as one of the two species included by Curtis under *Caenis (Caenis).*

The genotype of *Caenis* was validly designated by Curtis in 1837 as *macrura.* Campion called the generic unit, supposedly in need of a name, *Ordella,* giving *macrura* as genotype; this name must therefore be regarded as a strict synonym of *Caenis* Stephens. Lestage (1931d) reaffirmed the synonymy of *Eurycaenis* with *Brachycercus.* North American students of the Ephemeroptera have not accepted *Ordella,* though *Brachycercus* has been substituted for *Eurycaenis.* The present status of these names is therefore as follows:

Brachycercus Curtis 1834, genotype by monotypy *B. harrisella* Curtis.
- Syn. *Oxycypha* Burmeister 1839, genotype here selected *Oxycypha luctuosa* Burmeister (= *Brachycercus harrisella* Curtis, according to Eaton and Campion).
- Syn. *Eurycaenis* Bengtsson 1917, genotype by original designation *Brachycercus harrisella* Curtis.

Caenis Stephens 1835, genotype *Caenis (Caenis) macrura* Stephens, designated by Stephens 1837.
- Syn. *Ordella* Campion 1923, genotype by original designation *Caenis macrura* Stephens.

McDunnough (1931) treated the North American species of the family Caenidae, but apparently did not know of Campion's and Lestage's placement of *Eurycaenis* in synonymy, since he described *prudens* as *Eurycaenis prudens.* Traver (1932) likewise overlooked these papers, but in 1935 she used the genus *Brachycercus* to include those species which had previously been placed in *Eurycaenis.* It is obvious from a study of adults that the species of *Caenis* and those of *Brachycercus* form a closely knit phylogenetic unit, which is apparent both in wing characters and in genitalia. Ulmer (1933: 206) and Traver (1935: 630) have included the genus *Tricorythodes* in the subfamily Caeninae with the other genera mentioned above. Spieth (1933: 355), in disagreeing with this arrangement, stated that the resemblances are merely superficial.

Brachycercus nymphs are rather unusual in appearance, belonging to that large assemblage of mayflies which have operculate gill covers to act as protectors of the more delicate, highly tracheated, posterior gills. These nymphs are easily separated from those of *Caenis* by the presence of tubercles on the head and by the depressed body form. They differ from the nymphs of *Tricorythodes* in having tubercles on the head, a depressed body, and rather square, operculate gill covers. *Oreianthus* nymphs are differentiated from *Brachycercus* nymphs by the presence of hind wing pads. Finally, *Brachycercus* nymphs differ from those of *Ephemerella* in that the operculate gill covers of *Brachycercus* are on the second abdominal segment rather than on the third or fourth segment as in *Ephemerella.*

The following key, modified from that of Traver (1935: 639), separates the North American species of *Brachycercus:*

1	Small species (wings 3.0 mm. in length)	(2)
1′	Larger species (wings 3.8-5.0 mm. in length)	(3)
2 (1)	Meso- and metanotum straw to canary yellow; femur of foreleg light purplish brown; abdominal tergites creamy white, 8-10 with a purplish-black middorsal streak	*flavus*
2′	Meso- and metanotum light brown; femur of foreleg pale smoky; abdominal tergites yellowish white, immaculate	*prudens*
3 (1′)	Mesonotum dark red-brown, metanotum purplish; tergites 1-6 purplish gray, margins lavender; posterior margins of sternites purplish	*nitidus*
3′	Mesonotum paler brown; metanotum similar	(4)
4 (3′)	Tergites 1-6 pale, posterior margins blackish; dark lateral spots on some of the sternites	*idei*
4′	Tergites 1-6 prominently mottled; middorsal line evident on tergites 6-9, sternites without spots; anterior margins of sternites 1-9; blackish	*maculatus*

The species of *Brachycercus* are recorded from widely scattered localities. *B. flavus* was described from Louisiana at the Texas state line; *idei* is known from Ontario; *nitidus* occurs in the Appalachian region of North Carolina; *prudens* has been listed from Saskatchewan and Kansas; and *lacustris* from Michigan and New York. The species, *maculatus,* is known to occur only in Florida.

It appears that the genus, which is also Holarctic and African, is distributed throughout North America, although no species are known to occur in the Rocky Mountains and westward. *Flavus, maculatus,* and *Brachycercus* sp. A are Coastal Plain forms, and, as such, will probably be found to have a wide distribution in this faunal province.

Brachycercus maculatus Berner

TAXONOMY.—The colorational characters mentioned in the above key will distinguish adults of *B. maculatus* from other known species of the genus. I hesitate to place too much faith in such color differences, because in many insects coloration is a rather variable "constant." The few North American species of *Brachycercus* can be separated on the basis of coloration, but many of the distinctions may prove unreliable, particularly if additional species are found. The entire group needs to be re-examined on the basis of morphology. In male adults, characters that are worthy of further study include both the length of the forelegs in relation to the wing and body length, and the proportions of the parts of the forceps. There may prove to be some variability in venation and distance between compound eyes. In other genera of mayflies these characters have been used with some success, but they have not been employed in the study of this genus.

To date, the nymphs of only two North American species have been described. The following key separates the known nymphs of *Brachycercus:*

1	Tubercles present on lateral margins of pronotum .... (2)
1′	No prothoracic tubercles .... (3)
2 (1)	Third leg $2\frac{2}{3}$ to 3 times as long as foreleg; tubercles on lateral margins of pronotum anterior to center .... *nitidus*
2′	Third leg $1\frac{1}{2}$ times as long as foreleg; tubercles reduced to small projections from anterolateral corners of pronotum .... species A
3 (1′)	Prominent pyramidal horns on head; legs distinctly dark-banded, each segment with a single dark submedian band .... *lacustris*
3′	Tubercles on head less prominent, somewhat rounded; legs not banded .... *maculatus*

DISTRIBUTION AND ECOLOGY.—*B. maculatus* is known from Silver Springs and from the Santa Fe River at Poe Springs in Florida (map 15). The adults of related genera are strongly phototropic, but no specimen of *Brachycercus* has ever been attracted to my lighting sheet. The few male adults in my collection, together with numerous adults of *Tricorythodes,* were found floating on the surface of the Santa Fe River in one of the quieter parts. They were dead, but well preserved.

Almost nothing has been written about the nymphs of *Brachycercus* in North America. Traver (1932: 142) recorded *B. nitidus* from small rivers and mountain streams. Needham (1918: 249) pointed out that in America, *B. lacustris,* unlike its European counterpart, *B. harrisellus,* apparently inhabits only lakes. Spieth (1938: 3) stated that

> The paratypes *[Oligoneuria ammophila]* were all found on the clean sand of a bar in about one foot of water. The current was moderately swift. In addition to *O. ammophila* a species of *Brachycerus [Brachycercus]* was found in abundance on the same bar. Both species occupied distinct areas and the ranges barely overlapped. *O. ammophila* was found in the swifter current and the bottom was not smooth but rather filled with a great many small depressions. The sand was clean and lacked any trace of silt. *Brachycerus* selected calmer, smoother areas where a small amount of silt was mixed with the sand.

The indications are that the nymphs of *B. maculatus* prefer basic streams. Both Silver Springs and the Santa Fe River are somewhat alkaline, although the latter is not noticeably so. Nymphs were collected at the Santa Fe River in May, and these were observed inhabiting the occasional sandy marginal strips and living in water not more than three or four inches deep. They selected those stretches where silt tended to be mixed slightly with the sand.

SEASONS.—Adults were collected on February 28 and April 6, a cast nymphal skin on May 7, and nymphs on May 8. These specimens indicate a late winter and spring emergence. Comparison with the related *Caenis* suggests that emergence occurs throughout the year.

LIFE HISTORY.—Three males of *Brachycercus idei* were captured by Ide (1930: 218) as they swarmed at the edge of a small lake in Ontario early in the morning of September 5. His data and Eaton's statement (1884: 146) that *B. harrisellus* probably flies at night appear to be the only published references to the habits of the adults. The adult males which I collected were taken before noon; their fine state of preservation makes it appear likely that *B. maculatus* flies early in the morning.

Adult life probably lasts only a few hours—a characteristic common to all recognized members of this subfamily. Little is known about nymphal life of any of the species.

LOCALITY RECORDS.—Alachua County: Santa Fe River at Poe Springs, February 28, 1939, adults; April 6, 1940, adults; May 8, 1947, nymphs. Marion County: Silver Springs, May 7, 1934, nymphal exuviae. Putnam County: Orange Creek, April 10, 1948, nymphs.

Brachycercus sp. A

TAXONOMY.—This species is known only in the nymphal stage and, although it is believed to be new, the description will be withheld until the adult stage has been taken. The nymphs can be distinguished from those of *B. maculatus* by the presence of small anterolateral projections from the pronotum; the ways in which it differs from other known nymphs of the genus are shown in the key on page 190.

DISTRIBUTION.—*Brachycercus* sp. A nymphs have been collected from two tributaries of the Apalachicola River. These are the only known records for the species (map 15).

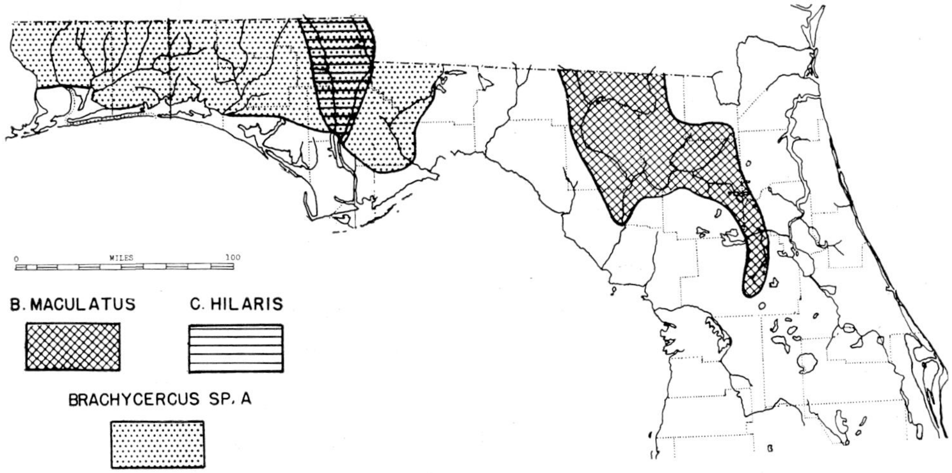

Map 15.—The distribution of *Brachycercus maculatus*, *Brachycercus* sp. A, and *Caenis hilaris* in Florida.

PLATE XVIII

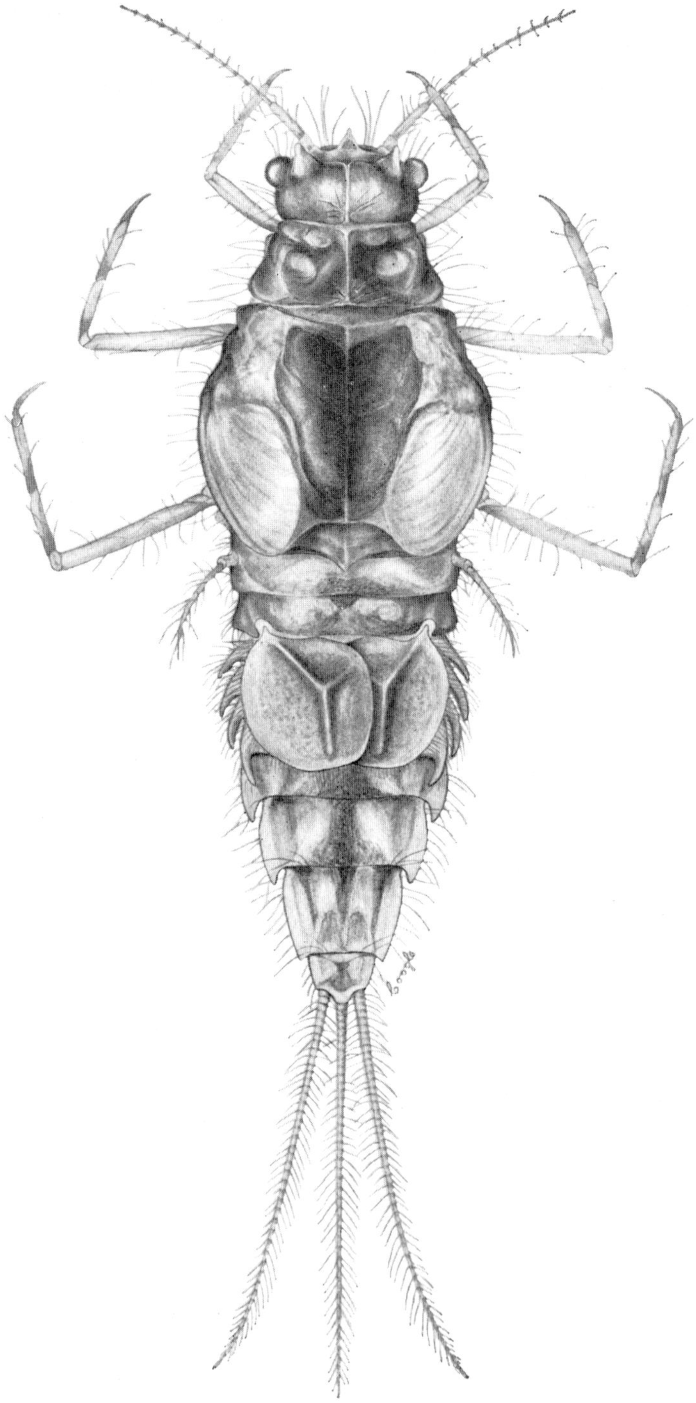

Brachycercus sp. A, nymph.

ECOLOGY.—While collecting in Sweetwater Creek, Liberty County, I found numerous nymphs of *Brachycercus* sprawling on the sand within a few inches of the water line. They normally chose sand which had a slight amount of silt mixed in it and which was partially protected from the main force of the current by projecting vegetation. They could be taken in greatest numbers just beyond the area where silt had accumulated in sufficient amounts for *Hexagenia* nymphs to be common. Frequently, when scraping the sand for *Brachycercus,* the strainer would also pick up small nymphs of *Hexagenia munda marilandica* and the dragonfly *Progomphus obscurus.*

The streams from which the nymphs were taken are typical of the sand-bottomed creeks with little vegetation which are described in the section on habitats.

SEASONS.—Nymphs in all stages of development were collected on April 29. From this evidence, it is likely that the species emerges throughout the year, but the rate of emergence slows down during the winter and completely stops during cold periods.

HABITS.—Little is known of the habits of this species. The nymphs sprawl on the sand, and when taken from the water, they can scarcely move—probably because of their long, spindly legs and attenuated claws. The slow, wriggling motions of the nymphs, the flattened abdomen, and the broad thorax immediately identify *Brachycercus* nymphs to the naked eye.

LOCALITY RECORDS.—Gadsden County: 4½ miles south of River Junction, March 17, 1939, nymphs. Liberty County: Sweetwater Creek, April 30, 1946, nymphs; February 3, 1949, nymphs.

Genus CALLIBAETIS Eaton

When Eaton treated the North American species of *Callibaetis* in his monograph, he described only four, three of which were already named. Between 1900 and 1918, Banks added eight additional species, among which were included *floridanus* and *pretiosus,* the two Florida forms. Since the publication of *The Biology of Mayflies* in 1935, three additional species of *Callibaetis* have been described, making a total of twenty-one known forms from that part of North America north of Mexico. Though adults as well as nymphs of the genus are quite distinct and easily separable from other members of the Baetinae, within the genus itself taxonomic characters are not so well defined and differentiation of many of the species is difficult.

The genus is unique among North American forms in being ovoviviparous. *Callibaetis vivipara* from South America was first found to exhibit this condition. In 1935 a specimen of an unidentified species from New York was reported to contain first-instar nymphs. In 1941 I reported the phenomenon of ovoviviparity in *C. floridanus, C. pretiosus,* and another species from Michigan, and noted that apparently many, if not all, of the species of this genus are normally ovoviviparous. Subsequently Edmunds (1945: 169-71) has confirmed my findings and has described the phenomenon in two species from Utah.

Callibaetis is both a Neotropical and Nearctic genus being generally distributed over both continents. In North America, it is found from coast to coast in mountain regions as well as on the plains. According to published records, the range of the genus does not extend very far north into Canada. Taxonomically, *Callibaetis* is one of the most difficult of all mayflies with which to deal; hence distributional records are few and far between.

Spieth (1933: 341) considered *Callibaetis* to be a separate stock of the Baetinae, and stated: " . . . within this compact group of genera, it is possible to distinguish three distinct lines of evolution. *Callibaetis* represents one line, which is the most primitive of the three; the other two branches are highly specialized and about equal in position. *Baetis* and *Pseudocloeon* make up one line and *Centroptilum* and *Cloeon* the other." I am quite in accord with his opinion concerning the primitiveness of *Callibaetis*. The structure most clearly indicating phylogenetic relationships in the Baetinae is the hind wing; in *Callibaetis*, it is the best developed of the subfamily and retains many cross veins.

Callibaetis floridanus Banks

TAXONOMY.—Some fifty years ago, Mrs. A. T. Slosson collected *Callibaetis floridanus* from Biscayne Bay and submitted her specimens to Dr. Banks for description. Her series consisted entirely of females, and the species was known only from this sex until 1940, when I wrote a description of the male and nymph. My specimens from north Florida, which I am calling *C. floridanus* (form A), may prove to be a new species. For the time being, however, I am not considering them as such. The north Florida forms differ in certain minor respects from the south Florida insects as follows:

North Florida *Callibaetis floridanus* (form A)	South Florida *Callibaetis floridanus* (form B)
Forewings colorless or only slightly colored.	Forewings tinged with brown.
Spots covering body are brown or reddish brown.	Spots covering body are red.
Coloration somewhat dulled.	Coloration intense.

Although in south Florida there are no mayflies which show the characteristics of the north Florida *Callibaetis*, I have taken insects in the Gainesville area which are very similar to the south Florida forms. From an examination of the specimens at hand, it would appear that two subspecies of *C. floridanus* inhabit Florida, with their region of intergradation probably in central Florida. As I have only nymphs from the region between Highlands and Marion counties, the separation must be tentative; when adults are secured from the intervening

PLATE XIX

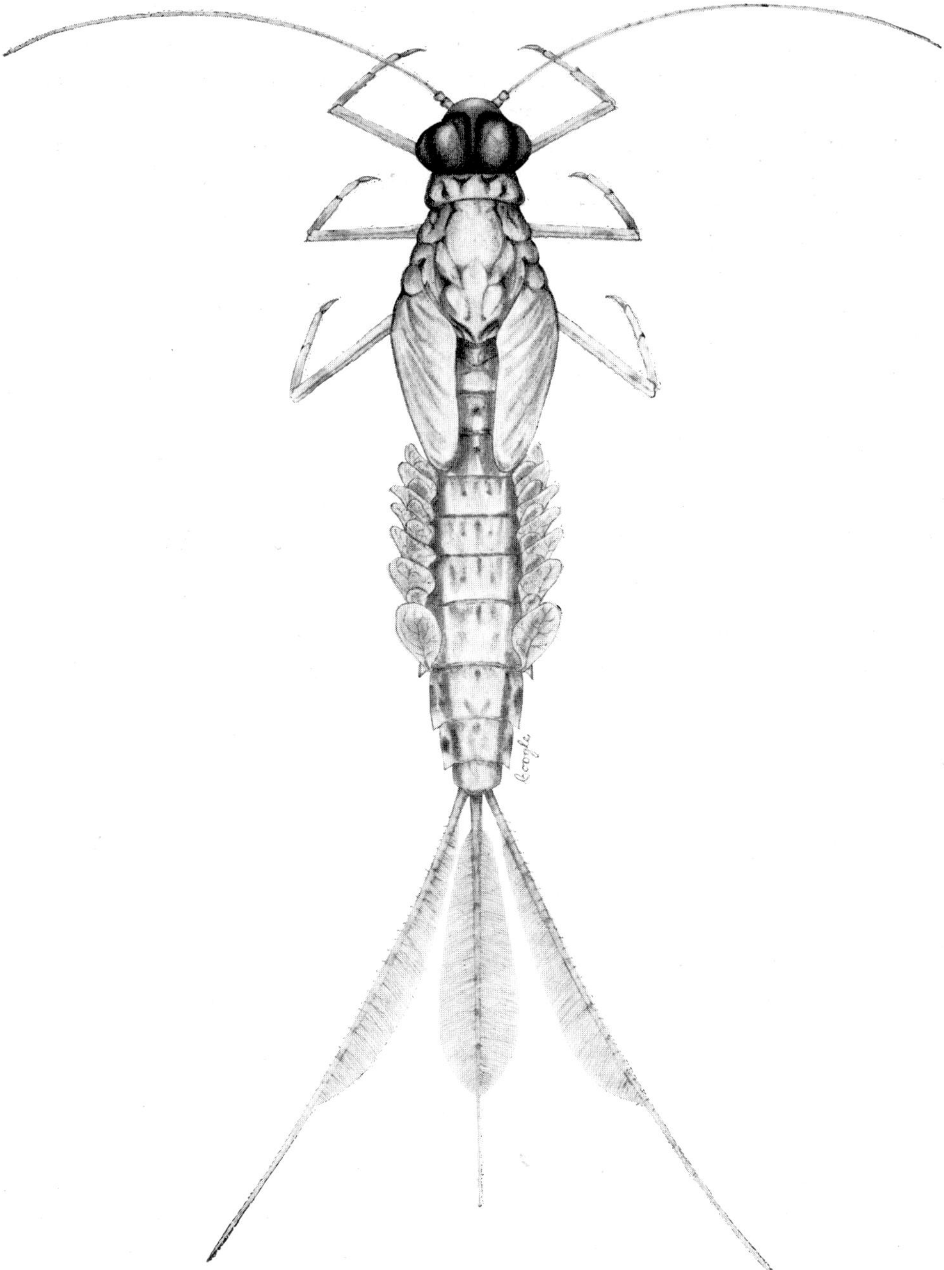

Callibaetis floridanus Banks, nymph.

area, *C. floridanus* (form A) and *C. floridanus* (form B) may prove to be distinct species. The appearance of *C. floridanus* (form B) in the Gainesville area may be only a further indication of the racial character of the two populations; the region of intergradation may be in this more northern locale rather than in central Florida, and the true *C. floridanus* (form A) may occur farther north and west.

In the taxonomy of the genus *Callibaetis*, the males and nymphs are much more difficult to identify than are the females. In the Florida species, the males and nymphs are not exceptional, and unless females are available, identification must be tentative.

DISTRIBUTION.—*C. floridanus* (form B) is distributed throughout the southern tip of Florida and is known to spread as far north as Highlands County. Probably with further collecting in central Florida, the species will be shown to have a much wider range. *Floridanus* (form B) has been collected throughout the Everglades and as far south as Key Largo.

C. floridanus (form A) occupies a range extending from Lake County north into Georgia and northwest to Jackson County. The species most likely spreads farther west, but collections in the northwestern region have been almost entirely from streams. The few specimens of *Callibaetis* that were taken were nymphs.

Probably *C. floridanus* (form A) is not confined to Florida, but may extend into Georgia as far as the Piedmont Province. Westward, the species probably spreads throughout the panhandle of Florida, but here, too, no final conclusions can be reached until adults are taken.

ECOLOGY.—*Callibaetis floridanus* nymphs have the widest limits of toleration of any mayfly in Florida.

The Everglades, which have been described many times and in many places, are broad open expanses of saw grass, partially submerged by water accumulating on the surface of the exposed pitted limestone (Miami oolite) during the rainy season. An occasional hammock indicates a slightly higher mass of rock on which water does not stand. Almost stagnant drainage canals that are filled with gar and other fish, turtles, and vegetation penetrate much of the Everglades. During the dry season, mayfly nymphs develop in the canals, but in the rainy season, when most of the land is under water, mayfly nymphs can be found in almost any area where the water has been standing for a few weeks. At that time, not only the mayflies but also crayfish, minnows, beetles, water bugs, and other members of the canal association become widely disseminated. The water in the "glades" is shallow, one to three feet in depth, fully exposed to the sun, and consequently quite warm. The population of the mayfly nymphs tends to thin out over this region and the mayfly fauna in a particular area would be smaller than the fauna in an area of equal size in the canals.

In south Florida, nymphs have also been taken from temporary ponds, roadside ditches, and from small pools in abandoned rock pits. The species has even been found breeding in a small, artificial fish pond. A collection taken at Key Largo included nymphs that were living in water where mangrove trees were growing. The nymphs were collected not more than ten feet from the roots of the trees. Although the water from which the nymphs came tasted brackish, analysis showed that it had a salinity of only 1.7 per cent.

In north Florida, the habitat of *C. floridanus* is similar to that of the south Florida form. The nymphs are most common in roadside ditches where they live among the stems and leaves of submerged vegetation, clinging to the plants, darting swiftly from one stem or leaf to another, or moving slowly and gracefully over the surface of the plant. In the roadside ditches the water is usually not more than three to four feet deep, choked with vegetation, and stagnant. The plants most commonly include *Pontederia, Persicaria, Globifera, Saururus, Isnardia, Websteria,* and many algae. The nymphs are seldom found in the algae where they form mats, for the insects' movements would be much too restricted by the entangling strands. Normally, the nymphs can be found where the vascular plants are most dense.

Numerous *Callibaetis* nymphs are usually found in temporary woods ponds, sinkhole ponds with open surfaces, and lakes with marginal vegetation. Cypress swamps, marshes of various types, and margins of slow-flowing streams where there is vegetation suitable for harboring nymphs also maintain their populations of *C. floridanus.* The immatures cannot withstand a strong flow of water, and I have never taken them from rapid-flowing creeks unless these streams formed pools or had thick growths of shore vegetation which helped to slow the current in this zone.

Those sinkhole ponds in which the surface is covered with duckweed usually do not have mayfly nymphs. Several factors may account for their absence: the covering of plants which makes it impossible for adults to lay eggs or for nymphs to emerge; an absence of submergent vegetation; an insufficiency of oxygen due to the absence of submergent vegetation; lack of food materials; or lack of suitable refuges. In this sort of place, mayflies find it difficult to live and the normal pond-margin inhabitants are also absent. Sinkhole ponds and lakes in which the margins are either permanently or intermittently covered by water hyacinths are usually uninhabited by *Callibaetis* nymphs, probably for the same reasons that they are absent from the duckweed-covered ponds.

The ditches, ponds, swamps, and other bodies of water which *Callibaetis* nymphs inhabit may range from a very acid condition to a very basic one. Nymphs have been collected from a *Pontederia* marsh in which the pH (colorimetric measurement) was lower than 4.0 which was below the lowest range of the instrument. In Biven's Arm, a lake in Alachua County which is choked with vegetation, principally *Myriophyllum,* nymphs were common at a pH above 10.0 (glass-electrode measurement).

Not only are the nymphs very tolerant to the acidity or basicity of the water, but also to the degree of freshness. In the summer of 1939, Mr. Jack Russell, while working on salt-marsh mosquitoes near New Smyrna, frequently took mayfly nymphs along with mosquito larvae. He neglected to save the nymphs, but he did notice that they occurred in water which ranged from fresh to definitely brackish. On June 30, one adult came to his light at Coronado Beach; since that island is some distance from the mainland and its surface water is brackish, the nymphs must have developed in the saline water. Moreover, I collected immatures from canals in the city of Miami not more than one or two miles from Biscayne Bay, and subsequently I have learned that some of these almost stagnant canals are brackish, particularly at high tide. I have been told that during the early part of the twentieth century Biscayne Bay, described by Banks as the type locality of *C. floridanus,* was brackish in places. *C. floridanus* is the only mayfly in North America which is known to inhabit brackish water. Eaton (1895: 144) recorded a Cingalese *Palingenia* which inhabits an estuary where the water "occasionally must be brackish." He also records *Cloeon dipterum* and *Caenis halterata* from brackish-water streamlets of Hamman-es-Salakin, Biskra.

In the standing fresh waters of Florida, the only ephemerid associates of *C. floridanus* are *C. pretiosus* and *Caenis diminuta.* Along the margins of slow-flowing streams where *C. floridanus* (form A) may occasionally be found are also *Caenis diminuta, C. pretiosus, Ephemerella trilineata, Paraleptophlebia bradleyi, Blasturus intermedius, Baetis spiethi, B. spinosus,* and *Stenonema smithae.* The usual associates other than mayflies in the ponds and ditches include Odonata larvae, Hemiptera, various water beetles, caddisflies, and other still-water forms.

The adults of *C. floridanus* remain in a fairly moist situation, although I did collect a fair series from a burnt-over field adjacent to one of the canals in the Everglades. In the rock pits of south Florida, there are few or no bushes, but there are relatively dense growths of grasses where the mayfly adults remain hidden in the low vegetation until time for larviposition. In one section of the Everglades, which is in part a cypress swamp with emergent grasses scattered throughout the area, I collected some ninety mayflies by sweeping the low plants. Apparently the adults preferred to remain close to the water rather than to fly into the trees.

SEASONS.—Throughout Florida, emergence of *Callibaetis floridanus* occurs over the entire year. Records for adults of *C. floridanus* (form B) include only the summer months, November, and February. There are imagoes or mature nymphs of *C. floridanus* (form A) in my collection from all months of the year.

HABITS.—*Callibaetis* nymphs are the most graceful of all the mayfly immatures found in Florida waters. Its body arched, the green insect hangs from a plant stem and blends almost perfectly with its background. Only the delicate movements of the antennae and the shuttling motion of the large, foliaceous gills draws one's eyes to the nymph. If disturbed, the insect darts rapidly away from the disagreeable area by rapid flicks of the hairy caudal filaments. Its streamlined shape, hairy caudal filaments, and ability to become motionless in an instant assure *Callibaetis* maximum efficiency in its darting actions. Dr. Needham (1935: 13) has described the actions of *Callibaetis* nymphs as follows:

> . . . The ordinary activities of *Callibaetis* nymphs may readily be seen by watching specimens in a small aquarium. Their clambering about among the water plants is strikingly intermittent: a run here, a dash yonder, a sprint across the bottom, and then a sudden halt, in an attitude of alertness, body gracefully arched, head and tail up, antennae extended, and caudal plumes outspread. Such sudden starting and stopping is common to the behavior of many defenseless herbivores and is well calculated to throw the eye of a pursuer off the track. The stopping is so sudden, and the nymph sits so tight among the protective greens and grays and browns of its environment, that the eye with difficulty locates it while at rest. So it sits still for a time, lightly poised, facing an opening between the stems, ready for another instant spring, its banded caudal filaments gracefully arched and drooping, displayed like the plumes in a proudly decorated rooster's tail. Very slowly the antennae, long and slender, and so transparent as to be well nigh invisible, are set swinging up and down. They are moved alternately, one being lowered while the other is lifted.

When collected and placed in a white-enameled pan in which the water has been drained away, the nymphs hop about very much like small minnows if one attempts to lift them. This action is particularly evident in larger nymphs, especially those in the penultimate or last instar. Frequently the nymphs can be detected in the pan only by draining the water away from them and watching for their kicking and hopping as they attempt to get back into the water.

Examination of the contents of the alimentary tract of nymphs showed that they feed predominantly on algae of the filamentous type, but plant material of other sorts was present. The nymphs graze on the algae covering the plant stems and leaves in the ponds and ditches, rather than move into algal mats which sometimes form in these places. The algae most commonly found in the enteron included *Ulothrix, Mougetia,* and *Oedogonium.*

LIFE HISTORY.—Needham reported that *Callibaetis* nymphs develop in from five to six weeks from the laying of the egg. I believe that this statement is probably based more on estimate than on actual rearing, for the species of this genus are ovoviviparous and the nymph develops in the body of its mother for nearly a week. There is some rather interesting information at hand concerning the rate of development of *C. floridanus* (form A) in the Gainesville region. Mr. J. C. Dickinson, while investigating the rate of regeneration of life in intermittent ponds, discovered that within five to six weeks after the first water appeared in the ponds, mature *Callibaetis* nymphs could be found.

C. floridanus is an ovoviviparous species. The female contains from 450 to 500 eggs, which completely fill the abdomen and occupy a large part of the thorax. When the eggs are ready to hatch, the young nymphs can be clearly seen. They appear as small, whitish ovals with five black spots at one end, and are coiled within the very thin, transparent chorion, their legs folded beneath the thorax, the abdomen bent double so that the legs are concealed, the cerci held beneath, and the antennae pressed close against the head. The head is the most prominent feature of the nymph and the developing eyes, five dark spots, stand out clearly against the white body. Abdominal segments, mouth parts, and the rather long legs are clearly discernible when a specimen is uncoiled. Gills have not yet formed, and if they behave in this species as in others in which the postembryonic development has been studied, these structures will probably develop in the second or third instar.

The subimago emerges in the late afternoon. On February 21 I noted many subimagoes rising from the surface of a sinkhole pond between 3:45 P.M. and 4:00 P.M. Other field observations indicate that emergence occurs about one to two hours before sunset and occasionally afterward. In the laboratory emergence time is now and then upset as indicated by transformations after dark and during early morning.

The subimaginal molt takes place about seven to nine hours after the winged stage is assumed. As adults, females have been kept alive for a maximum of thirteen days. In natural habitats, however, the imaginal period is undoubtedly several days shorter. The male dies within two days after emergence. Longevity is probably correlated with ovoviviparity in mayflies. Normally the life span of adult mayflies is from a few hours to two or three days, but a female *Cloeon dipterum,* the European ovoviviparous mayfly, survived for twenty-one days. After mating, the female becomes quiescent until the eggs are ready to hatch. When the time arrives, the female takes to the wing, flies to a body of somewhat stagnant water, and releases the eggs, which hatch at the moment of laying.

On several occasions, I have been able to obtain living nymphs from adult females. The imagoes were captured at light, placed in a cage with a container of water, and left alone. Within five to seven days, the females dropped to the water surface, released the young nymphs, and, apparently completely spent, remained behind on the surface film with wings outstretched. As yet, I have not succeeded in rearing the nymphs from the time of larviposition.

Before I discovered ovoviviparity in *C. floridanus,* I secured eggs from a female which I was reasonably certain had mated, placed them in a dish of water, and observed them daily until decomposition set in. Normally mayfly eggs when placed in water will form a protective jelly-like layer around themselves or release attaching strings. *C. floridanus,* however, did not form such protective devices. Also, I tried artificial fertilization of eggs, and as before, they decomposed.

I have not observed the mating flight of this species, although I have found as many as ninety females among the vegetation in a small area at the same time. Large groups of females have also been seen at lights and other observations of large numbers of females have been reported to me.

LOCALITY RECORDS.—Alachua County: Gainesville, October 24, 1937, adults; August 20, 1938, nymphs and adults; February 7, 1939, adults; April 13, 1939, nymphs and adults; March 4, 1940, adults; other records numerous from April, 1937 through March, 1940 and from February, 1946 through June, 1949. Hatchet Creek, February 26, 1938, adults; April 18, 1938, adults; April 1, 1939, adults; May 6, 1939, adults; February 16, 1940, nymphs and adults; March 2, 1941, nymphs. Santa Fe River, March 25, 1939, adults. 3 miles northeast of Gainesville, July 26, 1938, adults. Glen Springs, December 3, 1938, adults. Citrus County: 1½ miles south of the Withlacoochee River, March 25, 1938, nymphs. Chassahowitzka Springs, May 10, 1947, adults. Collier County: Pinecrest, August 10, 1937, adults; August 19, 1937, adults. 15 miles east of Munroe, September 6, 1940, adults. Columbia County: 6 miles east of Lake City, May 12, 1937, nymphs. Dade County: Miami, August 10, 1937, nymphs; August 13, nymphs and adults; August 14 through September 7, 1937, adults; November 21, 1937, adults; November 29, 1939, adults. 12 miles north of Miami, July 5, July 13, August 2, and September 1, 1937, nymphs and adults. 5 miles west of Miami, August 14, 1937, nymphs and adults; September 2, 1937, adults. 45 miles west of Miami, August 3, 1937, adults. Royal Palm State Park, July 18, 1938, nymphs; September 7, 1940, adults. 3 miles west of Royal Palm State Park, August 27, 1937, nymphs and adults. 8 miles south of Homestead, November 19, 1948, nymphs. Gilchrist County: Suwannee River, April 5, 1938, adults. Glades County: Moore Haven, June 10, 1939, nymphs. Hendry County: Clewiston, February 1, 1941, adults. Highlands County: Child's Crossing, August 11, 1938, adults. Highlands Hammock State Park, May 13, 1939, nymphs. Hillsborough County: Tampa, March 21, 1937, nymphs. Little Fish-hawk Creek, March 26, 1938, nymphs. Six-mile Creek, March 26, 1938, nymphs. 2 miles east of Tampa, March 26, 1938, nymphs. Jackson County: Blue Springs Creek, July 1, 1939, adults; December 1, 1939, adults; June 5, 1940, adults; May 3, 1941, adults. Florida Caverns State Park, December 2, 1939, nymphs. Lake County: St. Johns River at Crow's Bluff, September 12, 1938, nymphs and adults. Lee County: Bonita Springs, February 8, 1939, nymphs. Levy County: 22.3 miles northeast of Cedar Keys, April 9, 1937, nymphs. 6 miles northeast of Cedar Keys, April 9, 1937, nymphs. 4 miles south of Bronson, November 14, 1937, nymphs. Marion County: Juniper Springs, November 21, 1937, nymphs. Withlacoochee River, March 25, 1938, nymphs. Munroe County: Pinecrest, July 19, 1935, adults; August 3 and 24, 1937, nymphs and adults; December 26, 1937, adults. Turner's River, December 24, 1935, nymphs. 4.6 miles south of Dade County line on U. S. Highway 1, November 20, 1948, nymphs. Palm Beach County: Lake Worth, February 17, 1941, adults; March 15, 1941, adults. Polk County: Polk-Lake county line, May 13, 1939, nymphs. Putnam County: Palatka, December 29, 1938, nymphs. Welaka, July 5, 1939, adults; October 23, 1939, adults; July 22, 1940, adults; October 19, 1940, adults. St. Johns County: April 23, 1938, nymphs. Seminole County: Big Econlockhatchee River, February 22, 1941, nymphs. Sumter County: 1 mile north of county line, March 27, 1938, nymphs. Suwannee County: Suwannee River at Branford, June 6, 1940, adults. Taylor County: 6.6 miles east of Perry, February 5, 1938, nymphs. 7.3 miles northwest of Perry, April 1, 1938, nymphs. Volusia County: near Benson Springs, August 30, 1938, adults. Daytona, August 1, 1939, nymphs.

Callibaetis pretiosus Banks

TAXONOMY.—The taxonomic position of the Florida specimens which I am referring to *C. pretiosus* is still somewhat in doubt, but they seem to fit the description of this species better than that of any of the other described forms. Dr. Nathan Banks kindly presented me with two adults from Massachusetts, and examined a Florida specimen, which I sent to the Museum of Comparative Zoology for comparison with the types. He suggested that the Florida mayfly was *pretiosus,* but did not give a definite identification. Further comparison of my specimens (females) with the Northern forms leads me to believe that, at most, the Florida insects are of subspecific rank, but until adults from other localities can be examined, I consider it best to use only a specific name.

Since Banks described the species from Virginia in 1914, there have been no published records of its occurrence in other regions until I reported (1941) that the species was ovoviviparous. In Traver's presentation of *Callibaetis* (1935) the species was redescribed, but as only the female was known and as Traver had no specimens of even this sex, *C. pretiosus* still remains very poorly known. If the Florida species is *pretiosus,* then all stages in its life history are represented in my collection.

In *Callibaetis* imagoes, there is a very marked sexual dimorphism. The female *pretiosus,* as is usual for the genus, is strongly marked over the body and has a brown vitta in the forewing, as well as numerous cross veins behind this vitta and double intercalaries along the outer margin of the forewing. The male has the normal darkened thorax and darkened distal abdominal segments with the pale intermediate area. In the forewings behind R_1, there are, as in the male of *C. floridanus,* relatively few cross veins, and the marginal intercalaries are usually single. Even though the females are very easily distinguished from one another, the males of the Florida species are so similar that they can be differentiated only on the basis of coloration of the abdominal tergites. The nymphs, on the other hand, are easily separated from those of *floridanus.* Because so few *Callibaetis* nymphs have been described, it is impossible to state whether the type of flap found on the seventh gill of *pretiosus* is unique.

DISTRIBUTION.—The distribution of *pretiosus* in Florida parallels that of *C. floridanus* (form A). I have records from as far south as Hillsborough County, and as far east as Volusia County, northward into Echols County, Georgia, and westward into Baldwin County, Alabama. As mentioned above, the only other record is from Virginia, which is the type locality of the species.

ECOLOGY.—*Callibaetis pretiosus* is much the same as *C. floridanus* (form A) ecologically, and frequently the two species can be found side by side in the same roadside ditches, ponds, or lakes. However, there are no indications that the nymphs inhabit brackish water, and the range of acidity-alkalinity seems to be somewhat narrower than for the former species.

SEASONS.—The species emerges throughout the year in Florida. In its more northern environs, however, the season of emergence is doubtless much shortened.

LIFE HISTORY.—The life history of *pretiosus* is very much the same as that of *floridanus.*

LOCALITY RECORDS.—Alachua County: Gainesville, March 22, 1937, nymphs and adults; May 11, 1937, adults; November 4, 1937, adults; January 23, 1938, adults; August 17, 1938, adults; November 21, 1938, adults; March 29, 1939, adults; April 13, 1939, adults; June 30, 1939, adults; September 3, 1939, adults;

PLATE XX

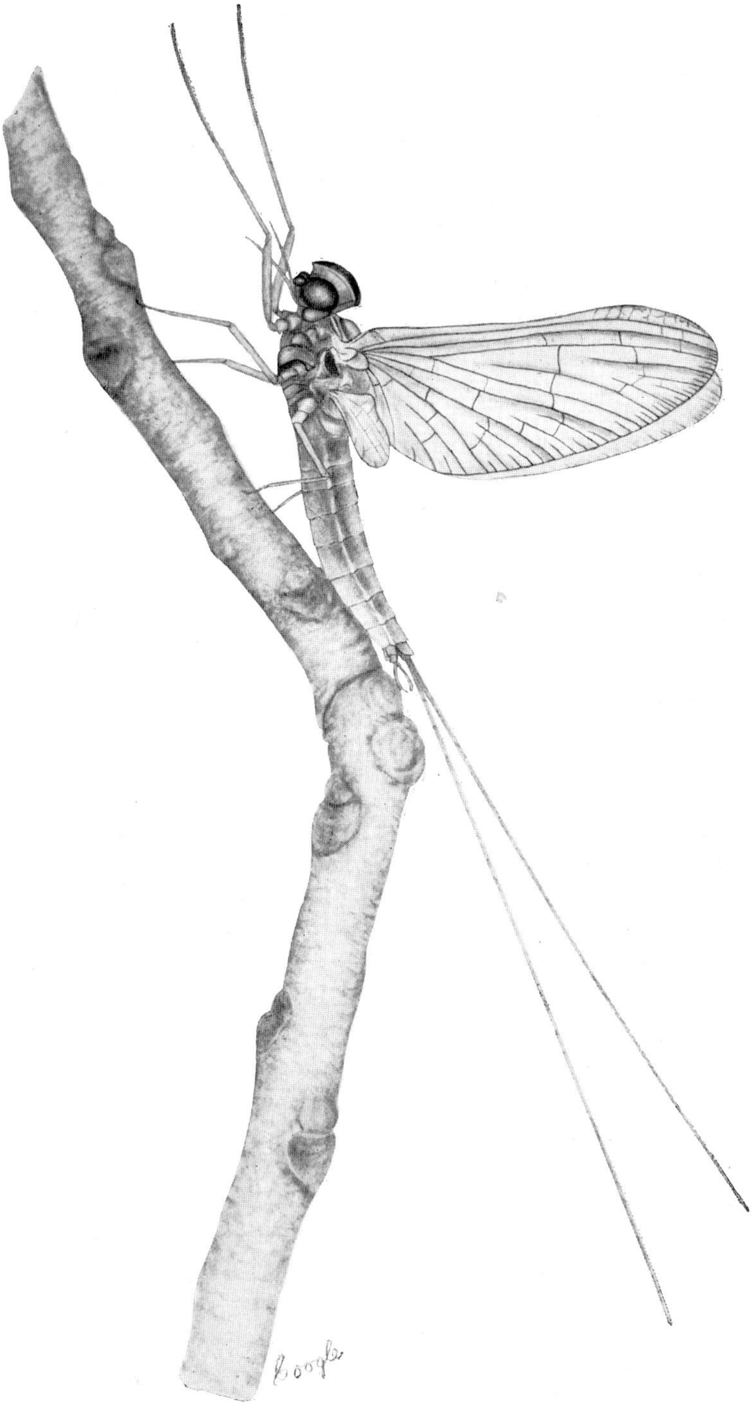

Callibaetis pretiosus Banks, male imago.

October 5, 1939, nymphs and adults; March 18, 1940, adults; March 30, 1940, adults; April 19, 1940, adults, July 28, 1947, adults. Swamp west of Newnan's Lake, January 15, 1938, nymphs and adults. Hatchet Creek, April 2, 1938, adults; February 16, 1940, nymphs. Santa Fe River, October 25, 1939, adults. San Felasco Hammock, October 26, 1937, adults. 2½ miles northwest of Gainesville, November 6, 1937, nymphs and adults. Bay County: 28.8 miles north of Panama City, June 8, 1938, nymphs. 3 miles west of county line, May 30, 1940, nymphs. Charlotte County: 12 miles southeast of Punta Gorda, September 12, 1947, adults. Citrus County: 1½ miles south of Withlacoochee River, March 25, 1938, nymphs. Columbia County: 2.7 miles north of Lake City, October 27, 1938, nymphs. 19.6 miles north of Lake City, October 27, 1938, nymphs. 1 mile northwest of Ocean Pond, May 12, 1937, nymphs. Franklin County: 4.5 miles west of Carabelle, June 6, 1938, nymphs. Hamilton County: 0.6 miles north of Live Oak road at U. S. Highway 41, February 4, 1938, nymphs. 4.1 miles northwest of White Springs, February 4, 1938, nymphs. Hernando County: Bayport, November 1, 1947, adults. Hillsborough County: 2 miles east of Tampa, March 26, 1938, nymphs. Hillsborough River State Park, August 16, 1938, adults. Holmes County: 1.2 miles west of county line, June 8, 1938, adults. Jefferson County: Aucilla River, May 2, 1946, adults. Leon County: 13.8 miles west of Tallahassee, June 5, 1938, nymphs. 20.3 miles west of Tallahassee, June 5, 1938, nymphs. Madison County: 5.9 miles west of Madison, February 5, 1938, nymphs. Marion County: Orange Springs, October 2, 1937, nymphs. Near Oklawaha River, February 12, 1938, nymphs. Ocala National Forest, February 12, 1938, nymphs and adults; March 19, 1938, nymphs; July 24, 1938, adults. Okeechobee County: 5 miles south of county line at Highway 29, August 18, 1940, nymph. Pasco County: 20 miles north of Tampa, March 25, 1938, nymphs. Putnam County: Welaka, September 11, 1939, adults; October 12, 1939, adults. 30 miles east of Gainesville, October 2, 1937, nymphs and adults. 1.2 miles north of Palatka, October 2, 1937, nymphs. 11 miles north of Palatka, October 2, 1937, nymphs and adults. Taylor County: 6.6 miles east of Perry, February 5, 1938, nymphs. Volusia County: De Leon Springs, August 5, 1939, adults. Walton County: 1.4 miles west of Portland, June 7, 1938, nymphs.

Genus ACENTRELLA Bengtsson

In a paper (1940c) dealing with the Florida Baetinae, I discussed the taxonomic status of the genus *Acentrella*. I expressed doubt as to the validity of the genus, and at the present time, I am even more doubtful. Until more work is done on the Northern forms, it is perhaps best to retain the name. The characters used by Traver in separating this group of species from *Baetis* proper include the presence of a penis cover, absence of a costal projection on the metathoracic wings in the adult male, and a two-tailed nymph. It is now definitely known that at least one of the Florida species, *ephippiatus,* has a three-tailed nymph, and most likely the nymph of the other Florida species also exhibits a three-tailed condition.

Although *Acentrella* is poorly known, published records indicate that it ranges widely over North America. *A. insignificans* has been recorded from British Columbia and Idaho, *frivolus* from Quebec, and *ephippiatus* from Florida. No species are known to occur west of Alabama in the South, and none are listed from the central United States, although *dardanus* is described from Manitoba.

Spieth (1933: 339) did not accept either *Acentrella* or *Heterocloeon* as good genera, but treated them as elements of *Baetis.* Without doubt, *Acentrella* is derived from *Baetis.* The *Acentrella* group seems to have split off from a species similar to *B. spinosus* in which the costal projection is so minute as to be almost negligible. Of the described species of *Baetis,* there are several which might have given rise to *Acentrella,* and it seems certain that *Acentrella* merely represents a slightly more advanced type than those species which are included in *Baetis* proper. From the standpoint of the nymph (of *ephippiatus*), there are no differentiating characters that might warrant a generic separation from *Baetis,* and in them, phylogenetic indicators point to no advances over *Baetis.* The two-tailed nymphs might indicate that this group of species in which they occur is intermediate between *Pseudocloeon* and *Baetis,* even though *B. bicaudatus* has only two caudal filaments. Spieth concluded that "it is possible that even *Pseudocloeon* should be considered part of the genus *Baetis,* comparable with the short winged forms known among Drosophila, leaf hoppers, beetles, parasitic hymenoptera, gall wasps, etc."

Acentrella ephippiatus (Traver)

TAXONOMY.—Traver described *A. ephippiatus* in *The Biology of Mayflies* (1935) as *Baetis ephippiatus,* but based her description on males only. In 1937 she re-erected *Acentrella,* and by this action removed *ephippiatus* to Bengtsson's genus. I described the nymph of this species in 1940, and at the time pointed out that this species, on the basis of the nymph, formed an intermediate between *Baetis* and *Acentrella.*

The adult male is easily distinguished from the various *Baetis* species by the absence of the costal process from the metathoracic wings; the adult female is less easily separated. I have found that the best character for separating these females from those of *B. spinosus* and *B. spiethi* is a distal femoral band on each leg of *ephippiatus.* The nymphs are distinct from all other Florida species in having a strongly colored seventh pair of gills; without use of this characteristic, many of the nymphs might be confused with *B. spinosus,* for the maxillary palpi of the two species are almost identical. In well-marked nymphs, the prominent red-brown tergites of segments 2, 5, and 8, and the reddish markings on the venter easily separate *A. ephippiatus* from other species. *A. ephippiatus* adults may be separated from *A. propinquus* by the coloration of the middle abdominal tergites of the male of the former species; the female adult and nymph of the latter have not yet been collected in Florida.

DISTRIBUTION.—*A. ephippiatus* is widely distributed in northwestern Florida and in the two southernmost counties of Alabama; in the remainder of Florida, however, the species is poorly known. *A. ephippiatus* was described from Fort Valley and Rome, Georgia, which are located in distinct physiographic provinces. Rome is located near the southern tip of the Valley and Ridge Province; Fort Valley is on a plateau just below the Fall Line Hills. The Florida and Alabama records are from typical Coastal Plain country (map 16).

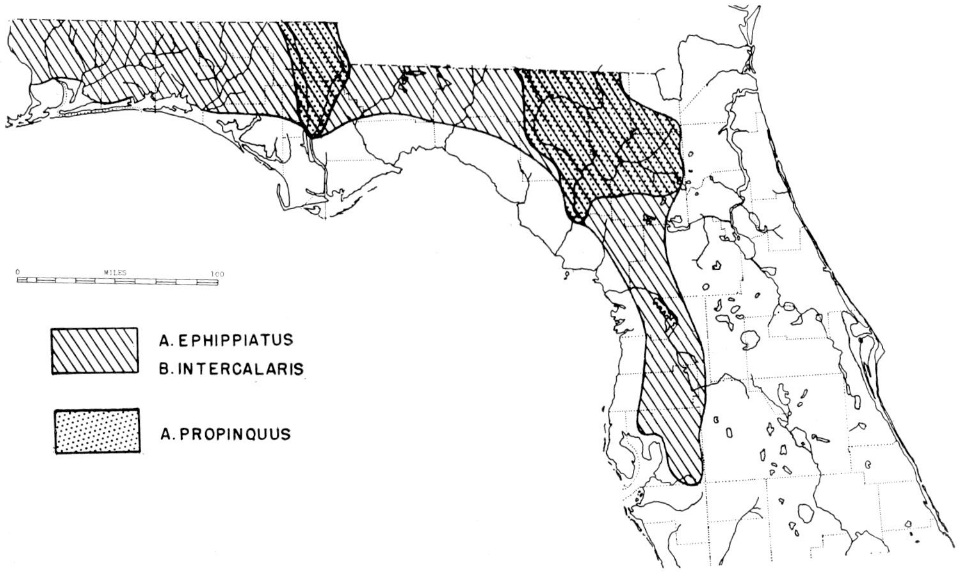

Map 16.—The distribution of *Acentrella ephippiatus*, *A. propinquus*, and *Baetis intercalaris* in Florida.

ECOLOGY.—Nymphs inhabit sand-bottomed streams where they cling to the vegetation. I have found them most frequently in those creeks emptying into the Choctawhatchee Bay. The streams have dense growths of *Vallisneria*, *Potamogeton*, *Sagittaria*, and algae in all parts except the deepest, and in them, *ephippiatus* nymphs are the most frequently encountered of the ephemerids. The nymphs cling to the vegetation in the swiftest, as well as the slowest, parts of the creek, but are seldom found on submerged logs and boards or in the very slow-flowing water near shore. Many of the creeks in the panhandle region of Florida are small, shallow, and have very little submergent vegetation, but nymphs manage to live in them by clinging to the small bits of plant material and to the accumulated debris. In these streams there is a paucity of habitats and this is reflected in the number of nymphs present, for there are far fewer insects per cubic foot of area than in the streams first described.

In Alachua County, the nymphs and adults have been taken only from Hatchet Creek, which has one of the richest mayfly faunas in central Florida. In the shallower parts of the creek, there are dense growths of *Potamogeton*, *Fontanalis*, and other plants among which the nymphs live. They seldom venture into the quieter water nearer shore where the kindred nymph, *B. spinosus*, is often found. Even the more rapid water of this stream is fairly slow when compared to Northern streams, but the flow is continuous and the current is quite noticeable, particularly in the shallower portions.

Subimagoes only are known from the Hillsborough River; they were collected where it is fairly shallow, rapid, and filled with vegetation, principally *Vallisneria*.

Every stream from which I have collected nymphs of *A. ephippiatus* and at which I have taken the pH showed a definite acidic reaction. In no case did the acidity range below 5.8 nor above 6.5; all readings were by the colorimetric method. The Hillsborough River, from which I have only adults, is slightly basic. The acidic streams drain flatwoods or swampy areas and usually are slightly to strongly tinted.

Ephemeropteran nymphs found to be associated with *A. ephippiatus* were *Pseudocloeon bimaculatus, P. alachua, Baetis spiethi, B. australis, B. spinosus, Paraleptophlebia volitans, Tricorythodes albilineatus, Stenonema exiguum, S. smithae, S. proximum, Cloeon rubropictum, Blasturus intermedius, Habrophlebiodes brunneipennis, Choroterpes hubbelli, Hexagenia munda marilandica, Baetisca rogersi, Ephemerella* sp. A, *E. choctawhatchee, E. trilineata, Caenis hilaris, Centroptilum viridocularis,* and *Isonychia* sp.

SEASONS.—*A. ephippiatus* occurs in Hatchet Creek near Gainesville, but the species has been taken only intermittently on many collecting trips to the stream; consequently, seasonal data are not very complete for it. Though a study of the specimens from west Florida and from Hatchet Creek indicates that the species emerges throughout the year, adults in my collection were taken only in March, April, May, June, and October. Nymphs in their last instar were collected in August and September, whereas half-grown specimens are known for April, May, and June; and very immature nymphs, for June only. These latter specimens seem to indicate that emergence must occur during the winter and fall. However, I have no definite information for the remainder of the year owing primarily to lack of collecting in the other months. Professor P. W. Fattig collected the type specimens in Georgia on June 1 and August 3, 1931.

HABITS.—The nymphs of *A. ephippiatus* are rapid swimmers, which cling gracefully to vegetation, the head facing upstream and the tails waving as the current swings the body from side to side. As they move slowly over their supports they graze on the materials covering the plants and occasionally dart to another plant, but always face upstream. The nymphs, which seldom go deep into the clumps of vegetation, remain mostly near the outer leaves and stems away from accumulations of silt and detritus.

The body of the nymph is quite rounded, and when a specimen is placed in a pan or dish, it usually falls on its side and can move only by flicking the abdomen. If the insect falls on its belly surface, it can crawl, but not so rapidly as the Leptophlebines.

Examination of the alimentary tract of west Florida specimens indicates that the nymphs feed predominantly on filamentous algae, but an occasional diatom is picked up along with the filaments.

LIFE HISTORY.—Life as a nymph probably lasts between six and nine months. When the subimago is ready to emerge, the nymph swims to the surface, the thoracic covering splits, and in an instant the subimago rises from the water. This speed is particularly necessary in streams, for if the adult is thrown off balance and the wings become wet, they crumple when freed from the water. The insect is then doomed, for it cannot again straighten its wings. When the subimago rises from the water, it flies upward and may reach a support close to the stream. Usually, however, the flight is upward until the insect is out of sight some twenty-five feet in the air. It then probably flies to trees which line the stream margin and there sits quietly and awaits the ultimate molt. The shedding occurs within eight or nine hours. In the field, adults have been observed emerging just after sunset, and in the laboratory the performance is the same. Nothing is known of the mating flight or oviposition.

LOCALITY RECORDS.—Alachua County: Hatchet Creek, April 2, 1938, adults; September 14, 1938, nymphs; March 22, 1939, adults; June 24, 1939, adults; October 11, 1939, nymphs. Little Hatchet Creek, September 13, 1940, adults. Bay County: Pine Log Creek, May 31, 1940, nymphs. 27.4 miles north of St. Andrews, May 30, 1940, nymphs. 24.7 miles northwest of Panama City, June 6, 1938, nymphs. 23.7 miles northwest of Panama City, June 8, 1938, nymphs. Columbia County: Falling Creek, June 30, 1939, nymphs. Escambia County: 5 miles southwest of county line, June 1, 1940, nymphs. Bayou Marquis, June 1, 1940, nymphs. Jackson County: 3.6 miles north of Altha, July 1, 1939, nymphs. Hillsborough County: Hillsborough River, October 21, 1940, adults. Holmes County: Sandy Creek, May 1, 1946, nymphs. Okaloosa County: 5.1 miles west of county line, June 7, 1938, nymphs; May 31, 1940, nymphs. Niceville, June 7, 1938, nymphs. Santa Rosa County: 4.8 miles north of Navarre, June 1, 1940, nymphs. Walton County: 7.3 miles west of Ebro, June 7, 1938, nymphs. 15.8 miles west of Ebro, June 7, 1938, nymphs. 13.8 miles west of Freeport, June 7, 1938, nymphs and adults. 16.6 miles west of Freeport, June 7, 1938, nymphs. 5.4 miles west of Washington County line, May 31, 1940, nymphs. 6.7 miles west of Portland, May 31, 1940, nymphs. 9.5 miles west of Portland, May 31, 1940, nymphs and adults.

Acentrella propinquus (Walsh)

TAXONOMY.—In my paper on the Baetinae of Florida (1940c), I recorded *Acentrella propinquus* from Florida and at that time expressed some doubt as to the validity of the identification. I have been unable to obtain authentically identified specimens, and have seen none which seem to resemble the description of *propinquus* more closely than do the Florida insects.

In 1862 Walsh described "*Cloe vicina* ? Hagen"; however, this proved not to be the *vicina* of Hagen, and in 1863 Walsh substituted the name *propinquus* for *vicina*. In 1871 Eaton transferred *propinquus* to *Baetis*, and in 1885 he redescribed the species briefly in his monograph. McDunnough, in 1925, examined the type series in the Museum of Comparative Zoology, selected a lectotype, and suggested that *dardanus* might be a synonym of *propinquus*. The former species was treated by Traver (1935), who mentioned the possibility of synonymy. Traver recorded the species as *Acentrella propinquus* (?) Walsh in her paper on "Mayflies of the Southeastern States" (1937: 83) and briefly described the male and female imagoes.

A. propinquus is very similar to *A. ephippiatus* and can be differentiated from the latter only by the fact that *ephippiatus* has some dorsal markings on the pale abdominal segments. I have found several specimens of *ephippiatus* on which these markings are obsolescent, and because of this and the very similar venation and genitalia, the insects which I am calling *propinquus* may be only variants of *ephippiatus*.

DISTRIBUTION.—*Acentrella propinquus* has been recorded from three localities in North America: Rock Island, Illinois (type locality); Sheffield, Alabama; and Marianna, Florida. Each of these localities is in a different physiographic province, and the character of the streams must certainly be very different. As mentioned elsewhere in this paper, many Northern forms find ingress into Florida by means of the Apalachicola River drainage system. In Florida the species is now known from places as widely separated as Blue Springs Creek near Marianna and from the Hillsborough River near Tampa (map 16).

ECOLOGY.—The adults, which were all taken at streams that are distinctly basic, are the only representatives of *propinquus* in my collection. Because of the scarcity of the species in Florida, I believe that the nymphs must be confined to water in which the pH ranges above 7.0. The imagoes from Blue Springs Creek were taken at a lighting sheet. At the Hillsborough River, Dr. L. J. Marchand found adults clinging to the wall of an old dam in company with numerous subimagoes of several other Baetinae and *Stenonema*.

SEASONS.—This species has a year-round emergence, and in this respect probably does not differ from the other Baetinae. Adults are known only for July 1 and October 21. No other information concerning the Florida insect is available.

LOCALITY RECORDS.—Alachua County: Santa Fe River, October 25, 1939, adults. Hillsborough County: Hillsborough River, October 21, 1940, adults. Jackson County: Blue Springs Creek near Marianna, July 1, 1939, adults.

Genus BAETIS Leach

Baetis is one of the first-known mayfly genera, having been described by Leach in 1815. *Brachyphlebia,* which Westwood described in 1820, proved to be synonymous with *Baetis*. Later, in 1843, a portion of *Baetis* was synonymized by Pictet. The genus *Acentrella* was split off in 1912 by Bengtsson, and since then has caused some dissension among taxonomic Ephemeropterists. Many species of mayflies were described under the genus *Baetis* before this all-inclusive category was divided into several smaller genera. At the time of publication of *The Biology of Mayflies* in 1935, there were forty-one species of *Baetis* listed from North America north of Mexico, and since that date there have been five additional species described in the genus: *B. foenima* McD., *B. jesmondensis* McD., *B. persecuta* McD., *B. macdunnoughi* Ide, all from Canada, and *B. spiethi* Berner from Florida. The re-erection of *Acentrella* (Traver, 1937) removed six of the species from *Baetis,* thus reducing to forty the number known at present. Of this group, which includes *Acentrella* and *Baetis,* twenty-three have been described by McDunnough, eight by Traver, seven by Dodds, and the remainder by other workers.

The treatment of *Baetis* as given by Traver (1935) was severely critized by McDunnough (1938), but the sections of the genus treated by him are all Western and Northern and are, therefore, quite distinct from the Florida species. Hence, his criticisms need not be discussed here.

On the basis of genitalia, Traver divides the genus into two main groups—the *moffati* and *intercalaris* types. Each of these is then subdivided. The former includes the modified *moffati* type, the *spinosus* type, and the true *moffati* type; the latter comprises the true *intercalaris* type and the *quilleri* type. McDunnough, even in his caustic review of *Baetis,* follows Traver's divisions, but he states that several of her species placements are erroneous. The Florida species of *Baetis* fall into the typical *moffati,* the *spinosus,* and the *intercalaris* types.

The species of *Baetis* are rather generally distributed throughout the world, occurring in the Holarctic, Neotropical, and Indo-Australian faunal regions. In the Nearctic and Neotropical regions, the genus is widely dispersed from southern South America to the Arctic region in North America, its only limitation apparently being the presence of permanently moving fresh water.

Because of the overlapping of many generic characters, Spieth (1933: 339) considers it likely that the "classification [of certain of the Baetinae] may be and probably is an artificial one, and that it cannot be said with certainty that it represents a picture of the phylogenetic history of the group." He considers *Acentrella* and *Heterocloeon* as merely subdivisions of *Baetis* and also suggests that *Pseudocloeon,* even though it completely lacks hind wings, should also be included as one of the group under *Baetis.* I cannot agree with this latter conclusion, for, even though the female of *Baetis spiethi* possesses hind wings of microscopic proportions, I believe that the complete absence of wings should be sufficient to segregate these groups into higher categories than subgenera.

Baetis is probably derived from some stock which was basically similar to *Callibaetis.* The author quoted above distinguishes three distinct lines of evolution in the Baetinae: *Callibaetis; Baetis* and *Pseudocloeon; Centroptilum* and *Cloeon.*

From a consideration of the hind wings of *Baetis* there appear to be at least three lines of development: enlarged wings with three or more longitudinal veins and a small costal process; medium-sized wings, usually with three longitudinal veins and a rather large, hooklike costal process; and smaller hind wings with two longitudinal veins and usually a small costal process. From the standpoint of nymphs, other lines of development may be evolved: those nymphs with three caudal filaments, and those with only two. The former group can be further subdivided into those possessing rounded seventh gills, and those with lanceolate seventh gills. Again, the former group can then be subdivided into those species with an expanded second segment of the maxillary palp and those in which this segment is not expanded.

Ecologically, the nymphs are confined to water in which there is constant motion. In Florida, *Baetis* nymphs occur only in streams, but in more northern localities the immatures can tolerate lake shores where there is continuous wave action.

Baetis spinosus McDunnough

TAXONOMY.—McDunnough described *Baetis spinosus* in 1925 from specimens collected in Manitoba, Canada. The species remained unnoticed until 1935, when Traver recorded adults from New York and the Chattahoochee River at Atlanta, Georgia. My descriptions (1940c) of females and nymphs from Florida completed the knowledge of the stages of *B. spinosus.*

The genitalia of *B. spinosus* are distinct from those of all other species of the genus. McDunnough (1925b: 174) says that they "are, however, very characteristic, the second joint of the forceps having a strong pointed projection on the inner apical margin, a feature which is unique in our North American *Baetis* species." However, when the nymphs of *spinosus* are compared with those of *B. australis,* the two seem structurally almost identical. My only criteria for distinguishing between them are the slightly longer caudal filaments and the more or less unicolorous dorsum in the last-instar male nymphs of *australis.* Neither of these characters offers a good basis for separating the species when only one or a few nymphs are involved; if a series is examined, the immatures can usually be divided into two distinct groups.

According to McDunnough, *B. spinosus* is allied to *frondalis* McD. and in general appearance is close to *dardanus* McD. In Florida, *australis* is the only species which appears to be closely related to *spinosus.*

B. spinosus is one of the more advanced species of *Baetis,* for in this species there are only two longitudinal veins in the hind wings and the costal process is reduced to a small hump. Judged by the genitalia, its phylogenetic relationship is not so clear, for these organs are distinctive, particularly the forceps, which possess a very prominent tubercle at the juncture of the second and third segments.

DISTRIBUTION.—Apparently *Baetis spinosus* is distributed over the entire eastern portion of the United States and part of Canada, ranging from Manitoba to south-central Florida. In Florida, *B. spinosus* is one of the most widely dispersed of all the mayflies. It ranges from the southwestern border of Alabama to the northeast corner of Florida, and from this latter locality as far south as Hillsborough County. In the panhandle of Florida, *B. spinosus* nymphs can be found in every stream in which there is permanently flowing water, and in north and central Florida this is true almost without exception.

The distribution of *spinosus* is rather interesting in that four physiographic provinces are involved: the Canadian Shield, the Adirondack, the Coastal Plain, and the Piedmont. Apparently, geographic barriers are of little importance in determining the range of this species (map 4).

ECOLOGY.—Without doubt, this is one of the most adaptable of all Florida mayflies. The following instance is a fairly good example of its ubiquity. Nymphs were observed in a flume leading to a ram at the origin of Little Sweetwater Branch in Liberty County. The flume, constructed of wood, had

PLATE XXI

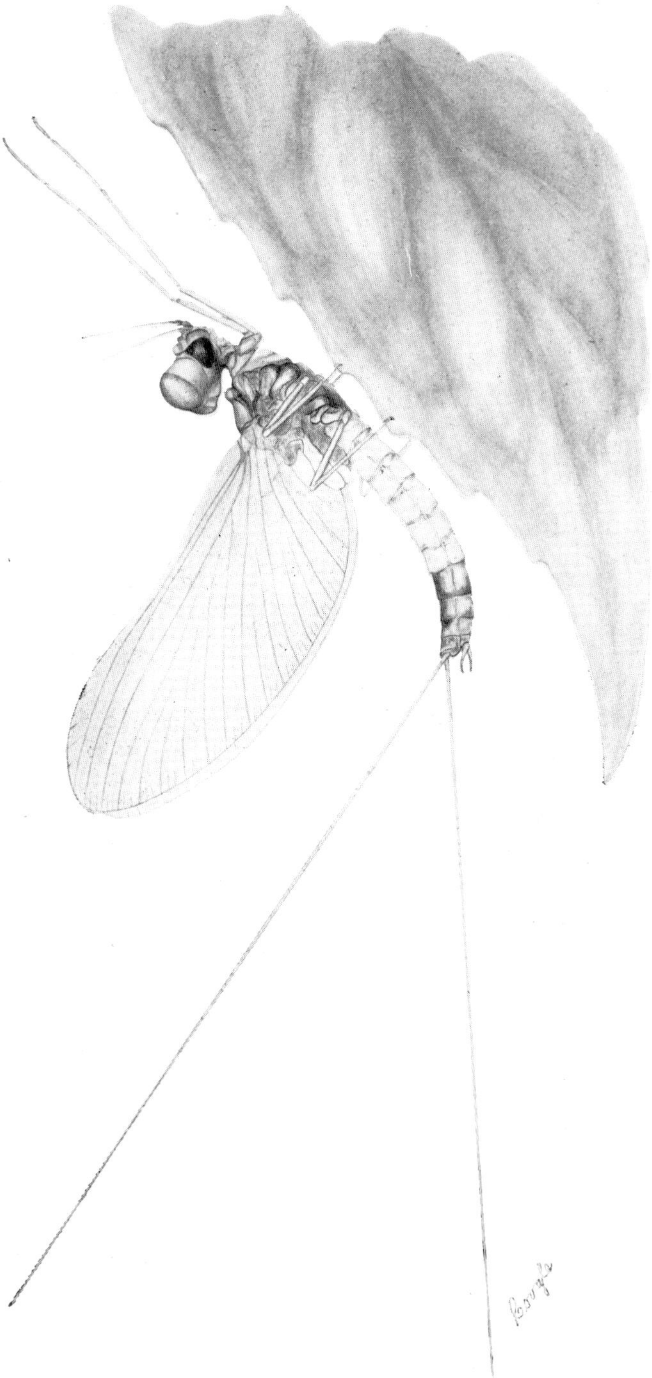

Baetis spinosus McDunnough, male imago.

its sides covered with a thin growth of algae, and there was a continuous flow of water through the trough. The level of the water was not much over two or three inches, yet it supported a rather large population of *B. spinosus* nymphs of all sizes. I have found immatures along the edges of creeks where the water was practically stagnant and the nymphs were living under accumulated masses of dead leaves and silt. Conversely, the nymphs have been found in the most rapid parts of the most rapid Florida streams, in which they can be found in all regions ranging from the slowest to the swiftest.

Those streams in which there are dense growths of *Vallisneria, Sagittaria,* and *Potamogeton* support the largest populations of *B. spinosus* nymphs. The reason for this abundance is seen in the tremendous amount of surface presented by the thick masses of this submerged vegetation. An abundance of food materials in the form of algae and diatoms covers the surfaces of the leaves. An excellent example of this type of stream is the Santa Fe River in the vicinity of Poe Springs. There, the *Vallisneria* and *Sagittaria* are so thick that in wading the stream one might easily walk almost across it without touching the stream bed. During the late winter, spring, or summer, fifty to one hundred nymphs may be easily collected with a few strokes of the dip net through this tangle of vegetation. In this mass, the nymphs seem to prefer the more distal or free portion of the leaves and become less numerous near the base of the plants.

The spring runs in the west-central and the north-central part of Florida are usually similar to the Santa Fe River in having great growths of eelgrass; however, near the head of the runs, where the springs emerge, mayflies are very rarely found. Within a short distance downstream they become noticeable. Their absence from the head of these runs may possibly be explained by the fact that the water emerging from the ground carries in solution large amounts of bicarbonate which are deposited as carbonate when the water reaches the surface. By forming a covering over the surface of the plants this salt prevents the mayflies from securing sufficient food to survive. Another explanation of the absence of mayflies from the springheads might be the lack of sufficient oxygen in the water just after it emerges from its subterranean channels. As surface water, it has not yet had time to become sufficiently oxygenated for the maintenance of mayfly populations. At the head of the springs the snail *Goniabasis* normally occurs in almost unbelievable numbers. I have almost filled a dip net with them after taking a few strokes through the eelgrass; these snails may also be a factor in restraining the introduction of mayfly nymphs into springheads.

Any slightly acid stream containing detritus, submerged logs, leaf drift, or vegetation in midstream or near shore will ordinarily have nymphs, even though they may occur there in very limited numbers. Many of these streams have so little vegetation and other suitable habitats for development of nymphs that only small numbers can be supported. Only rarely do nymphs enter such dense mats of algae as those formed by *Spirogira* and *Batrachospermum.*

Physically, the water inhabited by the nymphs of *B. spinosus* shows a wide range. The immatures are found in streams with a pH as low as 5.0, as well as in those which are alkaline, with the pH as high as 7.8. In general, it seems that the nymphs thrive better in slightly alkaline water with a pH about 7.3. The temperature factor is not particularly important in Florida, although it is of the greatest importance farther north, for during a cold spell the degree of emergence is greatly lessened and may completely stop. As the water gradually warms, the nymphs again become active enough to attain the adult state.

Nymphs can apparently tolerate standing water for a time, because I have been able to keep them alive in unaerated aquaria for as long as two days, and in shallow pans of water for as long as three. If the water is aerated, however, the life of the insect may be prolonged as much as two weeks in the laboratory. This ability to tolerate standing water is probably one of the factors which accounts for the wide distribution of the species. If the nymphs can live in intermittent streams and in those which flow most of the time but which do occasionally become stagnant, it would greatly increase the flexibility of the species. There is some field evidence which indicates such tolerance. I have collected nymphs from several streams which connect two bodies of standing water. Normally there is a slight movement of water from one of these bogs to the other, but during low water the flow ceases. If the flow ceases for too long a time, the nymphs die off; but as soon as the water begins moving again, these intermittent creeks become repopulated from nearby permanently flowing streams.

Even though *B. spinosus* is so tolerant, it has not yet become adapted to lake margins in Florida as have *Ephemerella trilineata, Choroterpes hubbelli,* and *Stenonema proximum.* I fully expected to find the nymphs in this situation, but examination of the margins of many sand-bottomed lakes produced not a single immature, despite the presence of the other species mentioned above.

Every Florida species of mayfly, except *Callibaetis floridanus* (form B), *Hexagenia munda orlando, H. bilineata, Pentagenia vittigera,* and *Campsurus incertus,* has, at one time or another, been found along with the nymphs of *B. spinosus.*

SEASONS.—Among the materials at hand, there are adults collected or reared during nine months of the year; nymphs in their last instar are known for the other three months. *B. spinosus* is one of the many species of mayflies which emerges throughout the year in Florida. Traver does not state the time at which the Georgia specimens were taken, but McDunnough's adults were collected on July 13 and 16, and on August 16. There is no doubt that in the northern part of its range *spinosus* has a short seasonal emergence, and that the farther south it occurs, the longer the season, until it reaches Florida, where seasonal limits disappear.

HABITS.—The habits of posture and movement easily distinguish *B. spinosus* from other mayfly nymphs when they are examined in the laboratory. Usually the insect swims to some point of attachment, grasps it with the claws, and raises the abdomen well above the support. If there is nothing to which it can cling, the nymph swims to the perpendicular sides of the container and hangs there until motivated to move. The gills are held close to and over the abdomen in contrast to *B. intercalaris*, one of the species with which *spinosus* may be most easily confused. When stimulated, the nymphs dart in short spurts from one spot to another by rapidly moving the caudal filaments up and down. The hairs on the filaments overlap and form a very effective organ for rapid propulsion by vibration.

Certainly as swift or swifter than *Callibaetis* nymphs, those of *B. spinosus* are less graceful in their resting attitudes than the *Callibaetis* nymphs. They cling rather closely to their resting place, with their heads upstream and their abdomens swinging from side to side in the current. Slowly, the nymphs crawl over the surface of the leaves or rocks as they search for new and tender bits of food. During this slow movement, the orientation to the current is maintained no matter whether the nymph is on the upper or the underside of the leaf; in a current, however, the insect rarely moves to the upper side of its support, for there it is less protected from the force of the moving water and the food supply is more limited. While feeding, the nymph swings its head from side to side as the maxillary palpi, continually moving, aid it in bringing its food within reach of the maxillae and mandibles.

The food found in the alimentary canals of nymphs indicates that mastication is very thorough, for in every specimen studied, the diatoms were broken and the remainder of the food was destroyed almost beyond recognition, even though it was still in the anterior portion of the tract. The food of the Santa Fe River nymphs proved to be predominantly filamentous algae and plant epidermis, although an occasional diatom was noted.

Adults are only mildly phototropic, with the phototropism more pronounced in the subimaginal stage. Orientation is negatively geotropic, both the subimago and imago clinging with head upward.

LIFE HISTORY.—Repeated and unsuccessful attempts have been made to hatch eggs in the laboratory. Females which were forced to oviposit may not have copulated, for I took none during the mating flight. I tried artificial fertilization, but it, too, produced no results. Published data record the rearing of only a single species, *B. vagans*, which Dr. Helen Murphy (1922) raised from egg to adult both in the laboratory and in the field at Ithaca, New York. She found that under laboratory conditions the nymphs went through twenty-seven instars, and, according to temperature, this developmental period was six to nine months, and the definite broods that were produced emerged in May, August, and October. In Florida, such brooding has not been observed and, moreover, would not be expected of *B. spinosus*.

Emergence occurs in late afternoon, usually just after sundown or as the sun is setting. The period of transformation lasts fifteen to thirty minutes, with the peak occurring about ten minutes after the first subimago rises. As the subimago lifts itself from the stream, it can be seen as a slow-moving grayish body rising upward. If it encounters a tree or bush, the insect stops; if not, it continues its upward flight until it is lost to sight. When sufficiently high, it probably flies horizontally to a nearby tree or bush. I have seen the subimagoes fly almost straight up for thirty or forty feet before they were lost to sight. When the insects begin to emerge, one can easily spot the area of transformation by observing the path of the giant dragonflies which begin to patrol the stream as the sun sets. The slow flight of the mayflies makes them easy and choice prey for these Odonata. One day during the spring, as I stood in midstream netting subimagoes, a large dragonfly darted in front of me, grasped a mayfly which was at the mouth of the net, and darted away without even entering the moving bag. Another excellent indicator of emergence is the increased activity of the insectivorous birds along the stream margins.

The spiders which are numerous along the streams are among the most important decimators of *Baetis spinosus* subimagoes and adults. Many adult mayflies and subimaginal skins have been found entangled in the meshes of spider webs under bridges. If the subimago does not entangle itself too badly, and if the spider does not kill the insect at once, there is some chance of its becoming free at its final molt; this is, however, a comparatively rare occurrence.

If the subimago escapes predators, it alights and becomes motionless and inconspicuous, hardly moving during the entire period of its life as an immature adult. After a period varying from seven to ten hours, the final shedding occurs. The imago remains in a resting state until time for the mating flight. While resting, adults of *Baetis*, both male and female, frequently twitch the abdomen and swing the caudal filaments from side to side. At Ithaca, New York, *Baetis* performs the mating flight in the forenoon; no observer has been fortunate enough to note the phenomenon in *B. spinosus*.

Emergence of the subimago of *B. spinosus* has frequently been observed in the laboratory. When ready for emergence to occur, the nymph floats freely at the surface of the water and almost immediately the subimago pops free, the whole process taking not more than five to ten seconds. Naturally, in such rheocolous forms, rapid emergence is an absolute necessity unless the nymphs migrate into slow water; even those *Baetis* nymphs which happen to be in almost stagnant water emerge very rapidly.

Numerous attempts to rear the species in quiet water in the laboratory failed. Individuals were, however, successfully reared in a cage placed to allow a slow but steady stream of tap water to flow through it. The use of an aquarium equipped with an aerator gave an even greater degree of success. The aerator emitted a steady stream of bubbles which disturbed the surface. Apparently the agitating of the water aids the subimago in releasing itself from the nymphal skin, perhaps by battering, or perhaps because the increased amount of dissolved gases in the water aids the nymph in obtaining the air necessary for flotation and loosening of the skin.

With respect to the longevity of the adult, the male shows almost no signs of life about thirty-eight to forty hours after emergence. The female adult in the subimaginal and imaginal stages may live as long as forty-eight to fifty hours. Mating obviously must occur very soon after the subimaginal molt, and oviposition shortly thereafter.

LOCALITY RECORDS.—Alachua County: Hatchet Creek, May 11, 1937, nymphs; February 8 and 26, 1938, nymphs; March 23, April 2, April 18, May 5, July 9, 1938, adults; March 22, April 1, May 6, June 24, October 11, 1939, adults; August 10, 1940, adults; October 5, 1940, nymphs; March 2, 1941, adults. 2½ miles west of Gainesville, April 4, 1937, nymphs; October 25, 1937, nymphs; March 5, 1938, nymphs; August 9, 1938, adults; March 10, 1939, adults; April 21, 1939, adults. 3 miles north of Paradise, February 12, 1938, nymphs. 1 mile west of Newnan's Lake, May 11, October 4, 1937, nymphs; January 8, 1938, nymphs; January 25, 1938, adults; August 13, 1938, adults. Hogtown Creek, March 14, 1933, nymphs; May 10, May 30, September 25, October 30, 1937, nymphs; July 17, August 24, 1938, adults. Campus of the University of Florida, July 30, 1938, nymphs and adults; February 7, March 10, April 12, April 20, November 11, November 22, nymphs and adults. Santa Fe River at Poe Springs, May 21, 1934, nymphs; March, 1935, nymphs; March 24, 1937, nymphs; March 12, 1938, nymphs; March 18, 1938, adults; February 11, 1939, nymphs and adults; February 18, 1939, nymphs and adults; March 4, 1939, adults; March 25, 1939, nymphs; October 3, 1939, nymphs and adults; October 25, 1939, adults; March 1, 1940, nymphs; April 6, 1940, adults; June 13, 1940, nymphs and adults. Little Hatchet Creek, August 22, 1940, nymphs and adults. Bay County: 24.7 miles northwest of Panama City, June 6, 1938, nymphs. 14.1 miles north of Panama City, June 8, 1938, nymphs. 27.4 miles north of St. Andrews, May 30, 1940, nymphs. Pine Log Creek, May 31, 1940, nymphs. Citrus County: 1½ miles south of Withlacoochee River, March 25, 1938, nymphs. Chassahowitzka Springs, May 10, 1941, nymphs and adults. Clay County: Green Cove Springs, October 15, 1938, nymphs. Columbia County: 11.5 miles north of Lake City, October 27, 1938, nymphs. Falling Creek, February 4, 1938, nymphs; November 13, 1938, nymphs and adults; June 30, 1939, nymphs. Escambia County: Bayou Marquis, June 1, 1940, nymphs. Gadsden County: River Junction, March 17, 1939, nymphs; June 30, 1930, adults. 4½ miles south of River Junction, July 1, 1939, nymphs. 5 miles south of River Junction, June 6, 1940, nymphs. Gilchrist County: Suwannee River near Old Town, March 4, 1939, nymphs. Hamilton County: 0.6 miles north of Live Oak road at U. S. Highway 41, February 4, 1938, nymphs. 1 mile north of White Springs, February 4, 1938, nymphs. Hernando County: Southern county line, March 27, 1938, nymphs. Hillsborough County: 2 miles west of Alafia, March 26, 1938, nymphs. Little Fish-hawk Creek, March 26, 1938, nymphs. Bell Creek, March 26, 1938, nymphs. Six-mile Creek, March 26, 1938, nymphs and adults. Hillsborough River, October 21, 1940, adults. Holmes County: Sandy Creek, December 11, 1937, nymphs; July 2, 1939, adults; December 14, 1939, nymphs; May 1, 1946, nymphs. Jackson County: 3.6 miles north of Altha, December 10, 1937, nymphs; July 1, 1939, nymphs. Florida Caverns State Park, December 2, 1939, nymphs. Jefferson County: Drifton, February 5, 1938, nymphs. Leon County: 11.2 miles west of Tallahassee, March 16, 1939, nymphs. 16.9 miles west of Tallahassee, March 17, 1939, nymphs. Liberty County: 10 miles south of River Junction, March

17, 1939, nymphs. Near Ochlocknee River, March 16, 1939, nymphs. 4.5 miles from the turnoff to Torreya State Park, June 10, 1938, nymphs. Hosford, March 17, 1939, nymphs. Little Sweetwater Branch, December 10, 1937, nymphs; June 10, 1938, adults. Sweetwater Creek, November 4, 1938, nymphs; July 1, 1939, nymphs; May 2, 1941, nymphs. Madison County: 4.3 miles east of Jefferson County line, February 5, 1938, nymphs. Marion County: Withlacoochee River, March 25, 1938, nymphs. Nassau County: 19 miles north of Duval County line, August 28, 1938, nymphs. Okaloosa County: 5.1 miles west of county line, June 7, 1938, nymphs; May 31, 1940, nymphs. Niceville, June 7, 1938, nymphs. Crestview, December 12, 1937, nymphs. Shoal River, December 12, 1937, nymphs. Putnam County: Johnson, April 10, 1948, nymphs. Santa Rosa County: 4.8 miles north of Navarre, June 1, 1940, nymphs. Sumter County: 2 miles north of Bushnell, March 27, 1938, nymphs. 1 mile north of Sumter County line, March 27, 1938, nymphs. Taylor County: Fenholloway River, March 18, 1939, nymphs. Wakulla Springs run, May 29, 1940, nymphs. Walton County: 1 mile west of Walton County line, June 7, 1938, nymphs. 5.4 miles west of Walton County line, May 31, 1940, nymphs. 10.6 miles west of Walton County line, May 31, 1940, nymphs. 15.8 miles west of Ebro, June 7, 1938, nymphs. 2.6 miles west of Freeport, June 7, 1938, nymphs; May 2, 1946, nymphs. 0.8 miles west of Portland, April 3, 1938, nymphs. 6.7 miles west of Portland, June 7, 1938, nymphs; May 31, 1940, nymphs. 9.5 miles west of Portland, May 31, 1940, nymphs; June 7, 1938, nymphs. Washington County: Holmes Creek, December 11, 1937, nymphs; April 2, 1938, nymphs; June 9, 1938, nymphs; July 2, 1939, nymphs.

Baetis australis Traver

TAXONOMY.—Traver described *Baetis australis* from North Carolina in 1932 and redescribed it in *The Biology of Mayflies* in 1935. I recorded the occurrence of the species in Florida in 1940, but nothing was added to the taxonomic knowledge of the species. Morphologically, the male is distinct from all other *Baetis* adults, and can be easily separated from the remaining Florida species by the coloration of abdominal segments 2 through 6. The genitalia of *australis* are quite similar to those of *frondalis* and resemble those of *spinosus*. The metathoracic wings of the three species are all very similar in that they are long and narrow, the longitudinal veins are reduced to two (the third being variable), and the costal process is much reduced.

I have seen no females which could be assigned to *australis*, although some are doubtless included among the specimens which have been identified as *spinosus*. The differences are probably so slight that the females of these two species are indistinguishable. The nymphs are likewise very difficult to separate, and I have found no characters which are easily used. However, examination of a fairly large series of specimens shows that the lateral caudal filaments of *spinosus* are shorter than those of *australis*. In the last-instar male nymphs, moreover, the abdomen of the *australis* is unicolorous whereas that of *spinosus* is more variegated. Gills and mouth parts are very similar.

DISTRIBUTION.—Present knowledge of the distribution seems to indicate that *B. australis* is a Coastal Plain species. Traver described the insect from specimens taken at Goshen Swamp, at Burncoat Swamp, and in Lenoir County,

North Carolina. Later, she recorded *B. australis* from Fort Valley, Georgia, which is in the more hilly section of the Coastal Plain. I have found the species ranging from Alachua County in Florida all the way to the western limits of southern Alabama. Probably its range extends as far south as Hillsborough County, Florida, but I have no adults from this area, and the nymphs seem to show only *spinosus* characters. *Baetis frondalis* McD., to which *australis* seems to be most closely related, is known from Ontario and Quebec; the other relative of *australis*, namely, *spinosus*—distribution of which has been discussed elsewhere in this paper—inhabits identical situations with it.

An interesting note is introduced when the distribution of *spinosus* is compared with that of *australis*. The latter species is confined to the Coastal Plain, but the former, apparently with no better adaptations, is known to occur from Canada to Florida. This would seem to indicate that both *australis* and *frondalis* are derived from *spinosus* or some similar species—*australis* arising at one end of the range, *frondalis* at the other. The only fundamental difference in this group of species seems to be in the presence of the enlarged projections on the medial surface of the claspers of *spinosus*. The dark color of the abdomen may be linked with a reduction in size of the projections in *frondalis* and *australis* (map 14).

ECOLOGY.—Identifiable nymphs and adults have been collected only from acid streams. All of these creeks fall into the sand- and silt-bottomed categories. In the former streams, the nymphs can be found living on the vegetation in the most swiftly flowing portion as well as in the slowest areas. Like *spinosus*, the nymphs tend to remain near the distal parts of the plants where there is the greatest flow and the smallest deposit of silt. Nymphs may occur on all sorts of stream debris. Submerged logs, in particular, are sought by the nymphs where they hide in the cracks and crevices on the downstream side. In the Alachua County creeks, the nymphs were found principally on *Juncus repans* and on a moss which grows profusely in certain parts of some of the streams where the flow is at least perceptible. The west Florida nymphs were chiefly inhabitants of the dense mats of *Vallisneria* and *Potamogeton* which choke the streams.

The silt-bottomed streams from which nymphs were taken had luxuriant growths of *Isnardia, Persicaria,* and eelgrass. Despite this plentiful vegetation, the mayfly fauna was not excessively large.

Associated with *B. australis* in Alachua County streams are *Pseudocloeon alachua, Baetis spinosus, B. spiethi, Stenonema smithae,* and *Paraleptophlebia volitans.* In the west Florida streams, *B. spinosus, B. spiethi, Acentrella ephippiatus, Paraleptophlebia volitans, Stenonema smithae, S. proximum, S. exiguum, Ephemerella trilineata, E. choctawhatchee, Ephemerella* sp. A, *Choroterpes hubbelli, Tricorythodes albilineatus, Caenis hilaris, Pseudocloeon bimaculatus,* and *Centroptilum viridocularis* occur along with *B. australis.*

SEASONS.—Adults have been reared or collected only during March, May, September, and October; emergence, however, is not seasonal in Florida.

Nymphs which appear to be *australis* were collected in June, and ranged in size from very immature to full grown. Very likely the young specimens would have emerged sometime during the winter.

HABITS.—There seem to be no essential differences between the habits of *B. australis* and *B. spinosus* nymphs.

LIFE HISTORY.—The immature stages probably last from six to nine months, but there is no substantiating evidence for this assumption. The process of emergence is like that of *B. spinosus.* The subimago emerges during the afternoon about two or two and one-half hours before sunset. All subimagoes which were observed in the field emerged from very small streams, and in every case the insect flew to the nearest support and settled about three or four feet above the surface of the water. The subimaginal molt takes place after seven to ten hours, and the imago remains quiescent until time for the mating flight. I have not observed the flight nor has it been reported by other workers. Morgan (1913: 392) described the flight of an unidentified species of *Baetis.* Her observations were made in May, 1911, as she watched a swarm of *Baetis* which was flying near Cascadilla Creek at Ithaca, New York, just after a shower. The insects flew at eye level and so were easily observed. The swarm was composed of males and females in about equal numbers. Morgan was able to observe the actual mating and described the process as follows:

> . . . The male of one of the couples flew up and attached himself beneath a female, pressed the dorsal side of his head against the ventral side of her thorax and extended his forelegs upward, in order to clutch her prothorax. The setae of the female extended straight out posteriorly, but those of the male were pointed forward over his back so that their tips projected between the heads of the two insects. The position of the abdomen could not be clearly seen, but judging from that of the setae, it must have been recurved in order to insert the penes inside the egg valve. Copulation did not last more than half a minute. When in copula, each pair was borne diagonally downward to the ground, but always separated immediately upon touching it.

From my observations of numbers of *B. australis* in Florida streams, it does not seem reasonable to believe that large swarms are formed. Though I have visited the creeks at all times of day, as well as on days when the presence of imagoes on bridges indicated they were on the wing, I have never encountered a swarm. The swarms in this southern part of the country are small, consisting perhaps of only a score of individuals. Mating takes place at all times of the year and the species has not telescoped this phenomenon into such a short time that huge swarms are necessary.

LOCALITY RECORDS.—Alachua County: Hatchet Creek, October 28, 1939, adults. Little Hatchet Creek, August 10, 1940, nymphs and adults; August 22, 1940, nymphs and adults; September 13, 1940, adults. Campus, University of Florida, March 23, 1939, adults; October 6, 1939, adults. Gainesville, November 22, 1940, adults. Escambia County: 5 miles southwest of county line, June 1, 1940, nymphs. Bayou Marquis, June 1, 1940, nymphs. Jackson

County: Florida Caverns State Park, December 2, 1939, nymphs. Liberty County: Sweetwater Creek, May 2, 1941, nymphs. Marion County: 9.6 miles southwest of Salt Springs, April 10, 1948, nymphs. Okaloosa County: 5.1 miles west of Walton County line, May 31, 1940. Walton County: 6.7 miles west of Portland, June 1, 1940, nymphs. 9.5 miles west of Portland, May 31, 1940, nymphs and adults. Freeport, May 2, 1946, nymphs.

Baetis intercalaris McDunnough

TAXONOMY.—McDunnough's taxonomic work on the Ephemeroptera began with his description of *Baetis intercalaris* in 1921. *B. intercalaris* was again described by McDunnough in 1923 when he examined living insects in order to clarify the differences existing in the eyes of certain species of *Baetis*. Traver redescribed the male adult in the taxonomic portion of *The Biology of Mayflies*, and Ide completed the life history with his description of the nymph (1937: 227, 229). In 1940 I reported that the species was present in Florida. The identification was tentative, as only nymphs and females were available for study.

The shape of the labial palp, which is not expanded, easily separates the nymphs of *intercalaris* from other Florida *Baetis*. The relatively long caudal filaments and the mode of banding of these structures is also characteristic; and the rounded seventh gills immediately separate the nymphs from *B. spiethi*. The female imagoes, likewise, are absolutely distinct from all other Florida *Baetis* in having an enlarged hind wing and a third vein in this wing.

In her discussion of *B. intercalaris*, Traver (1935: 692) intimates that the species is most closely related to *B. flavistriga* and *B. phoebus* among the known species of the genus. Ide (1937: 320) considers the nymph of *intercalaris* to be close to the *cingulatus* group of nymphs *(B. phoebus, cingulatus, levitans,* and *flavistriga),* but distinct from it. I have seen few specimens of these species and can add nothing further concerning the relationship of *intercalaris* to them.

DISTRIBUTION.—If the Florida form is *intercalaris*, the distribution of the species is certainly most interesting, as it is known only from Ontario and from Florida. Because there are no records of the species from the intervening area between these Northern and Southern regions, such distribution introduces some doubt as to the validity of the identification of the Florida specimens as *intercalaris*, though the very close correspondence of the insects to the published descriptions cannot be denied. In June, 1940, a single male imago was taken from a bridge near Wetumpka, Alabama, and it fits the description of *intercalaris* fairly well, except that the wing length falls far short of the measurements given for that species. If this male is *intercalaris*, it would offer a connecting link between the Northern and Southern forms, for it was collected just at the edge of the Fall Line on the Coastal Plain side.

Southern nymphs or adults of *intercalaris* are known from Hillsborough County, Florida, to Elmore County, Alabama. In Florida, the species is rather widespread, but not so widespread as *spinosus*. The check on the range is apparently ecological, as the nymphs seem to be less tolerant than are those of *spinosus* (map 16).

ECOLOGY.—Basic streams, where the flow is swiftest, support the largest populations of *B. intercalaris* nymphs. Although the species does inhabit acid water, the creeks are usually only slightly acidic and have sand bottoms. In the central portion of Florida, nymphs have been collected in greatest numbers from the Santa Fe River and from the Hillsborough River, both of which are basic streams. The nymphs are very rare in the sand-bottomed streams of peninsular Florida, and, in Alachua county where collecting has been most intensive, nymphs have been taken only at the Santa Fe River.

In the Santa Fe River, the number of immatures fluctuates greatly from one time to another, but the population is never so great as that of *spinosus*, with which *intercalaris* is associated. Both species inhabit eelgrass in the swiftest water, but in more rapid currents the *intercalaris* nymphs congregate in greater numbers than do those of *spinosus*. In the slower areas the ratio of *intercalaris* to *spinosus* is reversed.

There are never large concentrations of nymphs in the sand-bottomed streams and frequently, even though other species of *Baetis* were fairly common, only one or two specimens of *intercalaris* were taken. The nymphs inhabit the vegetation and debris wherever the water is permanently flowing. If there is debris in the middle of the stream, the nymphs will more likely be found there in greater numbers than near the shore. In the creeks of Gadsden County there are often masses of exposed roots of terrestrial plants which are washed by rather rapidly flowing water. *B. intercalaris* nymphs can usually be collected from these mats. Submerged logs in the swifter water are also productive situations, but examination of such logs is a rather arduous task and less profitable than root collecting.

The size of the stream evidently has no bearing on the presence or absence of nymphs, as the insects have been found in brooks no wider than three feet and no deeper than three inches, as well as in creeks and in rivers. In the smallest streams there was no vegetation, and the nymphs inhabited debris lodged in the stream bed. One small creek, about three or four inches in depth, had its bottom covered with a thin growth of algae, and several nymphs were found in this material.

The principal mayfly associates of *intercalaris* are *B. spinosus, B. spiethi, Pseudocloeon alachua, P. punctiventris, P. parvulum, Centroptilum viridocularis, C. hobbsi, Stenonema smithae, S. exiguum, S. proximum, Tricorythodes albilineatus, Isonychia* spp., *Caenis hilaris,* and *Oreianthus* sp. No. 1.

SEASONS.—*Baetis intercalaris* emerges throughout the year, but there is evidence of a tendency to form broods within a particular stream. The collections from the Santa Fe River seem to indicate that March and October are important months for emergence because the mature nymphs were very common during these periods; however, mature nymphs were also found on other dates but not in such numbers. Nymphs collected from Sweetwater Creek in west Florida were predominantly in the last nymphal stage during December, but mature nymphs were also present in all other months in which collections were made there.

The type specimens of *B. intercalaris* were taken by McDunnough at the Rideau River, Ottawa, from June 11 through June 14. In 1922 he collected again at the type locality and noted that there were apparently two generations, one emerging in early June and the other from the middle of August until early in September. McDunnough also noted that individuals of the second generation were somewhat smaller than those of the first.

Ide (1935) working in Ontario on the effect of temperature on the distribution of mayflies in a stream, found that in the lower, warmer reaches of the stream, *intercalaris* gave definite evidence of the occurrence of two generations in the same season, the first during the early part of June, the second starting in July and proceeding until the middle of August. Collections from the upper part of the stream indicated that there is probably only one generation produced at the upper stations in a season.

HABITS.—The living nymphs can usually be distinguished from those of *B. spinosus* and *australis* without microscopic examination, although they are superficially similar and agree in body size and in length of caudal filaments. When nymphs of *B. intercalaris* are placed in a small amount of water in a white-enameled pan, the gills are spread outward from the abdomen so that they overlap its borders considerably. Furthermore, the gills, as seen against the white background, frequently have a greenish tinge. *Spinosus* and *australis,* on the other hand, usually hold the gills, which are relatively colorless, much closer to the body. In addition, the caudal filaments of *intercalaris* are usually less intensely banded than are the filaments of *spinosus* and *australis.* Swimming and clinging activities of *intercalaris* are very similar to those of *spinosus.*

The nymphs live on vegetation growing in the swift water and cling near the upper, free ends of the leaves where the full force of the current can strike them. Remaining on the undersurface of the leaves, they feed on the materials covering the plants. The food consists chiefly of algae, but diatoms, desmids, and plant epidermis are frequently eaten.

Although the upper surfaces of the rocks in the Santa Fe River are completely covered with the moss *Fontinalis,* this moss is not a habitat of *B. intercalaris,* probably on account of the slowness of the current and the heavy silting where the growths occur. *Intercalaris* is one of the most rheocolous of all the mayflies inhabiting Florida streams.

LIFE HISTORY.—Because the limits of toleration of *B. intercalaris* are not very great, rearing in the laboratory presents some difficulties. Many times I have attempted to bring live specimens to Gainesville from the Santa Fe River, but most of them died en route. However, a few were successfully transported to the laboratory, and several lived for slightly more than two days in an aerated aquarium. Of these, only females emerged. Essentially, emergence is like that of *spinosus.* The nymph makes a few rapid dashes that end at the surface, the thoracic covering splits, and the fully emerged subimago immediately appears and at once flies to the nearest support. Emergence has also been observed out of doors, and it was seen that the insects do not at once head for shore, but fly directly upward until lost to sight. They probably seek the protecting leaves of the higher trees.

The subimagoes begin rising just about sundown and continue for approximately thirty minutes, the peak being about fifteen minutes after emergence starts. In the laboratory a subimago emerged an hour after dark, at 7:30 P.M., and underwent its imaginal molt at 8:07 A.M. This is an excessively long subimaginal period and, I believe, atypical, because most specimens that were observed in the laboratory molted only eight to ten hours after transformation.

The mating flight of *B. intercalaris* has not been described, nor have I ever observed it.

LOCALITY RECORDS.—Alachua County: Santa Fe River at Poe Springs, May 21, 1934, nymphs; March 24, 1937, nymphs; March 12, 1938, nymphs; February 18, 1939, nymphs; March 25, 1939, nymphs; October 3, 1939, adults; October 25, 1939, nymphs and adults; March 1, 1940, nymphs; June 13, 1940, nymphs. Gadsden County: River Junction, March 17, 1939, nymphs. 4½ miles south of River Junction, July 1, 1939, nymphs. Hillsborough County: Bell Creek, March 26, 1938, nymphs. Hillsborough River, October 21, 1940, adults. Jackson County: Florida Caverns State Park, December 2, 1939, nymphs. Jefferson County: April 1, 1938, nymphs. Leon County: 11.2 miles west of Tallahassee, March 16, 1939, nymphs. Liberty County: Sweetwater Creek, November 4, 1938, nymphs; July 1, 1939, nymphs; December 1, 1939, nymphs; May 2, 1941, nymphs; May 29, 1946, nymphs. 10 miles south of River Junction, March 17, 1939, nymphs. Sumter County: 2 miles north of Bushnell, March 27, 1938, nymphs. Walton County: 1.2 miles east of Okaloosa County line, November 6, 1938, nymphs. Washington County: Holmes Creek, July 2, 1939, nymphs.

Baetis spiethi Berner

TAXONOMY.—*Baetis spiethi* is the first species of the *pygmaeus* group to be reported in the South. The insect was described from a fairly large series of both reared and wild specimens in 1940. The male is easily distinguished from all other Florida species of *Baetis* by the shape of the genitalia, the coloration of the abdomen, and, most readily, by the structure of the hind wings, which have only two longitudinal veins and a much more prominent costal projection than does *spinosus*. The differentiation of the females is a more difficult matter. Compared with those of *spinosus* (with which they can most easily be confused), the metathoracic wings of *spiethi* females are minute—being barely visible against the dark background of the thorax with the highest magnification of the dissecting microscope—and the head of *spiethi* is more yellowish. The females of *Acentrella ephippiatus* have a distal brownish band on the femora which distinguishes them from the females of *B. spiethi*. Nymphs are immediately distinguishable from all other known Southern Baetid mayflies by having lanceolate gills on abdominal segment 7.

PLATE XXII

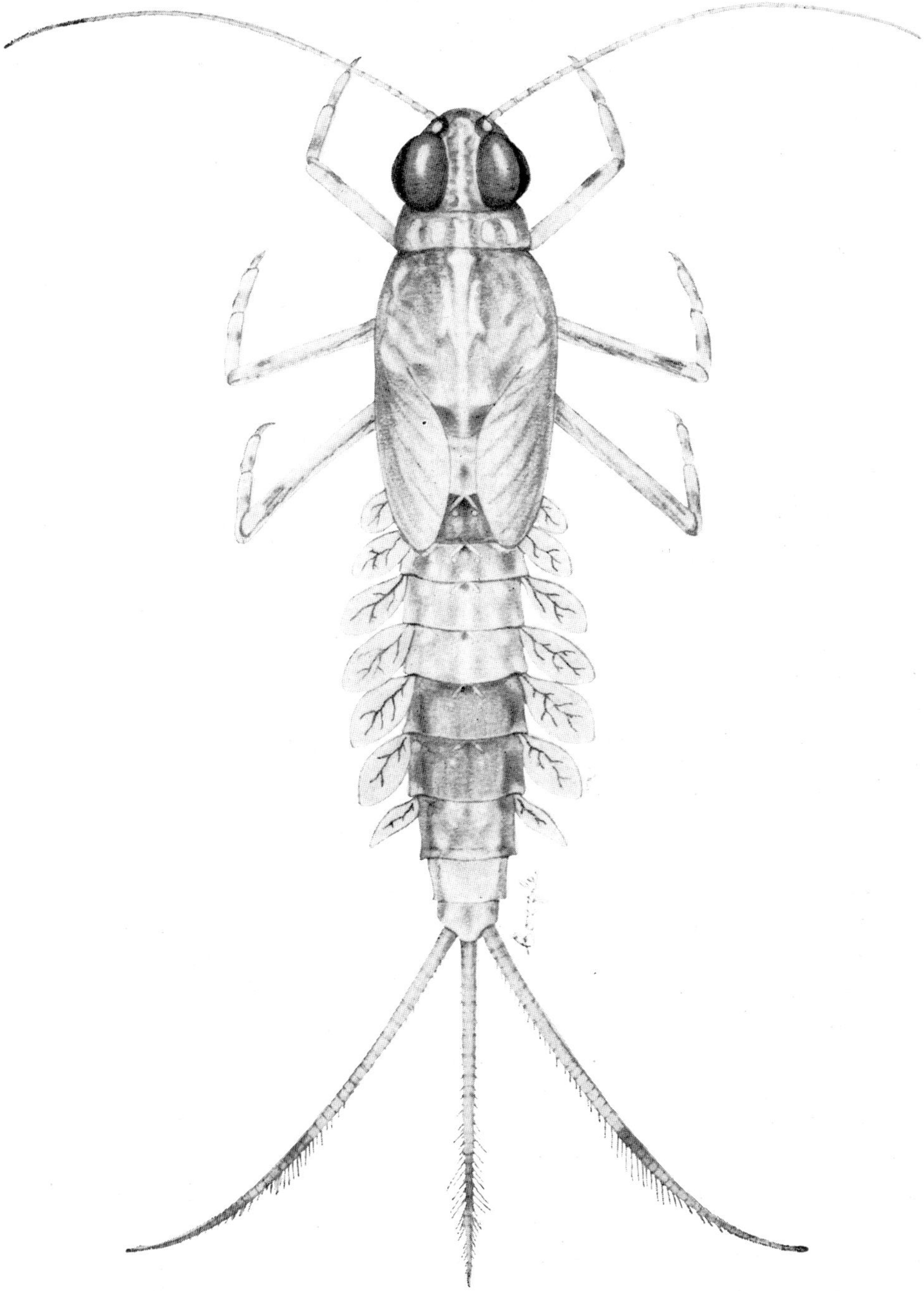

Baetis spiethi Berner, nymph.

The taxonomic relationships of this species were discussed in the original description, and I will merely summarize them here. *B. spiethi* is obviously very closely related to *B. pygmaeus* (Hagen) and *B. macdunnoughi* Ide. The males of *B. pygmaeus* and *B. spiethi* are so very similar that only by careful examination can the two species be separated. In *spiethi,* marginal veinlets are in the first interspace of the mesothoracic wing, the costal projection of the metathoracic wing is small and not curved, the first forceps segment has no tubercle, and the basal part of the turbinate eyes is red-brown. All these characters are apparently of only minor import, but they show that *spiethi* is derived from an ancestral form close to *pygmaeus.*

This conclusion is based not only upon morphology but also upon the distribution of the *pygmaeus* group. According to Ide (1935), Kennedy has proposed the theory that "the older or more primitive species of groups are northern in distribution or, if not northern, are active in the cooler seasons of the year or cooler parts of the day." Further evidence in support of this generalization has been added by Ide in his study on the effect of temperature on mayflies. This concept is in conformity with morphological evidence that *spiethi* is more advanced than its closest relative, the more Northern *pygmaeus,* which has been recorded no farther south than the mountains of Maryland and West Virginia.

DISTRIBUTION.—*B. spiethi* is known to occur from Hillsborough County in west-central Florida northward to the Georgia line and westward to Mobile County, Alabama. Its distribution in Georgia is unknown, and there are few Alabama records. It seems, however, that in the eastern part of the latter state the species does not extend far to the north, for no specimens were taken above Conecuh County. *B. spiethi* is probably as widespread in Florida as *spinosus,* in spite of the fact that it has not been collected from the northeastern part of the state.

Baetis spiethi is almost certainly confined to the Coastal Plain, and very likely to its southern portion, although it may range northward into the Carolinas. While Traver's work in North Carolina did not bring to light any of the *pygmaeus* group of species, it is probable that either *pygmaeus* and/or *spiethi* occur in this area* (map 14).

ECOLOGY.—This insect has been found most frequently in the smaller, more slow-flowing creeks. Within a particular stream, they are very similar in habitat preference to *B. spinosus.* In general, the nymphs seem to remain in quieter and shallower water than do those of *spinosus,* but almost always the two live side by side. They are most frequently found in sand-bottomed streams where the vegetation is scattered along the edge of the stream and where the flow is still evident and constant. The nymphs thrive among these plants, and with other Baetinae form the major faunal element. Hatchet Creek, which has been described elsewhere in this paper, is a typical example of a stream in which

* I have examined nymphs from Greene County, Tennessee, which were collected by Dr. M. Wright and which fall into the *pygmaeus* group. I have also collected nymphs of this group from the mountain streams of North Carolina.

spiethi prospers. In this creek the small beds of *Potamogeton, Globifera umbrosa,* and mosses make up the predominant vegetation from which the nymphs have been collected. There they may be found in the swiftest as well as in the slowest part of the current. In the Torreya State Park region, the nymphs are also abundant among the exposed roots of terrestrial plants in the streams which drain into the Apalachicola River.

In every stream from which *spiethi* nymphs were collected, *spinosus* or *australis* nymphs were also collected or seen, but not all streams from which *spinosus* was taken contained *spiethi* nymphs. The size of the stream does not affect the presence or absence of nymphs, as they have been collected from brooks which were merely trickles, as well as from streams as large as the run of Wakulla Springs. This river, like other west Florida streams from which collections were made, has dense growths of *Vallisneria.* Although the nymphs have not been found to be exceedingly numerous in the eelgrass, they do occur there. Some of the streams near the western border of the state have thick mats of *Potamogeton* in which *spiethi* nymphs occur as commonly as in Hatchet Creek near Gainesville.

The nymphs live on the undersides of the leaves near the free ends, where they are protected from the current and find food plentiful. Near shore among the vegetation, they may be found most often on the stems about halfway up from the stream bed. Occasionally nymphs are found in leaf drift, and somewhat more frequently on submerged logs in the current.

B. spiethi nymphs are the most tolerant of all the Florida species of the genus. I have kept nymphs alive in a shallow pan of water without aeration for as long as a week, and they have remained alive in an aerated aquarium for as long as fourteen days. The percentage of nymphs transforming in an aerated aquarium is usually greater than that in the other *Baetis* species, except perhaps *spinosus.*

Although nymphs of *spiethi* do inhabit basic streams, they seem to thrive far better in circumneutral or slightly acid creeks. These usually have a high content of humic and other organic acids that give the water a dark tinge.

The ephemerid nymphs associated with this species are the same as those listed for *B. spinosus.*

Adults are often encountered in spider webs under bridges. The bridges are also the most productive places for securing free adults or subimagoes, as the humidity over the streams is fairly high and shade is plentiful—two qualities which the subimagoes and adults seem to demand. Apparently the subimagoes of *Baetis* can tolerate a much dryer atmosphere than can the Leptophlebines, for they can molt where the Leptophlebines would perish in the process.

SEASONS.—Seasonally, this species is not restricted. While emergence is, of course, greater in the summer than during the colder months, no month is entirely free of adults. The evidence for this statement is conclusive, as I have adults for the ten-month period from February through November. Furthermore, I took last-instar nymphs during the early part of December and the early part of January, and these nymphs would undoubtedly have emerged sometime during the second week of these months. In the North, the other species of the *pygmaeus* group are strictly seasonal: *pygmaeus* emerges during the summer, and *macdunnoughi* adults are known for two dates, June 24 and July 18.

HABITS.—The habits of *spiethi* are not essentially different from those of *spinosus*. The food I examined was so finely masticated that it was difficult to determine its nature from a study of the contents of the alimentary canal. Some of the identifiable fragments included algae, parts of diatoms, small roots, and plant debris.

LIFE HISTORY.—Nymphal life probably occupies a period of about six months, but as I was not successful in rearing the species from egg to adult, the length of time required must remain conjectural. At all times of the year, nymphs of all ages can be found in Florida waters.

When the nymph is mature and ready to transform, it swims to the surface. As a subimago, it suddenly bursts free of the nymphal skin, floats on the discarded skin for a moment, and then flies away. Emergence takes place in the slower flowing parts of the stream, where upsetting is least likely to occur and where the insect may rest before testing its newly acquired wings. After this short rest, the subimago rises to nearby bushes or trees where it remains, unless disturbed, until the ultimate molt takes place. The subimaginal stage lasts from seven to ten hours, and the males may live about twenty-four hours longer.

The transformation to the subimago takes place from perhaps an hour before sunset until sunset, but before darkness has fallen. The height of emergence occurs just after sunset, and lasts for about fifteen minutes.

I have observed the mating flight of *spiethi* on only one occasion. While I was collecting nymphs at the Waccasassa River on October 13, 1946, a swarm of small mayflies was seen gathering at 10:15 A.M. in the warm sunlight. The males came together in small groups and then amalgamated into one large swarm of about two hundred individuals. They flew about six feet above the water and near the south shore of the stream, confining themselves to the bank and seldom moving out over the water. The swarm hovered over the vegetation, now and then rising to a height of about fifteen feet above the water, but rarely more than six or eight feet above the highest ground level. Only males were observed and collected from the swarm. The group would fly over the bank for a few minutes and then move away from the water only to return again within a short time. The presence of a black railroad bridge and the very dark water outlined the adults clearly so that they were easily observed as long as they remained in the vicinity of the stream.

The up-and-down flight varied greatly, having a rise and fall from six inches to three or four feet. After a period of about an hour, the swarm apparently dispersed. No females were seen joining the males and none was noted until 3:00 P.M. when a single individual was collected while it was flying close to the water.

From a distance, the mating flight resembled the swarming of midges so commonly observed along the margins of the lakes of Florida. It was only by approaching the swarm closely and noting the trailing caudal filaments that the members were identified as mayflies.

LOCALITY RECORDS.—Alachua County: Hatchet Creek, February 8, 1938, nymphs; February 26, 1938, adults; March 23, 1938, adults; April 2, 1938, adults; April 18, 1938, adults; May 5, 1938, adults; September 14, 1938, adults; November 13, 1938, adults; March 22, 1939, adults; April 5, 1939, adults; April 13, 1939, adults; May 6, 1939, adults; June 24, 1939, adults; October 11, 1939, adults; April 27, 1940, adults; March 2, 1941, adults; April 24, 1948, adults; January 15, 1949, adults. 5 miles northwest of Gainesville, April 4, 1937, nymphs; November 8, 1937, nymphs and adults; March 5, 1939, nymphs and adults; August 9, 1939, adults. Hogtown Creek, April 30, 1940, adults. 1 mile west of Newnan's Lake, January 8, 1938, nymphs; August 13, 1938, nymphs; January 16, 1938, nymphs; January 29, 1939, nymphs. 2½ miles west of Gainesville, January 29, 1938, nymphs; January 28, 1939, nymphs; March 18, 1940, adults. Worthington Springs, May 12, 1937, nymphs; February 12, 1938, nymphs; February 5, 1939, nymphs. Santa Fe River, February 5, 1939, nymphs. Bay County: 26 miles north of Panama City, June 8, 1938, nymphs. 5.6 miles north of Panama City, May 30, 1940, nymphs. 27.4 miles north of St. Andrews, May 30, 1940, nymphs. Pine Log Creek, May 31, 1940, nymphs. Calhoun County: Chipola River, November 5, 1938, nymphs. Columbia County: 11.5 miles north of Lake City, October 27, 1938, nymphs. Falling Creek, November 13, 1938, nymphs and adults; June 30, 1939, nymphs. Hamilton County: White Springs, February 4, 1938, nymphs. 8.3 miles south of Jasper, February 4, 1938, nymphs. Hernando County: Southern county line, March 27, 1938, nymphs. Hillsborough County: Bell Creek, November 20, 1937, nymphs; March 26, 1938, nymphs. Hurrah Creek at Picnic, December 27, 1937, nymphs; March 26, 1938, nymphs. Mullis City, December 31, 1937, nymphs. Little Fish-hawk Creek, March 26, 1938, nymphs. 2 miles west of Alafia, March 26, 1938, nymphs. Holmes County: Sandy Creek, December 14, 1939, nymphs. Jackson County: 3.6 miles north of Altha, December 10, 1937, nymphs; June 9, 1938, nymphs; July 1, 1939, nymphs and adults. Florida Caverns State Park, December 2, 1939, nymphs. Jefferson County: Drifton, February 5, 1938, nymphs. Leon County: 7 miles south of Highway 20 on Sopchoppy road, June 5, 1938, nymphs. Levy County: Waccasassa River, October 13, 1946, adults. Liberty County: Sweetwater Creek, November 4, 1938, nymphs; July 1, 1939, nymphs; December 1, 1939, nymphs; May 2, 1941, nymphs; April 29, 1946, nymphs. Madison County: Aucilla River, June 4, 1938, nymphs and adults. Marion County: Eureka, Oklawaha River, February 12, 1938, nymphs. Rainbow Springs, March 9, 1940, nymphs. Okaloosa County: 5.1 miles west of county line, June 7, 1938, nymphs; May 31, 1940, nymphs. Crestview, December 11, 1937, nymphs. Putnam County: Johnson, April 10, 1948, nymphs. Santa Rosa County: Pace, June 1, 1940, adults. Wakulla County: Smith Creek, June 5, 1938, nymphs. Wakulla Springs run, May 29, 1940, nymphs. Walton County: 7.3 miles west of Ebro, June 7, 1938, nymphs. 10.6 miles west of Washington County line, May 31, 1940, nymphs. 2.6 miles west of Freeport, June 7, 1938, nymphs; May 2, 1946, nymphs. 13.8 miles west of Freeport, June 7, 1938, nymphs. 9.5 miles west of Portland, May 31, 1940, nymphs. Washington County: Holmes Creek, December 11, 1937, nymphs; July 2, 1939, nymphs.

Genus CENTROPTILUM Eaton

Eaton (1869) established *Centroptilum* to accommodate those European Baetinae which possess metathoracic wings and in which the intercalaries of the forewings are single. In his taxonomic revision of the mayflies, he included in the genus a species from Cuba as well as a species close to *C. luteolum* (European) from the Hudson Bay region. Eaton (1892) again wrote of *Centroptilum* when he treated the Ephemeroptera of Central America, at which time he added to the list an undescribed species from north Sonora, Mexico. The status of the genus in the western hemisphere remained unchanged until McDunnough (1923) published descriptions of *C. fragile* and *C. curiosum*. Following this publication, descriptions of new species of *Centroptilum* began to appear frequently, and now there are twenty-three species known to occur north of Mexico, fourteen of which were described by McDunnough.

Prior to the work in Florida, the genus was entirely unknown from either the Coastal Plain or Piedmont provinces. Traver has recorded *C. album* (?) from the mountains of North Carolina and *Centroptilum* sp. from the mountains of Alabama. These records bring the genus as close to Florida as it was previously known to occur. *Centroptilum* occurs in the Holarctic area, in Africa, and generally in the Nearctic.

Ide (1930: 222-24) very briefly mentioned the habitat preference of *C. convexum* and *C. bellum*. Beyond this, nothing has been written about the ecology of the genus. Of the former species Ide reported that nymphs were abundant in the Mad River at Horning's Mills, where they lived among the lily pads and pond weeds. The nymphs of *C. bellum*, he continued, were numerous in moderately swift water, where they clung to the upper surfaces of stones.

Phylogenetically, all the Baetinae form a rather compact group. Spieth derived these genera from a common stock, considering those possessing well-developed hind wings to be more primitive than those in which these wings are absent. Based on the degree of reduction of hind wings, *Centroptilum* would be at approximately the same level of development as *Baetis* and *Acentrella*, with *Callibaetis* the most primitive, *Heterocloeon* an intermediate form, and *Pseudocloeon*, *Cloeon*, and *Neocloeon*, which lack hind wings, the most highly developed group. Genitalia do not seem to offer any indices to phylogenetic relations of the genera, nor does eye shape. Even nymphal characteristics do not appear to indicate true relationships.

Centroptilum viridocularis Berner

TAXONOMY.—*Centroptilum viridocularis*, one of the most recent additions to the genus, has been described both in the nymphal and adult stages. *C. viridocularis* may be distinguished from other species by the mesothoracic color, the red markings on tergites 2 through 6, the color of tergites 7 through 10, and the presence or absence of a projection on the inner margin of the second forceps segment. A comparison of Traver's drawings (1935) of the metathoracic wings of some species of *Centroptilum* with those of *C. viridocularis* leads me to associate the latter with *C. convexum, C. conturbatum,* or *C. rufostrigatum.* However, on the basis of Traver's drawings of genitalia, the relationships would seem to lie with *C. rufostrigatum* or *C. fragile.* Because both wings and genitalia are similar to those of *C. rufostrigatum,* these two species may possibly be the most closely related.

DISTRIBUTION.—The nymphs of this species are usually difficult to find, but the few specimens in my collection indicate that they occur in all parts of Florida where there are permanently flowing streams. Specimens have been taken in Hillsborough and Alachua counties and from the Suwannee River, as well as from a strip of counties in west Florida, the range extending even into Baldwin County, Alabama (map 17).

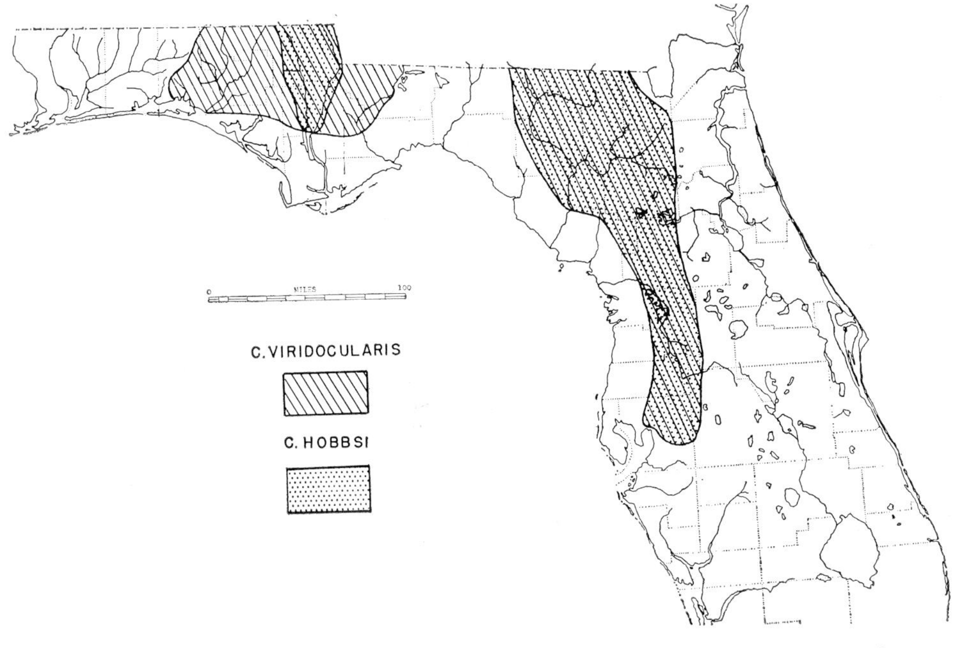

Map 17.—The distribution of *Centroptilum viridocularis* and *C. hobbsi* in Florida.

ECOLOGY.—The nymphs of *C. viridocularis* prefer moderately swift streams, the size of the channel seemingly being of no consequence. I have spent many hours searching for nymphs of this species in Hatchet Creek, near Gainesville. I considered two or three immatures a most unusual catch; yet examination of the stomach contents of several small bass *(Huro salmoides)* revealed that *C. viridocularis* was their choicest food and that each fish had eaten fifteen to twenty-five nymphs. After this discovery, I searched diligently, but without success, in the place from which the bass were seined. I then examined all other conceivable habitats, but still found only two nymphs. Subsequently, collections from a much smaller stream near Gainesville have yielded propor-

PLATE XXIII

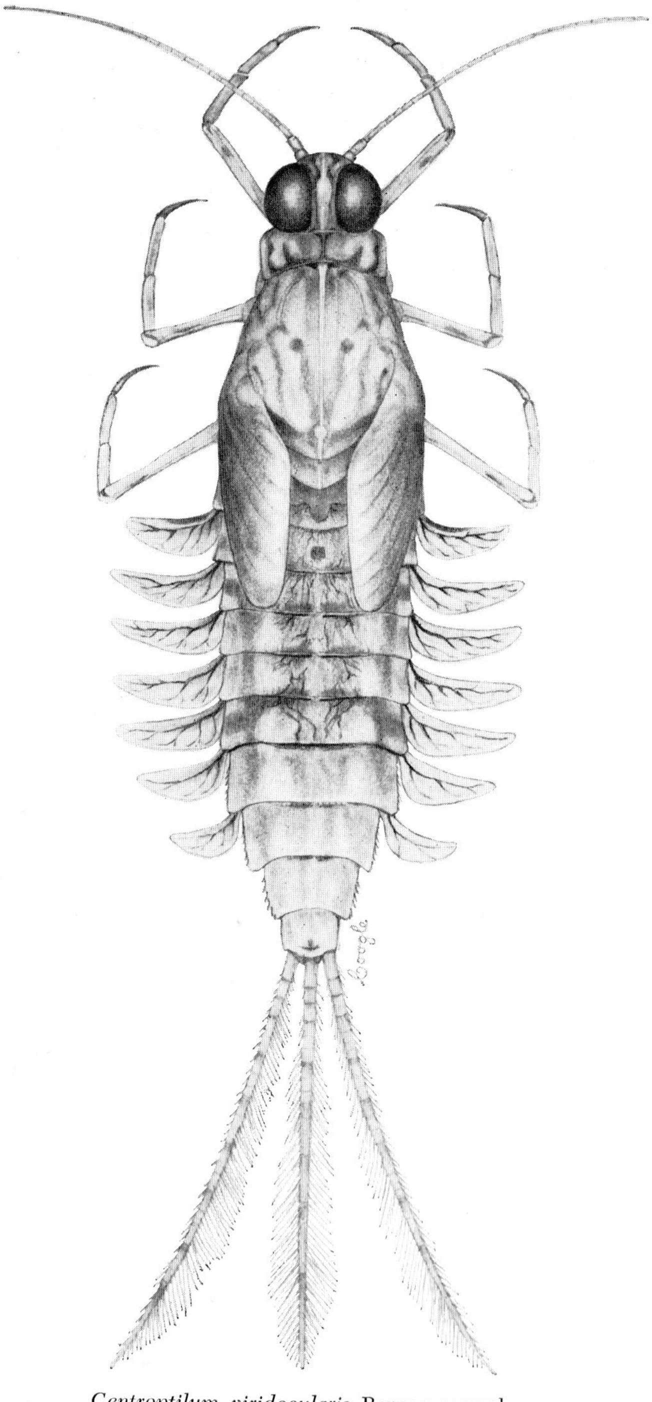

Centroptilum viridocularis Berner, nymph.

tionately many more nymphs than did Hatchet Creek, and some of these were reared to the adult stage.

Hatchet Creek is a sand-bottomed stream, rather deep in places, but, for the most part, a shallow, moderately rapid creek. The depth varies from six or seven feet to as little as six inches in the more swift-flowing portions. Near the shore there are dense growths of parrot's-feather (*Myriophyllum* sp.), and in the stream proper there are clumps of *Potamogeton, Juncus repans, Globifera umbrosa,* and some *Fontinalis.* Near the shore in quieter areas there are rather large accumulations of leaf drift, and the stream is margined with various grasses which are partially submerged. Though two nymphs were collected from the leaf drift, most of the insects lived near the base of the grasses where there was a silty accumulation and very little current.

Little Hatchet Creek, from which most of the nymphs were taken, definitely falls into the sand-bottomed category and is about three feet wide and one foot deep. In the shallower parts where the current is more rapid, there are growths of *Juncus repans;* in the more slow-flowing, deeper regions the mosses become the dominant element of the flora. Nymphs of *C. viridocularis* appeared frequently on the silt-covered mosses, and a few nymphs were also present in the detritus which accumulated between the projecting roots of large trees along the stream margin.

As stated previously, the Santa Fe River at Poe Springs is one of the swiftest of Florida streams. It is only in the slowest flowing parts of the river that nymphs cling to the vegetation, principally *Nais.* They never venture into the swifter waters where *C. hobbsi* occurs. In their choice of environment within the river itself the two species of *Centroptilum* afford one of the best examples of the theory that claw size is directly related to habitat. The claws of the quiet-water form, *C. viridocularis,* are thin and attenuated, being three-fourths as long as the tarsi, whereas the claws of *C. hobbsi,* from swift water, are shorter and thicker, being about one-half as long as the tarsi.

The habits of the species in other streams in Florida are very much like those just described, but in none has the species been found to be as abundant as in the small stream near Gainesville.

The streams inhabited by *C. viridocularis* nymphs may be either slightly acidic or basic. In Hatchet Creek, an excellent example of a swamp stream in which the water is heavily tinted with organic acids, the coloration is so intense at times that the water takes on the appearance of strong tea. On the other hand, the Santa Fe River, fed by springs, is normally colorless and slightly basic in reaction. Although the species is tolerant of very slow-flowing water, it does not inhabit ponds and other standing bodies of water. The nymphs live in places in Hatchet Creek where there is no perceptible current. Yet in standing unaerated water in the laboratory they can survive only a few days.

Associated with *C. viridocularis* in Hatchet Creek are a number of species of mayflies: *Baetis spinosus, B. spiethi, Acentrella ephippiatus, Pseudocloeon alachua, Callibaetis floridanus* (form A), *Caenis diminuta, Blasturus intermedius, Choroterpes hubbelli, Ephemerella trilineata, Paraleptophlebia volitans, P.*

bradleyi, Stenonema smithae, and *S. proximum.* Although all these species do not typically occupy the same habitats as *C. viridocularis,* individuals of any one of them can, at times, be found with this species. In the Santa Fe River, a different group of mayflies which forms part of the biocoenosis includes *Centroptilum hobbsi, Baetis spinosus, B. intercalaris, Pseudocloeon parvulum, P. punctiventris, P. alachua, Stenonema smithae, S. exiguum, S. proximum,* and *Tricorythodes albilineatus.* These species are, in general, also associated with *C. viridocularis* in other Florida streams.

SEASONS.—Evidence points clearly toward a year-round emergence of *C. viridocularis.* I reared or collected adults only during April, May, June, July, and August. My collection also includes a mature nymph taken in October. In addition, very young nymphs were collected in March, July, and December, and half-grown specimens are known for February, May, June, and July. The young nymphs of March and July, as well as the half-grown nymphs of May, June, and July may all emerge during the fall and winter. Related genera show a disregard for seasonal emergence, and with such strong evidence as given above, it does not seem unlikely that transformation occurs in all months of the year.

HABITS.—The nymphs of *Centroptilum* are very easy to distinguish from other mayflies (except *Cloeon*) in the field. When the live nymph is placed in water, the tails are not held straight out as in *Baetis,* but are depressed slightly at the tips. Moreover, a brown band is usually prominent on the caudal filaments of those mayflies likely to be confused with the nymphs of *Centroptilum.* In addition, the long hairs on the tails of *Centroptilum* are not evident, and the shape of the head is quite different from that of many of the other genera of the Baetinae. Its rather abrupt frontal margin gives the nymph the appearance of having pulled the lower part of its head close in to the sternum, and its frontal margin is not so rounded as that of *Baetis.* Finally, the body of *Centroptilum* is shorter and thicker than that of *Baetis.*

The species swims adeptly by rapidly flicking the abdomen and holding the tails stiffly out from the body. Its movement, however, is slower than that of *Baetis.*

When placed in a white-enameled pan without water, the nymphs, in common with the other Baetinae, usually fall on their sides where they remain helpless. If stimulated, they begin to hop about, flipping the abdomen like a small minnow, but with much less strength. The older and more rounded the nymph, the more helpless it becomes when stranded out of water. The young nymphs, because of their more flattened bodies, often fall with the dorsal side up and are then able to crawl over the surface.

In feeding, the nymphs do not often use the legs for bringing food to the mouth, but put the entire burden on the mouth parts. As the nymphs come upon food material, the heads are moved into a position best suited for seizing the food, which is taken into the mouth by means of the maxillary palpi as they shuttle rapidly inward and outward. The food is made up of plant materials of various sorts, principally epidermis, algae, and, to a slight extent, diatoms.

LIFE HISTORY.—The length of nymphal life has not been determined, but it probably extends over a period of six to nine months. Emergence of the subimago during the summer occurs in late afternoon, just before or at approximately the same time as sunset. The final molt takes place within ten to twelve hours. In the laboratory, a female emerged as late as 6:30 P.M. on August 23. Although it was still lively on the morning of August 25, by 9:00 P.M. of that day it had begun to shrivel and showed only the slightest signs of life. Under ideal conditions it would seem that maximum longevity for females of this species is approximately fifty hours.

Some of the nymphs, which were kept in the laboratory in an effort to rear adults, were rather immature and remained alive long enough to undergo ecdysis. The following is a summary of the data on the molting:

Nymph 1.	Two-thirds grown	
	Collected at 3:00 P.M.	8/13/38
	Molted about 1:00 P.M.	8/15/38
	Molted between 9:15 P.M. 8/19 and 9:30 A.M.	8/20/38
	Molted	8/21/38
	Died	8/24/38
Nymph 2.	One-third grown	
	Collected about 10:00 A.M.	8/10/40
	Molted between 10:00 A.M. and 4:00 P.M.	8/14/40
	Died	8/16/40
Nymph 3.	Collected about 10:00 A.M.	8/10/40
	Molted between 6:00 and 9:00 P.M. (in last instar)	8/10/40
	Wing pads blackened	8/13/40
	Died without emerging	8/13/40

The mating flight of the North American and the Florida species of the genus has never been described nor observed. However, as the sun was setting on October 21, I did note a single female flying over midstream about six to ten inches above the water and apparently ovipositing.

LOCALITY RECORDS.—Alachua County: Hatchet Creek, May 11, 1937, nymphs; July 6, 1938, nymphs (from fish stomachs); July 9, 1938, adults; April 1, 1939, adults; April 13, 1939, adults; April 27, 1939, nymphs; May 6, 1939, adults; June 24, 1939, adults; August 10, 1940, nymphs; October 5, 1940, adults; May 11, 1946, nymphs; April 24, 1948, adults. 1 mile west of Newnan's Lake, August 13, 1938, nymphs. Hogtown Creek, October 13, 1940, nymphs. Santa Fe River at Poe Springs, May 21, 1934, nymphs; February 11, 1939, nymphs; February 28, 1939, adults; March 25, 1939, nymphs. Little Hatchet Creek, August 10, 1940, nymphs and adults; August 22, 1940, nymphs and adults; September 13, 1940, nymphs and adults. Citrus County: Chassahowitzka

Springs, May 10, 1941, adults. Gilchrist County: Suwannee River, March 4, 1939, nymphs. Hillsborough County: Six-mile Creek, March 26, 1938, nymphs. Hillsborough River, October 21, 1940, adults. Jackson County: Blue Springs Creek near Marianna, July 1, 1939, adults. Leon County: 7 miles south of Highway 20 on Sopchoppy road, June 5, 1938, nymphs. Walton County: 7.3 miles west of Ebro, June 7, 1938, nymphs. 13.8 miles west of Freeport, June 7, 1938, nymphs and adults. Freeport, May 2, 1946, nymphs.

Centroptilum hobbsi Berner

TAXONOMY.—The species was described in 1946 from a few nymphs and three females which were sufficiently different from any described species to warrant their receiving a new name. Traver (1935: 716) described *C. rivulare* in the adult stage, but dismissed the nymph with the assertion that it has small dorsal flaps on the basal pairs of gills. *C. bellum* has double gills on the basal segments at least. *C. hobbsi,* which also has double gills on the first segment, is certainly different from *bellum* as drawn and described by Ide (1930: 224), and is probably just as distinct from *C. rivulare* and *C. ozburni.* All four of these have double gills. The uniqueness of the Florida species lies in the fact that only the first pair of gills is doubled and that these gills are markedly different in form from the other six pairs.

The nymphs of *C. hobbsi* are easily distinguished from the nymphs of *C. viridocularis* by the gill structure and by the unmarked venter of *C. hobbsi.*

DISTRIBUTION.—Distribution records are few and scattered. The species is known from Hillsborough, Alachua, Marion, and Jackson counties. *C. hobbsi* appears to be less widely distributed than *C. viridocularis,* and its limiting factors are ecological. The species is entirely unknown outside of Florida (map 17).

ECOLOGY.—Though the ecological factors limiting *C. hobbsi* are complex, two are rather obvious. In the first place, the species is confined to streams; and secondly, these streams must be definitely basic in reaction. In regard to the second characteristic, nymphs have been taken most commonly from the Santa Fe River at Poe Springs in which the pH ranges close to 7.3, but other streams in which the nymphs were found have measured up to and above 8.0. For the most part, the rivers and creeks in which *C. hobbsi* nymphs occur are spring fed, and, in some cases, are the main spring runs.

As noted elsewhere in this study, both the spring runs and the Santa Fe River have abundant growths of *Vallisneria* and other water plants in mid-stream. *C. hobbsi,* an inhabitant of this vegetation zone in streams, has rarely been taken from other regions. The nymphs are clingers that attach their claws to the long, trailing strands in moderately swift- to swift-flowing water. The immatures occur in as swift water as the *Vallisneria* can tolerate; they can also be found in the more slow-flowing portions where *Nais* and other aquatic vegetation predominate. The claws of this species are proportionately much shorter and stouter than those of *C. viridocularis*—a condition which is correlated with the fact that *C. hobbsi* inhabits swifter flowing water.

In Jackson County, two nymphs were taken from a stream with a pH of 7.8, and with no vegetation in the flowing portions. The immatures were found clinging to submerged logs in the swiftest part of the creek. Farther west in Florida, along the Choctawhatchee Bay region, there are a number of small creeks with dense mats of *Vallisneria, Sagittaria,* and other aquatic plants in midstream. I took nymphs of *C. viridocularis,* but found no *hobbsi.* These streams are all definitely acidic with a pH not much over 6.0. It would seem, therefore, that *C. hobbsi* is quite sensitive to conditions of alkalinity and that this sensitivity is probably the major limiting factor in the distribution of the species.

The ephemerid associates of *C. hobbsi* are the same as those listed for *C. viridocularis* in the Santa Fe River. In another stream they were taken in company with *Isonychia* sp. A and *I. pictipes.*

SEASONS.—From the paucity of records, little can be said about the seasonal occurrence of *C. hobbsi.* I have recorded adults in February, March, and April and have taken mature nymphs in February, March, May, and June. In March I collected one young nymph, which probably indicates a late summer emergence. Very likely, emergence occurs throughout the year, but owing to the scarcity of individuals and their absence from acidic streams, I have been unable to secure sufficient evidence to make this statement without reservations.

HABITS.—Examination of the contents of the alimentary tract of one of the few nymphs in the collection from the Santa Fe River showed the food to consist almost exclusively of diatoms with a few interspersed strands of algae. My observations of *C. viridocularis* indicate that both Florida species of this genus graze on the diatoms and algae covering the plants on which the mayflies live.

LIFE HISTORY.—Nothing is known of the life history, but it is probably not different from that of *C. viridocularis.*

LOCALITY RECORDS.—Alachua County: Santa Fe River at Poe Springs, May 21, 1934, nymphs; March 12, 1938, nymphs; May 14, 1938, nymphs; February 11, 1939, nymphs and adults; February 18, 1939, nymphs; March 25, 1939, nymphs and adults. Hillsborough County: Six-mile Creek, March 26, 1938, nymphs. Hillsborough River, October 21, 1940, adults. Jackson County: 12.2 miles southeast of Marianna, June 9, 1938, nymphs. Marion County: Rainbow Springs run, March 9, 1940, nymphs.

Genus CLOEON Leach

Cloeon rubropictum McDunnough

TAXONOMY.—A series of nymphs from Florida are being tentatively referred to *Cloeon rubropictum.* On collecting trips into its range, I made several attempts to secure adults of this species, but took only nymphs and one female. Specific identification of nymphs in the Baetinae is a rather risky undertaking, particularly since the nymphs are so poorly known. I communicated with

Dr. F. P. Ide concerning the identity of my specimens, but at the time I was able to send to him only drawings of pertinent structures. He suggested that the species might be *rubropictum;* however, careful study of my specimens leads me to believe that it is unlikely that the Florida nymphs are of this species. The distributional data presented below tend to bear out this conclusion, but for the time being, I consider it best to hold the nymphs under this name. *C. rubropictum* does not conform to the definition of the genus as given by Traver (1935: 733), for instead of having double gills on segments 1 through 6 or segments 1 through 7, they are double only on segment 1. *C. minor* and *C. triangulifera* also form exceptions.

In 1937 Ide published the descriptions of three nymphal forms in the genus *Cloeon.* Now that a number of species are known in the nymphal stage, they may be divided into three groups on the basis of gill structure: those with double lamellae on segments 1 through 6 or 1 through 7; those with single lamellae on all segments; and those with double lamellae on segment 1 only. *C. rubropictum* falls into the last group. Some of Ide's Northern specimens have double lamellae on the first segment only, whereas others have double lamellae on segments 1 and 2. Although the structure is still in a state of flux, the tendency is apparently toward the adoption of the double-gill condition.

DISTRIBUTION.—*C. rubropictum,* described from Ontario in 1923, has since been recorded from Quebec; Willoughby, Ohio; and Ithaca, New York. If the Florida specimens are *rubropictum,* the distribution is certainly peculiar unless it be explained by lack of collecting. The species apparently does not come east of the Valley and Ridge Province until it reaches the Coastal Plain in Florida where it is dispersed over the western region of the state. However, nymphs are known only from a rather restricted portion of Florida—that part in which the streams most closely resemble those found in the Appalachians. The eastern spread of the species in Florida may have been stopped by the wide area of dry, sandy country which separates the Aucilla River from the Suwannee drainage. This distribution would seem to support the theory that the species moved into the state along the western side of the Appalachians and gradually spread into western Florida (map 18).

ECOLOGY.—*C. rubropictum* is an inhabitant of slow-flowing water. Of the numerous Florida streams I have examined, only a few supported populations of the species in numbers sufficient to indicate that it was a highly successful form in the creek. The first of these, from which I have collected on three different occasions, is located 5.8 miles north of Panama City in the panhandle of Florida. The creek is about fifteen feet across and varies in depth from one to three feet. Attached to the pilings supporting one of the two bridges over the stream were large clumps of a densely growing moss, which was also attached to all other firmly anchored objects in the stream. When I first collected there, the water was somewhat contaminated with masses of tar that had been discarded during road construction. Wherever the moss occurred, the current was slow and the movement of the water scarcely sufficient to cause the plants to stream out in the direction of the flow. Farther upstream, the water was more shallow and there was much less moss. A species of *Persicaria,* which is common in very slow-flowing water where the movement is barely perceptible, now became evident. My first trip to the creek yielded only a few specimens; my second, about twenty-five nymphs; and my third, more than one hundred. On my last trip, taken on May 30, 1940, nymphs were much more numerous than on the previous collecting trips. In every case, I collected the immatures from the moss, and although I thoroughly examined the *Persicaria,* I could not find a single specimen. The density of the nymphal population in the stream is indicated by the fact that I took sixty-seven nymphs in one sweep of the dip net through the moss.

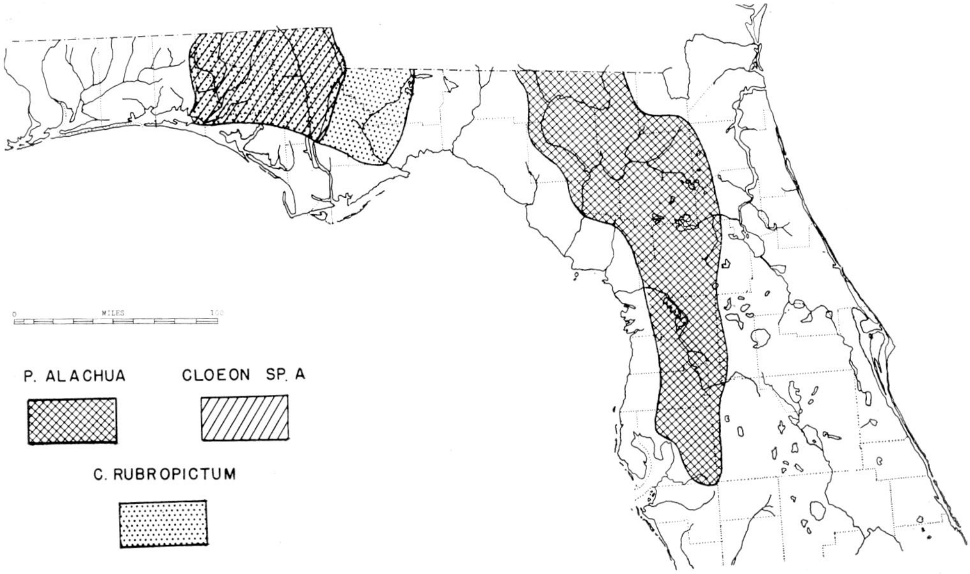

Map 18.—The distribution of *Pseudocloeon alachua, Cloeon rubropictum,* and *Cloeon* sp. A in Florida.

In May, 1940, I visited the Wakulla Springs run and collected ten nymphs in about one hour. This stream, as is usual in spring runs, has dense growths of *Vallisneria* both in the slow current of the shore region and in the deeper, more swiftly moving central part. The river is perhaps seventy-five yards in width and has a soft, sandy bottom that is overlaid with a deposit of calcium carbonate. The many *Goniobasis* taken in the dip net, as well as the numerous *Ampularia,* indicated a high alkalinity. I could collect only near shore because the river deepened rapidly, and even though I examined all types of habitats in the more shallow zones, I found *C. rubropictum* nymphs only on eelgrass.

Other streams from which nymphs of *rubropictum* have been collected are derived from swamps, are slow flowing, have tinted waters, and usually have much vegetation. Never have the immatures been found other than in protected places located in vegetation.

The kinds of streams in which the nymphs live do not seem to be limited by pH, as individuals have been found both in basic and acidic water. Temperatures in the western part of the state, which fall to a rather low level during the winter, would seem to interfere with emergence. My study of nymphal materials, however, suggests that such interference occurs only intermittently during sharp drops in temperature.

Mayfly nymphs associated with *C. rubropictum* in the stream north of Panama City are *Stenonema smithae, Habrophlebiodes brunneipennis, Choroterpes hubbelli, Baetis spiethi, Paraleptophlebia volitans, P. bradleyi,* and *Caenis* sp. In the Wakulla River, *Baetis spiethi, B. spinosus, Tricorythodes albilineatus,* and *Stenonema proximum* occur along with *C. rubropictum.*

SEASONS.—The specimens at hand indicate a year-round emergence for the species. My specimens include a single adult taken in July and mature nymphs collected in May, June, July, November, and December. The lack of material for other months is due only to the fact that collections were not made during those months in that part of western Florida where the species occurs. In the November collection, the presence of very immature, half-grown, and last-instar nymphs certainly indicates that emergence is continuous.

HABITS.—The collecting methods employed made it virtually impossible to keep the nymphs of *C. rubropictum* alive long enough to bring them to the laboratory at Gainesville. The living nymphs resemble *Centroptilum hobbsi* to such an extent that they cannot be differentiated without microscopic examination. The food of *C. rubropictum* consists of diatoms, algae, and plant epidermis.

LIFE HISTORY.—There is no definite seasonal emergence of the species, and the few references to it give little indication of its life history or mating habits. Available evidence would seem to indicate that the entire period from egg to adult probably requires about six months. Ide (1930: 226) has suggested that the species which Clemens (1913) called *Cloeon dubium* is, in reality, *C. rubropictum.* Clemens described the flight of the adults as it occurred at Station Island in July. The adults flew in small swarms along the shore at a height of from ten to fifteen feet about 7:45 P.M. Ide (1930) found the adults swarming in late evening at Lake Nipissing. The single female which I collected flew into the automobile as we stopped to collect at a stream in northwest Florida.

LOCALITY RECORDS.—Bay County: 5.6 miles north of Panama City, June 8, 1938, nymphs; November 5, 1938, nymphs; May 30, 1940, nymphs. Pine Log Creek, May 31, 1940, nymphs. Holmes County: Sandy Creek, July 2, 1939, adults. Jackson County: Florida Caverns State Park, December 2, 1939, nymphs. Wakulla County: Wakulla Springs run, May 29, 1940, nymphs. Walton County: 2.1 miles west of Walton County line, May 31, 1940, nymphs. Washington County: Holmes Creek, June 9, 1938, nymphs; July 2, 1939, nymphs.

Cloeon sp. A

TAXONOMY.—The four nymphs of this species in my collection clearly fall into the *Cloeon-Neocloeon* complex. Any final decision as to whether the species belongs in *Cloeon* or *Neocloeon* must await the discovery of the adult. On the basis of the characteristics set forth by Traver, *Cloeon* sp. A, which has single gills, should be included in *Neocloeon,* although Ide (1937: 241) places *Cloeon triangulifera* and *C. minor,* both species with single gills, in the genus *Cloeon.* As he was able to associate adults with the single-gilled nymphs which were definitely identified as *Cloeon,* it is reasonable to suppose that *Neocloeon* represents only a section of the genus *Cloeon.*

My discussion of *C. rubropictum* has indicated that there are probably three phylogenetic stocks within the genus based on gill structure. If the genus can logically be divided into three such stocks, then *Cloeon* sp. A would fall into the most primitive, namely, that in which all gills are single.

DISTRIBUTION.—*Cloeon* sp. A is known from only two localities, both in northwestern Florida. Three of the nymphs were taken from Holmes Creek which is a part of the Choctawhatchee drainage system; the other specimen was found in a small stream draining into the Chipola River, one of the Apalachicola River tributaries (map 18).

ECOLOGY.—From an ecological standpoint, the species is virtually unknown. In the center of one of the branches of Holmes Creek, from which the nymphs were collected, there was an old stump of a hydrophytic tree (*Nyssa* sp.) which on its downstream side had a small amount of *Persicaria* that grew in the sand and silt covering the roots. Intermingled in this mass was an accumulation of leaf drift, silt, and other trash. Behind the stump the water was about two feet deep and had no perceptible current. It was here that the three nymphs were collected. The single nymph from the Chipola drainage was taken from either the vegetation near the shore, or the leaf drift which had accumulated in this zone.

Both streams from which I collected nymphs are definitely acidic and are derived from strongly tinted swamp water. The mayfly associates behind the stump in Holmes Creek were *Baetis spinosus, Caenis hilaris, Tricorythodes albilineatus, Baetis spiethi, Hexagenia munda marilandica,* and *Cloeon rubropictum.* In the second creek, *Stenonema smithae, Blasturus intermedius, Caenis* sp., *H. munda marilandica, Habrophlebiodes brunneipennis,* and *Habrophlebia vibrans* were found in the same region from which *Cloeon* sp. A was collected.

SEASONS.—Four specimens taken on only two dates would hardly indicate the seasonal distribution of a species. However, in this instance, the data lead me to believe that emergence occurs during all months of the year. I took mature nymphs in July and a nymph in the penultimate instar in December. *C. rubropictum* emerges throughout the year, and, as the above evidence suggests, there is no apparent reason why *Cloeon* sp. A should not have a similar emergence.

LOCALITY RECORDS.—Jackson County: 3.6 miles north of Altha, December 1, 1939, nymphs. Washington County: Holmes Creek, July 2, 1939, nymphs.

Genus PSEUDOCLOEON Klapalek

From the standpoint of phylogeny, *Pseudocloeon* is one of the most interesting genera of the Baetinae. The lack of metathoracic wings in the adult and the vestigial median caudal filament of the nymph are characteristics which indicate a rather high degree of specialization. The genus was not erected until 1905, when Klapalek considered that the group of species lacking hind wings and possessing doubled intercalaries in the mesothoracic wings should, on the basis of these characteristics, be separated from the all-inclusive *Cloeon,* in which the intercalaries occur singly. Further, the nymphs of the two genera form compact assemblages, the nymphs of *Cloeon* being three-tailed and those of *Pseudocloeon* two-tailed. In the former genus the labial palpi are truncate; in the latter they are rounded. At present there are eighteen known species of *Pseudocloeon,* the latest of which I described in 1946. Of these eighteen species, the nine described by McDunnough are known chiefly from adults, especially males. Ide (1937) has cleared up some of the taxonomic difficulties concerning the immatures in his excellent paper on Baetinae mayflies, in which he describes the nymphs of six species.

The species of *Pseudocloeon* are differentiated on the basis of rather minor characteristics, of which the color pattern on the abdomen of the male adults is the most significant. I believe that when the presently described species of this genus are better known, color differences will be shown to be more variable than now supposed, and many of the species will be synonymous.

Published records showing the distribution of the species of *Pseudocloeon* are few and far between. Though some of the species are known only from the type localities, and though others have two or three locality records, the records show that the genus is widely distributed over the eastern portion of North America. There are also some records from the central United States and a few from eastern Alberta. The westernmost record south of Alberta is from Austin, Texas.

Previously the genus was unknown from the Coastal Plain (unless *P. veteris,* taken at the Fall Line in Texas, is considered to be from the Coastal Plain), but the genus has been found to be rather widely distributed through Florida wherever there are permanently flowing streams.

Spieth (1933: 338-39) concluded that *Baetis* and *Pseudocloeon* were closely related, and the mere fact that *Pseudocloeon* lacks the metathoracic wings did not indicate to him a close relationship to other genera which lack these wings. He stated:

> . . . The genus *Pseudocloeon* is like *Baetis* in every item of nymphal and adult structure considered here [venation, genitalia, mouth parts, and gills], except that the adult lacks hind wings and the nymph has only two caudal setae. McDunnough has established a genus *Heterocloeon* (of which the nymphs are

also unknown) for those species in which the hind wings are present but are reduced to a mere thread. . . . Considering wing characters alone, a graded series can be found which extends from the condition found in *Baetis parvus* to that found in *Pseudocloeon*. . . . It is possible that even *Pseudocloeon* [as well as *Acentrella* and *Heterocloeon*] should be considered part of the genus *Baetis*, comparable with the short winged forms known among *Drosophila*, leaf hoppers, beetles, parasitic hymenoptera, gall wasps, etc. Each of the types of reduced wings in these mayfly groups may have arisen by direct and independent mutation from a form such as *B. parvus*. It is not necessary that there has been a gradual decrease in the size of the hind wings. The *Pseudocloeon* species may be more closely related to a species of *Baetis* than are two species which are now unquestionably regarded as members of that genus.

Although Spieth's reasoning seems fairly sound, and although *Pseudocloeon* may represent only a subgenus of *Baetis*, I feel that the convenience of arrangement justifies the retention of the genus for the present. Further evidence to substantiate Spieth's argument is shown by the gradual diminution in size of the hind wings of females of certain Florida species of *Baetis*. Specifically, the metathoracic wings of *Baetis spiethi* females are so minute that in some specimens they are almost invisible and can be detected only by closest scrutiny of the lateral surfaces of the metathorax. Then, too, a series of *Baetis* nymphs of various species can be arranged to show all gradations in size of the median caudal filament.

In Florida the nymphs of *Pseudocloeon* are confined to running water, although in the North some species have been reported inhabiting lake margins.

Pseudocloeon alachua Berner

TAXONOMY.—*Pseudocloeon alachua* was described from Florida in 1940. Although similar to *P. parvulum*, it differs in having long, thin forceps, no intercalary veinlets in the first interspace of the forewing, and hyaline-whitish abdominal segments 2 through 6 marked with brownish-red patches. The nymphs are much like those of *P. parvulum*, but are easily separated by the faintly banded caudal filaments and the absence of coloration in the gills of *alachua*.

DISTRIBUTION.—The distribution of *P. alachua* is rather interesting, for the species is confined to that part of the state extending directly north from Hillsborough County into Madison and Hamilton counties. Although I have collected carefully in the areas to the west of these counties, I have taken only *parvulum* nymphs.

It is quite likely that *P. alachua* is one of the few endemic species of mayflies in Florida, but this conclusion must remain conjectural until specimens of *Pseudocloeon* from Georgia streams are available. *P. parvulum*, the most closely related species, is known only from Ontario and Alberta and, questionably, from Florida (map 18).

ECOLOGY.—Ecologically, *P. alachua* nymphs are confined either to streams of the sand-bottomed type or to clear, moderately slow-flowing streams such

PLATE XXIV

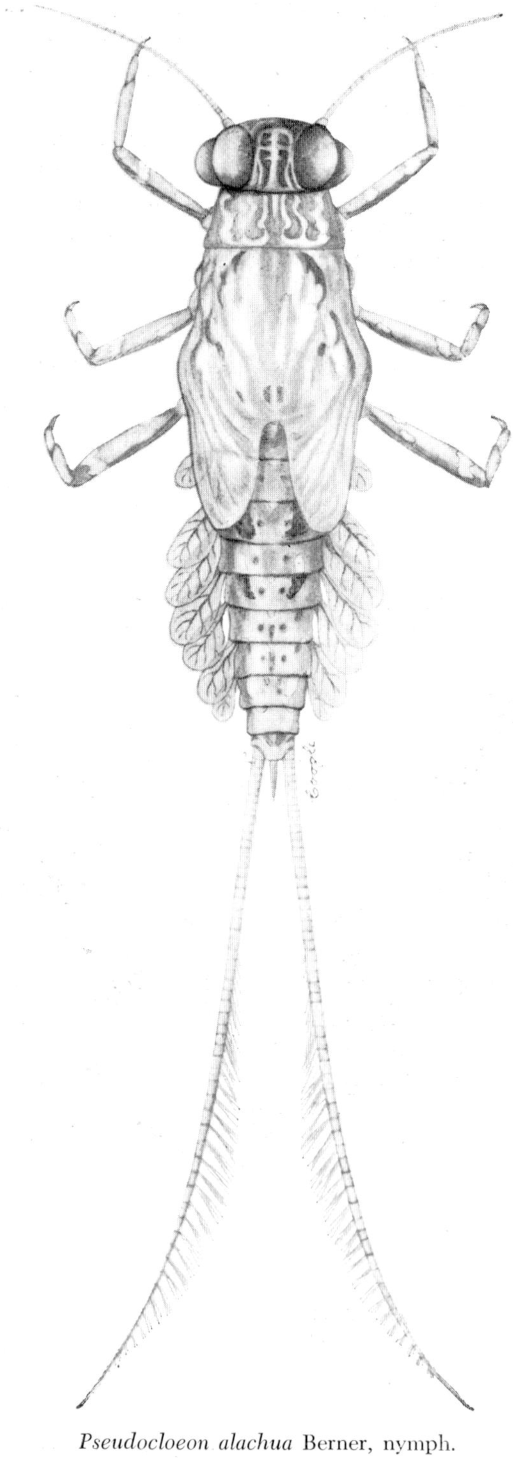

Pseudocloeon alachua Berner, nymph.

as Rainbow Springs run. I have taken nymphs from a creek of the former type that was not over a foot wide and three to four inches deep, as well as from Hatchet Creek, a considerably larger stream. The preferred habitats in these streams appear to be the upper sides of rocks or other solid structures in the swiftest current. The nymphs cling with their heads facing upstream and carry on their feeding activities with barely perceptible movements. Figure 88 shows the positions in midstream taken by the nymphs on the surface of a rock which was partially embedded in sand and which was covered with about one inch of water. Most of the *Pseudocloeon* nymphs on the rock were exposed on the upper surface and seldom moved into the crevices. *Baetis spinosus* nymphs, however, tended to remain in more sheltered areas on the downstream side of the rock, which measured twelve by six by four inches. The appearance

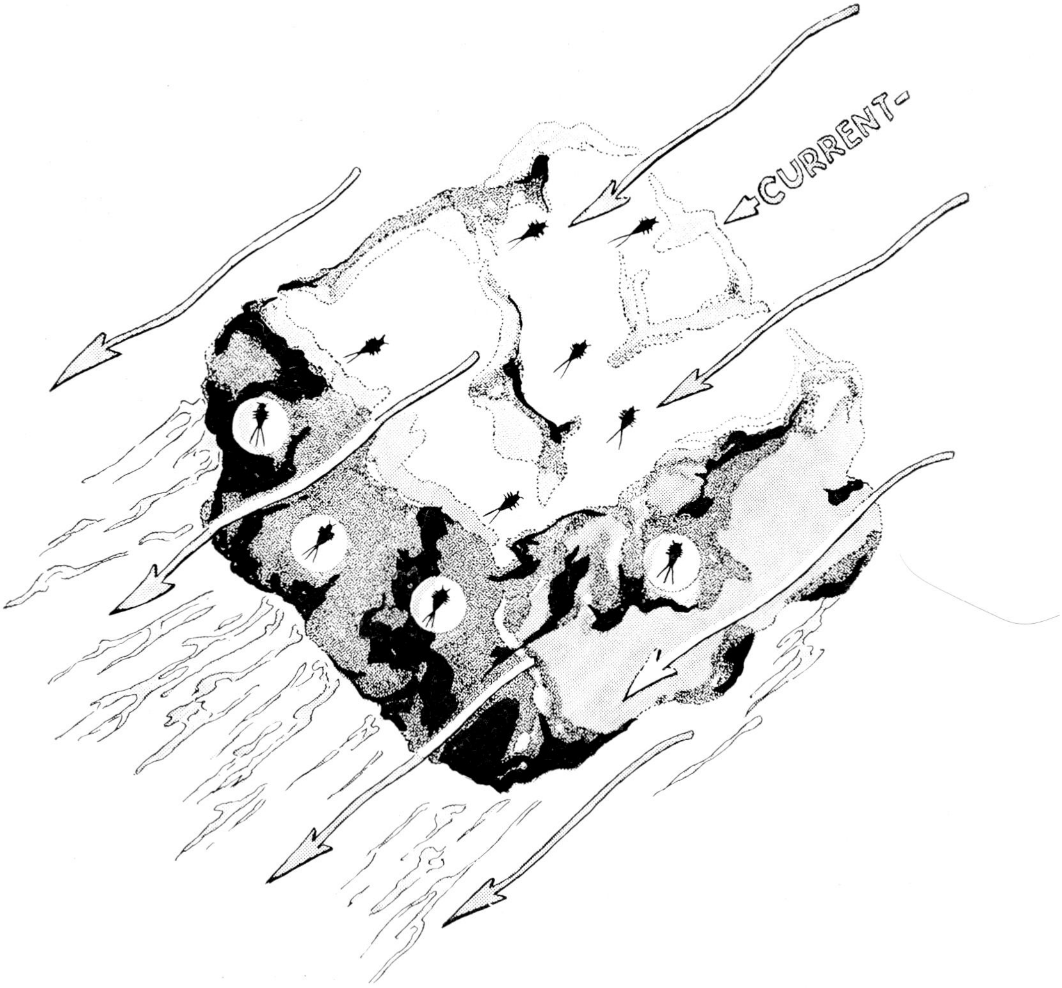

Fig. 88.—A diagrammatic representation of the distribution of the nymphs of *Pseudocloeon alachua* and *Baetis spinosus* on a rock in midstream. The two-tailed nymphs on the upper surface are *P. alachua;* the three-tailed forms on the sides of the rock are *B. spinosus.*

of such a large number of *Pseudocloeon* nymphs on this rock was the largest concentration that I have ever found at one time in such a small area in Florida. One or two nymphs are usually present on each of the smaller stones, and they seem to prefer the downstream side of these habitats.

Other sand-bottomed streams which lack loose stones have relatively few nymphs in comparison to the *Baetis* represented. In the swiftest water only an occasional specimen is collected from the vegetation, where the nymphs live on the distal parts of the plants which swing freely in the current. Specimens have rarely been collected from submerged logs, but the most productive areas are the small, pebbly riffles which are occasionally exposed in the sand bed.

The population of *P. alachua* nymphs is small in comparison with that of *B. spinosus* in the same stream. In every case where the two species were collected at the same time, there were many more individuals of *spinosus* than of *alachua,* even in the riffles and on the rocks.

The nymphs are more plentiful in the Santa Fe River than they are in the sand-bottomed streams. *P. alachua* nymphs seem to find conditions very suitable not only in that river, but also in the Rainbow Springs run, the Withlacoochee River, the Hillsborough River, and other streams which have rather dense growths of *Vallisneria* in their beds. The nymphs may be usually found on the free ends of the leaves of *Vallisneria* where it grows in the swiftest current of the streams, though they often inhabit slower waters in smaller numbers.

Acidity or alkalinity, which does not range below 6.0 or above 8.0 in pH, does not seem to affect nymphs materially. Both these extremes in pH values are rather rare in the streams of the section of Florida in which *P. alachua* occurs; nymphs have, however, been taken from streams with this lower reading and from spring runs which approach the higher pH. Though the immatures appear to thrive equally well in acidic or alkaline waters, they do not occur at the heads of springs—a peculiarity shared in common with all other species of Florida mayflies. The possible explanation for their absence from such habitats has been discussed under the ecology of *B. spinosus.*

The stream from which the types of *P. alachua* came is inhabited by two other species of mayflies, namely, *Baetis spinosus* and *B. australis,* the former being far more numerous. Other species taken along with *alachua* in the average sand-bottomed streams include *Baetis spiethi, Blasturus intermedius, Centroptilum viridocularis, Ephemerella trilineata, Habrophlebiodes brunneipennis, Paraleptophlebia volitans, Stenonema smithae,* and *S. exiguum.* In addition to these, *B. intercalaris, Pseudocloeon parvulum, P. punctiventris, Centroptilum hobbsi, Tricorythodes albilineatus,* and *S. proximum* may be found in such larger streams as the Santa Fe River.

SEASONS.—Adults of *P. alachua* emerge throughout the year. Although adults were taken or reared only in March, April, May, June, October, and November, nymphs in their last instar were collected in February, November, and December, and other nymphs ranging from very immature to last instar were secured in November. The evidence definitely indicates that disregard for seasons is characteristic of *P. alachua* in Florida.

HABITS.—The swimming habits of all members of this genus are similar, but because observations have been made more frequently on *alachua* nymphs than on any of the others, they will be described for this species. When the nymphs are placed in quiet water and stimulated, they move rapidly but very awkwardly. The tails, which merely trail out behind the insect and serve as sensory structures, have undergone such extensive changes that they are no longer efficient propelling organs like those of *Baetis* or *Callibaetis*. If a resting nymph be incited to move, it may crawl a short distance, then come to rest again. If the stimulus is stronger, the nymph may attempt to swim away from the source of annoyance. This act is accomplished by vigorous abdominal undulations and strong lashing of the almost useless caudal filaments. The whole process is very awkward, and the insect does not move far. As soon as swimming has ceased, the insect usually spreads its legs, separates and raises its tails, arches its back, and settles slowly to the bottom of the pan or dish.

Pseudocloeon nymphs have become highly specialized for living in flowing water. Dodds and Hisaw (1924: 142-43) have pointed out that such structural modifications as those exhibited by *Pseudocloeon* nymphs adapt them to the swiftest water. In their discussion of *Baetis bicaudatus* (this will certainly also apply to *P. alachua*, for it has undergone the same changes as *B. bicaudatus*) these authors stated:

> . . . The important changes [adaptations for inhabiting very swift streams and torrents] are a great increase in the relative size of thorax and legs, with a corresponding reduction of the abdomen, including an almost entire disappearance of the hairs of the caudal cerci. The decrease in absolute size of the body is probably also of significance, inasmuch as a small body offers less resistance to the current than a large one. Within the genus *Baetis*, as illustrated by our three species, the progressive adaptation to swift water follows the same line as that which differentiates this type from the swimming type, namely; decrease in the size of the body, without a corresponding decrease in the size of the legs; reduction of size of gill lamellae; reduction of middle cercus from a length two-thirds the other two, to a mere rudiment. The reduction of this cercus is, no doubt, of importance in reducing the pull of the water, by reducing the exposed surface in this part of the body, while the remaining two, when held close together, as they commonly are, serve to complete the pointed tip of the tapering body.

An examination of the alimentary canal of several nymphs indicated almost nothing concerning the food habits of the species, for in every case the food had been so thoroughly masticated that identification of the fragments was impossible. However, the diatoms and algae, which were abundant on the rocks and plants inhabited by the nymphs, were undoubtedly the food of the insects.

LIFE HISTORY.—I was unable to rear nymphs through the entire life cycle, but a study of the species in a very small stream on the campus of the University of Florida leads me to believe that *P. alachua* requires about six to eight months to complete its life history. An area near the source of the stream was thoroughly collected until no nymphs except very immature specimens could be found. About six months later the stream was again examined, and it was found that mature as well as very young nymphs were present. Although a few half-grown individuals were also discovered, these might well have been offspring of adults which emerged in other parts of the creek than those examined, and which had flown to the collected area to oviposit during the six-month period.

Emergence of the subimago occurs from about 2:00 P.M. to 5:00 P.M. This emergence time has been noted both in the spring and in the late fall, the only two periods during which specimens have been reared or observed emerging in the field. When ready to transform, the nymph swims to the surface, and immediately the subimago bursts free of the nymphal exuviae. It remains quietly resting on the floating skin for a moment, and then flies away. In a stream, this is a precarious time for the subimago, as it may be swept along at a rather rapid rate. At the slightest disturbance of the floating exuviae, the subimago immediately takes to the wing.

In the laboratory, the subimago molts after eight to ten hours, but the period is somewhat variable. The length of adult life, determined only for the female, has been found to be about forty hours.

The mating flight of *Pseudocloeon* has never been described, and it has not been observed for any of the Florida species.

LOCALITY RECORDS.—Alachua County: Campus, University of Florida, March 5, 1937, nymphs; July 31, 1938, adults; January 28, 1939, nymphs; February 7, 1939, nymphs; March 23, 1939, nymphs and adults; April 12 through 20, 1939, nymphs and adults; June 23, 1939, adults; November 8, 1940, adults; November 22, 1940, adults. Hatchet Creek, March 22, 1937, nymphs; April 1, 1939, nymphs and adults; April 13, 1939, nymphs and adults; May 6, 1939, adults; June 24, 1939, adults; October 11, 1939, adults; August 10, 1940, nymphs; March 2, 1941, nymphs. Hogtown Creek, April 3, 1937, nymphs; April 28, 1937, nymphs; September 25, 1937, nymphs; October 30, 1937, nymphs. Devil's Mill Hopper, October 25, 1937, nymphs; December 13, 1937, nymphs; March 5, 1938, nymphs. 2½ miles west of Gainesville, February 3, 1938, nymphs; March 18, 1940, nymphs and adults. Santa Fe River, March 25, 1939, nymphs and adults. 20 miles north of Gainesville, May 12, 1937, nymphs. Hamilton County: 8.3 miles south of Jasper, February 4, 1938, nymphs. Hernando County: Southern county line, March 27, 1938, nymphs. Hillsborough County: Six-mile Creek, March 26, 1938, nymphs; Bell Creek, March 26, 1938, nymphs and adults. Little Fish-hawk Creek, March 26, 1938, nymphs. Hillsborough River, February 11, 1939, nymphs; October 21, 1940, adults. Madison County: 4.3 miles east of Jefferson County line, February 5, 1938, nymphs. Marion County: Withlacoochee River, April 2, 1937, nymphs; March 25, 1938, nymphs. Rainbow Springs run, March 9, 1940, adults. 9.6 miles southwest of Salt Springs, April 10, 1948, nymphs. Taylor County: Fenholloway River, May 29, 1940, nymphs.

Pseudocloeon parvulum McDunnough

TAXONOMY.—There is still some doubt concerning the validity of the identification of the Florida specimens as *Pseudocloeon parvulum.* A series of nymphs closely resembles Ide's description and drawings of the species, and a single male adult in my collection is much like McDunnough's description of the true *P. parvulum.* The Florida specimens differ from the nymphs of the Northern *P. parvulum* only slightly in the abdominal maculation. The differences in the color patterns of the male adults hardly warrant the erection of a new species.

McDunnough described *P. parvulum* from a series of adults taken at Tillsonburg, Ontario, and from Kazabuzua, Quebec. In his description, he included the descriptions of male, female, and nymph, and at the same time indicated that the nymph of *parvulum* is at once distinguishable from all others of the genus by the fact that the caudal filaments are alternately banded with pale and dark colors as in the nymphs of *Stenonema.* Ide (1937: 236) added the maculation of the gills as a distinguishing feature of the nymphs. The Florida specimens all show very clearly the banded caudal filaments, the definitely maculate gills, and mouth parts which are almost identical with those figured by Ide. This banding, however, can no longer be considered unique, because *P. alachua* also shows faintly annulate caudal filaments.

DISTRIBUTION.—*Pseudocloeon parvulum* is known from the type locality listed above and from Milk River, Alberta, the Peche River, and Wakefield, Quebec. In the South, nymphs and adults have been collected from a limited area in Florida. The southern area of occupancy stretches from Alachua County to Liberty County, where it includes the tributaries of the Apalachicola River on the eastern side of the stream. There the distribution is somewhat spotty and rather difficult to explain, for, although the streams on the west side of the Apalachicola do not differ materially from those on the east side of the

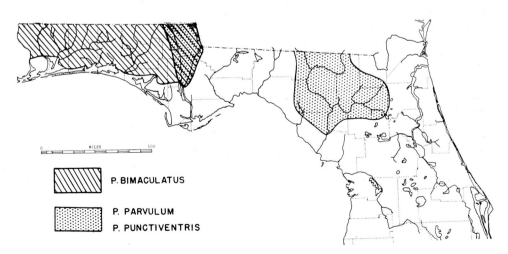

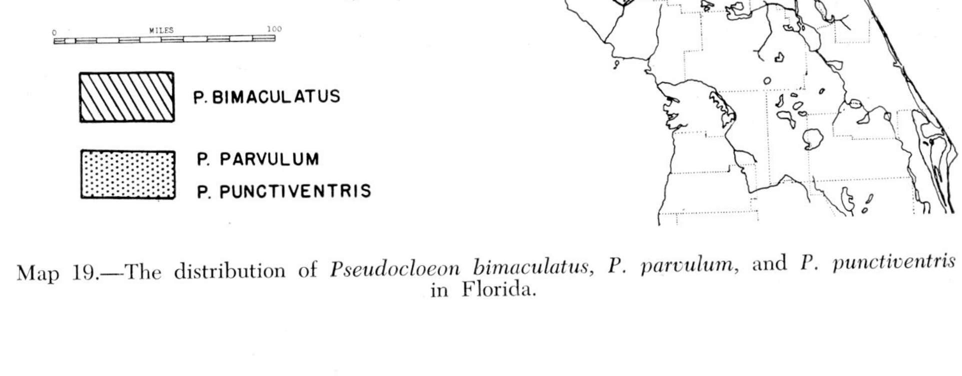

Map 19.—The distribution of *Pseudocloeon bimaculatus, P. parvulum,* and *P. punctiventris* in Florida.

river, the nymphs do not appear in any of my collections from the region. Likewise the absence of the nymphs from Hillsborough County seems unusual, as the conditions in the Hillsborough River are very similar to those in the Santa Fe (map 19).

ECOLOGY.—Nymphs have been found abundantly in the Santa Fe River. The species is known from only three other localities. In the Santa Fe the nymphs live in the swifter parts of the stream on the *Vallisneria* which grows there very thickly. On the submerged plants the nymphs inhabit the distal portions of the leaves which float freely in the current. Only rarely have the nymphs been found on the rocks which are strewn here and there over the stream bed, but this absence may be explained by the fact that mats of *Fontinalis* entirely cover the upper surfaces of the stones and render the conditions unsuitable for the nymphs. Occasional immatures may sometimes be found on the undersides or in crevices of the rocks, but they are far more numerous on the plants.

In the Suwannee River, of which the Santa Fe is a tributary, the nymphs must of necessity remain close to shore, as the water becomes deep very rapidly and suitable habitats for the nymphs are rather scarce. The river current is fairly rapid, but the limited aquatic vegetation consists chiefly of *Myriophyllum* and *Persicaria*. Mayflies were not common where my collections were made. Nonetheless, I caught nymphs of several species, among them *P. parvulum* taken from rocks along the river margin.

Two streams where nymphs were found have sand bottoms, are slightly acidic, and flow at a moderately swift rate. In these streams, the nymphs were on the vegetation and on exposed roots of terrestrial plants which projected into the water.

In the Santa Fe River I found *parvulum* nymphs together with those of *B. spinosus*, *B. intercalaris*, *Pseudocloeon alachua*, *Centroptilum viridocularis*, *C. hobbsi*, *Tricorythodes albilineatus*, *Stenonema smithae*, *S. proximum*, *S. exiguum*, and *Hexagenia munda elegans*. At the Suwannee River I took another species, *Isonychia pictipes*, along with *parvulum*. *Paraleptophlebia volitans* and *P. bradleyi* nymphs appeared with *parvulum* in the Fenholloway River; and in Sweetwater Creek *parvulum* occurred with *Oreianthus* sp. No. 1, *Caenis hilaris*, and the Santa Fe species referred to above.

SEASONS.—Although my collections of *P. parvulum* are not extensive, those at hand indicate that emergence occurs throughout the year. In the Northern reaches of its range, the species is seasonal; McDunnough has found that in the spring generation the individuals are larger than those of the summer generation, which are extremely minute.

HABITS.—The habits of this species are like those of *P. alachua*.

LIFE HISTORY.—Only one male and one female have been reared. These emerged during the afternoon about 3:00 and remained in the subimaginal stage for about nine or ten hours. No other data are available.

LOCALITY RECORDS.—Alachua County: Santa Fe River, March 24, 1937, nymphs; March 12, 1938, nymphs; February 11, 1939, nymphs; February 18, 1939, nymphs; March 4, 1939, adults; March 25, 1939, nymphs; October 25, 1939, adults. Gilchrist County: Suwannee River at Old Town, April 5, 1938, nymphs. Liberty County: Sweetwater Creek, June 10, 1938, nymphs; July 1, 1939, nymphs. Taylor County: Fenholloway River, March 18, 1939, nymphs.

Pseudocloeon punctiventris (McDunnough)

TAXONOMY.—There are a few nymphs in my collection which are doubtfully being referred to *Pseudocloeon punctiventris.* These immatures differ from Ide's description (1937: 237) in that the median pale band on the abdomen is quite distinct and the tracheae in the gills are prominent in preserved material. McDunnough (1923: 45) described the species from the Rideau River, Ottawa, and since then it has been recorded only in Ide's description of the immature.

The nymphs are distinct from all other Florida species in having a median dark band on each of the caudal filaments and in the absence of color from the seventh gills. If this species is *punctiventris,* the adults should be easily distinguishable by the absence of markings on the dorsum of the pale abdominal segments, as well as by the presence of a minute mid-ventral dot at the posterior margin of each abdominal sternite.

DISTRIBUTION.—The distribution of *P. punctiventris* is interesting as there are no records of its occurrence in the area between Canada and Florida. In Florida the nymphs have the same range as *P. parvulum* and have been taken from the same streams. During a short collecting trip in Georgia, I collected in a stream about ten miles north of Macon, and found nymphs there which are similar to those of Florida, although the ventral dark dots are not evident on the abdomen. As many of the Florida specimens do not have these dots, this characteristic does not seem of sufficient importance to separate the Georgia specimens from *P. punctiventris* (map 19).

ECOLOGY.—Ecologically, *P. punctiventris* is indistinguishable from *P. parvulum.*

SEASONS.—Nymphs in the last instar have been collected in February, March, April, and December. Even though there are wide gaps between collection dates, these specimens do indicate that *punctiventris* emerges throughout the year.

HABITS.—The habits are similar to those of *P. alachua.*

LIFE HISTORY.—Nothing is known of the life history, but it is probably very similar to that of *P. alachua.*

LOCALITY RECORDS.—Alachua County: Santa Fe River at Poe Springs, March 12, 1938, nymphs; February 11, 1939, nymphs; February 18, 1939, nymphs and adults; March 25, 1939, nymphs; March 1, 1940, nymphs. Gilchrist County: Suwannee River at Old Town, April 5, 1938, nymphs; March 4, 1939, nymphs. Liberty County: Little Sweetwater Creek, December 10, 1937, nymphs. Sweetwater Creek, April 29, 1946, nymphs. Putnam County: Johnson, April 10, 1948, nymphs.

Pseudocloeon bimaculatus Berner

TAXONOMY.—This species, known from three females, one male imago, and numerous nymphs, was described in 1946. The male adult may be distinguished from other species of *Pseudocloeon* by its color pattern, and particularly by the paired red spots on the abdominal tergites. These spots are also present on the abdomen of the female. Nymphs may be separated from others of the genus by the fact that the seventh pair of gills is deeply colored with reddish brown; moreover, the abdominal maculation is unique, and the banding of the caudal filaments and the length of the median filament in relation to the width of the laterals are distinctive. The nymphs are clearly marked and can be selected with the naked eye from unsorted Florida specimens, as it is the only species of *Pseudocloeon* found in this region in which the venter of the terminal abdominal segments is deep red-brown.

DISTRIBUTION.—*P. bimaculatus* is the only Florida species of *Pseudocloeon* known to occur west of the Apalachicola River. Both this fact and the similarity in banding of the caudal filaments suggest that *bimaculatus* may be the same species as those specimens I am calling *P. punctiventris*, which is limited to the east side of the river. Because the two areas are mutually exclusive and the two nymphs are so similar, except in maculation, the identification must remain in question until male adults of *punctiventris* are secured.

Nymphs are known from various streams throughout Florida west of the Apalachicola River, and from Mobile and Escambia counties in Alabama. The species is fairly common throughout west Florida and southern Alabama, and nymphs may be found in nearly all permanently flowing streams. A study of its distribution would seem to indicate that *bimaculatus* is endemic to the western portion of Florida and the southern part of Alabama, and is confined to the coastal area. Collections from more northerly localities in Alabama did not include a single specimen of *P. bimaculatus* (map 19).

ECOLOGY.—The sand-bottomed streams of the western part of Florida all support populations of *P. bimaculatus*, and it is much more common than are its kindred species of the more eastern regions in these streams. Particularly abundant in those streams with rather dense growth of eelgrass, the nymphs elsewhere are second in abundance only to *Baetis* species. In the creeks, the nymphs live on the *Vallisneria* where the water is swiftest, but they can also be found in shallower, more slowly flowing water. Some of the streams along the Choctawhatchee Bay region have populations of these nymphs living in *Vallisneria* not more than one hundred yards from the salt water. In other sand-bottomed streams, where there is no eelgrass, the nymphs live close to shore on the vegetation or on any other materials which are permanently anchored in the flowing water. In contradistinction to *P. alachua*, the immatures do not seem to occur as commonly on submerged logs and sticks as they do on the vegetation. Nymphs have not been found in the riffles of the west Florida creeks, probably because of the rarity of pebbly riffles, few of which have been examined. In one small creek, about three or four inches in depth, where the bottom was covered with an algal growth, I found several *P. bimaculatus* nymphs in midstream.

The creeks from which nymphs were collected ranged in pH from 6.0 to 7.6, but the immatures appeared to be more numerous in the acid streams. This pH relationship may not be significant, however, for vegetation is quite dense in many of the acidic and the basic streams of west Florida.

Mayflies which have been found associated with *P. bimaculatus* include *Baetis spinosus, B. australis, B. spiethi, B. intercalaris, Acentrella ephippiatus, A. propinquus, Caenis hilaris, Choroterpes hubbelli, Paraleptophlebia volitans, Ephemerella choctawhatchee, Ephemerella* sp. A, *E. trilineata, E. hirsuta, Baetisca rogersi, Habrophlebiodes brunneipennis, Tricorythodes albilineatus, Stenonema smithae, S. proximum,* and *S. exiguum.*

At a lighted sheet near streams, I caught one male and three females, but I did not collect nymphs at the stream where the adult male was taken. The females with which I am associating the male were collected at Perdido Creek, where the nymphs were plentiful in dense growths of *Vallisneria* and *Potamogeton.*

SEASONS.—Imagoes are known for June and July. Nymphs in the last instar are at hand for May, June, and December, and rather immature nymphs for June. Undoubtedly, emergence occurs throughout the year. The nymphs in my collection appear to fall into size groups and indicate that distinct broods form in this west Florida species. There are probably at least two winter and three summer broods. From available data it is impossible to estimate the number for the other times of the year.

HABITS.—Because of their low ecological valence, it was almost impossible to keep nymphs of this species alive for more than a couple of hours in quiet water. The species occurs only in west Florida, where no laboratory facilities were available for rearing. Because of these factors and of certain difficulties which prevented my remaining near one of the streams in which the nymphs live for a sufficient time to rear the species, little can be said concerning the life history, except that it probably does not differ materially from that of *P. alachua.*

LOCALITY RECORDS.—Bay County: 16.8 miles north of Panama City, June 8, 1938, nymphs. 26 miles north of Panama City, June 8, 1938, nymphs. 27.4 miles north of St. Andrews, May 30, 1940, nymphs. Jackson County: 12.2 miles southeast of Marianna, June 9, 1938, nymphs. Blue Springs Creek near Marianna, July 1, 1939, adults. Escambia County: Bayou Marquis, June 1, 1940, nymphs. Okaloosa County: 5.1 miles west of county line, June 7, 1938, nymphs. Crestview, December 12, 1937, nymphs. Niceville, June 7, 1938, nymphs. Walton County: 7.3 miles west of Ebro, June 7, 1938, nymphs. 5.4 miles west of Washington County line, May 31, 1940, nymphs.

REFERENCES

Banks, Nathan
1900. New genera and species of nearctic neuropteroid insects. *Trans. Amer. Ent. Soc.*, 26: 245-51.
1903. A new species of *Habrophlebia*. *Ent. News*, 14: 235.

Bartram, W.
1791. Travels through North and South Carolina, Georgia, East and West Florida. Philadelphia, pp. i-xxxiv + 1-522, 8 pls., 1 map.

Berner, Lewis
1940a. *Baetisca rogersi*, a new mayfly from northern Florida. *Canad. Ent.*, 62: 156-60, pl. 10.
1940b. Baetine mayflies from Florida (Ephemeroptera). *Fla. Ent.*, 23 (3 and 4): 33-45, 49-62, pls. 1 and 2.
1941. Ovoviviparous mayflies in Florida. *Fla. Ent.*, 24 (2): 31-34.
1946. New species of Florida mayflies (Ephemeroptera). *Fla. Ent.*, 28 (4): 60-82, pl. 1.

Byers, C. F.
1930. A contribution to the knowledge of Florida Odonata. *Univ. of Fla. Pub., Biol. Sci. Ser.*, 1 (1): 1-327, 19 figs., pls. 1-11.

Campion, H.
1923. On the use of the generic name *Brachycercus*. *Ann. Mag. Nat. Hist.*, 9: 515.

Carpenter, F. M.
1933. The lower Permian insects of Kansas. Part 6. *Proc. Amer. Acad. Arts and Sci.*, 68 (11): 487-503.

Carr, A. F.
1940. A contribution to the herpetology of Florida. *Univ. of Fla. Pub., Biol. Sci. Ser.*, 3 (1): 1-118.

Clemens, W. A.
1913. New species and new life histories of Ephemeridae or mayflies. *Canad. Ent.*, 45: 246-62, 329-41, pls. 5-7.
1915a. Rearing experiments and ecology of Georgian Bay Ephemeridae. *Contri. to Canad. Biology, Sessional Paper* 39b, pp. 113-28, 132-43.
1915b. Mayflies of the *Siphlonurus* group. *Canad. Ent.*, 47: 245-60, pls. 9-11.
1917. An ecological study of the mayfly *Chirotenetes*. *Univ. of Toronto Studies, Biol. Ser. 17*, pp. 1-43, 5 pls.
1922. A parthenogenetic mayfly (*Ameletus ludens* Needham). *Canad. Ent.*, 54: 77-78.

Cooke, C. W.
1939. Scenery of Florida. *State of Fla. Dept. of Conservation, Geol. Bull.* No. 17, pp. 1-118, 58 figs.

Cooke, C. W. and S. Mossom
1929. Geology of Florida. *Fla. State Geol. Survey, 20th Annual Report*, pp. 29-228, pls. 1-29.

Cooke, H. G.
1940. Observations on mating flights of the mayfly *Stenonema vicarium* (Ephemerida). *Ent. News*, 51: 12-14, 1 fig.

Cory, C. B.
1895. Hunting and fishing in Florida, including a key to the water birds known to occur in the state. Boston (Estes and Lauriat Co.), pp. 1-297.

Daggy, R. H.
1945. New species and previously undescribed naiads of some Minnesota mayflies (Ephemeroptera). *Annals Ent. Soc. Amer.*, 38 (3): 373-96, 12 figs.

DODDS, G. S. and F. L. HISAW

1924. Ecological studies of aquatic insects. I. Adaptation of mayfly nymphs to swift streams. *Ecology*, 5: 137-48, pls. 1 and 2.

EATON, A. E.

1883-1888. A revisional monograph of recent Ephemeridae or mayflies. *Trans. Linnean Soc. London, Second Series Zoology*, 3: 1-352, pls. 1-65.

1892. Family Ephemeridae (*in* Biologia Centrali-Americana). Pp. 1-16, 1 pl.

1896. Ephemeridae in brackish water. *Ent. Mag.*, p. 144.

EDMUNDS, G. F.

1945. Ovoviviparous mayflies of the genus *Callibaetis* (Ephemeroptera: Baetidae). *Ent. News*, 56 (7): 169-71.

FELT, E. P.

1901. Aquatic insects of the Saranac region. *New York State Mus. Bull. Rep. for 1900*, pp. 499-532, 14 figs.

FORBES, S. A.

1888. Studies of the food of fresh-water fishes. *Bull. Ill. State Lab. of Nat. Hist.*, 2 (Art. VII): 433-73.

FRISON, T. H.

1935. The stoneflies, or Plecoptera, of Illinois. *Ill. Nat. Hist. Sur. Bull.*, 20 (Art. IV): 281-471, 343 figs.

GLICK, P. A.

1939. The distribution of insects, spiders, and mites in the air. *U.S.D.A. Tech. Bull.*, No. 673, pp. 1-150, 13 figs., pls. 1-5.

HAGEN, H. A.

1861. Synopsis of the Neuroptera of North America with a list of South American species. *Smithsonian Misc. Coll.*, 4 (Art. I): 1-347.

HARKNESS, W. J. K. and E. L. PIERCE

1941. The limnology of Lake Mize, Florida. *Proc. Fla. Acad. Sci.*, 5: 96-116, 3 figs., pl. 1.

HOBBS, H. H.

1942. The crayfishes of Florida. *Univ. of Fla. Pub., Biol. Sci. Ser.*, 3 (2): 1-179, pls. 1-24, maps 1-11.

1944. Notes on the subterranean waters of the Florida peninsula with particular reference to their crustacean fauna. *The Biologist*, 26 (1 and 2): 6-8.

HUBBELL, T. H.

1936. A monographic revision of the genus *Ceuthophilus*. *Univ. of Fla. Pub., Biol. Sci. Ser.*, 2 (1): 1-551, pls. 1-38.

IDE, F. P.

1930. Contributions to the biology of Ontario mayflies with descriptions of new species. *Canad. Ent.*, 62: 204-13, pl. 17.

1935a. The effect of temperature on the distribution of the mayfly fauna of a stream. *Univ. of Toronto Studies, Biol. Ser. 39, Pub. Ont. Fish. Res. Lab.*, 50: 9-76, pls. 1-10.

1935b. Post embryological development of Ephemeroptera (mayflies). External characters only. *Canad. J. Res.*, 12: 433-78, pls. 1-13.

1935c. Life history notes on *Ephoron, Potamanthus, Leptophlebia*, and *Blasturus* with descriptions (Ephemeroptera). *Canad. Ent.*, 67: 113-25, pls. 4-5.

1937. Descriptions of eastern North American species of Baetine mayflies with particular reference to the nymphal stages. *Canad. Ent.*, 69: 219-31, 235-43, pls. 8-12.

1940. Quantitative determination of the insect fauna of rapid water. *Univ. of Toronto Studies, Biol. Ser. 47, Pub. Ont. Fish. Res. Lab.*, 59: 1-20, pls. 1-4.

1941. Mayflies of two tropical genera, *Lachlania* and *Campsurus*, from Canada with descriptions. *Canad. Ent.*, 73: 153-56, 3 figs.

IMMS, A. D.
1931. Recent advances in entomology. Philadelphia (P. Blakiston's Son and Co., Inc.), pp. i-viii + 374.

KRECKER, F. H.
1915. Phenomena of orientation exhibited by Ephemeridae. *Biol. Bull. Marine Biol. Lab.*, 29: 381-88, 2 figs.

LAFORGE, L., W. COOKE, A. KEITH, and M. R. CAMPBELL
1925. Physical geography of Georgia. *Geol. Surv. of Georgia*, Bull. 42, pp. 1-189, pls. 1-43.

LESTAGE, J. A.
1920. Nouvelles observations sur la ponte de *Cloeon dipterum* L. (Éphémère). *Bull. Soc. Ent. Brussels*, 2: 74.
1924a. Contribution à l'étude des larves Ephémères: série III—le groupe Éphémérelliden. *Ann. Biol. Lacustre*, 13: 227-302, 14 figs.
1924b. Notes sur les Ephémères de la monographical revision de Eaton. *Ann. de la Soc. Ent. de Belgique*, 64: 33-60.
1924c. A propos du genre *Caenis* Steph. = *Brachycercus* Curt. *Ann. de la Soc. Ent. de Belgique*, 64: 61-62
1931a. Note à propos de l'homonymie de deux Ephéméroptères. *Ann. de la Soc. Ent. de Belgique*, 71: 119.
1935. Contribution à l'étude des Ephéméroptères: IX—le groupe Siphlonuridien. *Bull. Ann. Soc. Ent. Belgique*, 75: 77-139, figs. 1-11.

LYMAN, F. E.
1943. Eye-color changes in mayflies of the genus *Stenonema* (Ephemeridae). *Ent. News*, 54 (10): 261-64.

MCDUNNOUGH, J.
1921. Two new Canadian mayflies. *Canad. Ent.*, 53: 117-20, 1 pl.
1923. New Canadian Ephemeridae with notes. *Canad. Ent.*, 55: 39-50, figs. 1-3.
1924. New Canadian Ephemeridae with notes, II. *Canad. Ent.*, 56: 90-98, 113-22, 128-33, pls. 1-3.
1925a. New *Ephemerella* species (Ephemeroptera). *Canad. Ent.*, 57: 41-43.
1925b. New Canadian Ephemeridae with notes, III. *Canad. Ent.*, 57: 168-76, 185-92, pls. 4-5.
1925c. The Ephemeroptera of Covey Hill, Quebec. *Trans. Roy. Soc. Canada*, 19: 207-24, pl. 1.
1927. Notes on the species of the genus *Hexagenia* with description of a new species (Ephemeroptera). *Canad. Ent.*, 59: 116-20, 1 fig.
1931a. The *bicolor* group of the genus *Ephemerella*, with particular reference to the nymphal stages. *Canad. Ent.*, 63: 30-42, 61-68, pls. 2-5.
1931b. The eastern North American species of the genus *Ephemerella* and their nymphs. *Canad. Ent.*, 63: 187-97, 201-16, pls. 11-14.
1931c. New North American Caeninae with notes. *Canad. Ent.*, 63: 254-68, pls. 17-18.
1936. A new Arctic Baetid (Ephemeroptera). *Canad. Ent.*, 68: 32-34, 1 pl.
1938. New species of North American Ephemeroptera with critical notes. *Canad. Ent.*, 70: 23-34, 1 fig., pl. 1.
1939. New British Columbian Ephemeroptera. *Canad. Ent.*, 61: 49-54, 1 pl.

MARTYNOV, V.
1922. The interpretation of the wing venation and tracheation of the Odonata and Agnatha. *Rev. Russe Ent.*, 18: 145-74, 3 figs., 1 pl. (Translated by F. M. Carpenter in *Psyche*, 37: 245-81, 1930).

MITCHELL, A. J. and M. R. ENSIGN
1928. The climate of Florida. *Univ. of Fla. Agri. Exp. Sta. Bull.*, No. 200, pp. 91-300, figs. 52-65.

MORGAN, A. H.
1911. Mayflies of Fall Creek. *Ann. Ent. Soc. Amer.*, 4: 93-119, pls. 6-12.
1912. Homologies in the wing-veins of mayflies. *Ann. Ent. Soc. Amer.*, 5: 89-106, figs. 1-6, pls. 5-9.
1913. A contribution to the biology of mayflies. *Ann. Ent. Soc. Amer.*, 6: 371-413, figs. 1-3, pls. 42-54.

1929. The mating flight and the vestigial structures of the stump-legged mayfly, *Campsurus segnis* Needham. *Ann. Ent. Soc. Amer.*, 22: 61-68, pl. I.

1932. The functions of the gills in burrowing mayflies *(Hexagenia recurvata). Physiological Zoology*, 5: 230-45, 1 pl.

MORGAN, A. H. and J. F. WILDER

1936. The oxygen consumption of *Hexagenia recurvata* during the winter and early spring. *Physiological Zoology*, 9: 153-69, 3 figs.

MORRISON, E. R.

1919. The may-fly ovipositor, with notes on *Leptophlebia* and *Hagenulus*. *Canad. Ent.*, 51: 139-46, pls. 10-11.

MURPHY, H. E.

1922. Notes on the biology of some of our North American species of mayflies. *Lloyd Library Bull. No. 22, Ent. Ser.* No. 2: 1-46, pls. 1-7.

NEAVE, FERRIS

1929. Reports of the Jasper Park Lake investigations, 1925-1926. IV. Aquatic insects. *Contri. Canad. Biol. and Fish. N. S.*, 4: 187-89.

1930. Migratory habits of the mayfly *Blasturus cupidus* Say. *Ecology*, 11: 568-76, 3 figs.

1932. A study of the mayflies *(Hexagenia)* of Lake Winnepeg. *Contri. Canad. Biol. and Fish. N. S.*, 7: 179-201, 10 figs.

1934. A contribution to the aquatic insect fauna of Lake Winnepeg. *Int. Rev. Hydrobiol.*, 31: 157-70, 3 figs.

NEEDHAM, J. G.

1905. Mayflies and midges of New York State. *N. Y. S. Mus. Bull.* 86: 17-62, figs. 1-14, pls. 4-12.

1907. Additional data concerning New York mayflies. *N. Y. S. Mus. Bull.* 124: 189-94, pl. 10.

1920. Burrowing mayflies of our larger lakes and streams. *Bull. Bur. Fish.*, Document No. 883, 36: 269-92, pls. 70-82.

1932. Three new American mayflies. *Canad. Ent.*, 64: 273-76, 1 fig.

NEEDHAM, J. G. and R. O. CHRISTENSON

1927. Economic insects in some streams of northern Utah. *Utah Agri. Exp. Sta. Bull.*, No. 201, pp. 1-36, 43 figs.

NEEDHAM, J. G. and H. E. MURPHY

1924. Neotropical mayflies. *Lloyd Library Bull. No. 24, Ent. Ser.* No. 4: 1-79, pls. I-XIII.

NEEDHAM, J. G., J. R. TRAVER, and YIN-CHI HSU

1935. The biology of mayflies with a systematic account of North American species. Ithaca (Comstock Publishing Co.), pp. i-xiv + 759, 168 figs., pls. I-XL, col. front.

ROGERS, J. S.

1933. The ecological distribution of the craneflies of Northern Florida. *Ecol. Monographs.*, 3: 1-74, 25 figs.

SCHOCK, G.

1887. *Ephemerella ignita*, Poda, eine Paedogentische Eintagsfleige. *M. T. Schw. Ent. Ges.*, 7: 48-50.

SMITH, O. R.

1933. The eggs and oviposition of mayflies. Abstract of Thesis Presented to Graduate School, Cornell University (enlarged *in* The Biology of Mayflies).

SNODGRASS, R. E.

1935. Principles of insect morphology. New York (McGraw-Hill Book Co., Inc.), pp. i-ix + 667, 319 figs.

SPIETH, H. T.

1933. The phylogeny of some mayfly genera. *Jour. N. Y. Ent. Soc.*, 41: 55-86, 327-90, pls. 16-29.

1936. The life history of *Ephemera simulans* in Lake Wawasee. *Canad. Ent.*, 68: 263-66, 1 pl.
1937. An oligoneurid from North America. *Jour. N. Y. Ent. Soc.*, 45: 139-45, 1 pl.
1938a. A method of rearing *Hexagenia* nymphs. *Ent. News*, 49: 29-32.
1938b. Two interesting mayfly nymphs with a description of a new species. *Amer. Mus. Nov.* No. 970, pp. 1-7, 2 pls.
1938c. Studies on the biology of the Ephemeroptera. I. Coloration and its relation to seasonal emergence. *Canad. Ent.*, 70: 210-18.
1938d. Taxonomic studies on Ephemerida. I. Description of New North American species. *Amer. Mus. Nov.* No. 1002, pp. 1-11, 1 fig., 1 pl.
1940. The North American ephemeropteran species of Francis Walker. *Ann. Ent. Soc. Amer.*, 33: 324-38.
1941. Taxonomic studies on the Ephemeroptera. II. The genus *Hexagenia*. *Amer. Mid. Nat.*, 26: 233-80, 50 figs., 12 maps.
1947. Taxonomic studies on the Ephemeroptera. IV. The genus *Stenonema*. *Ann. Ent. Soc. Amer.*, 40 (1): 87-122, 3 figs. pls. I-II.

STEGER, A. L.
1931. Some preliminary notes on the genus *Ephemerella*. *Psyche*, 38: 27-35, 1 fig.

TILLYARD, R. J.
1923. The wing-venation of the order Plectoptera or mayflies. *J. Linn. Soc. London Zool.*, 35: 143-62, figs. 1-10.

TRAVER, J. R.
1925. Observations on the ecology of the mayfly *Blasturus cupidus*. *Canad. Ent.*, 57: 211-18.
1931a. The ephemerid genus *Baetisca*. *J. N. Y. Ent. Soc.*, 39: 45-66, pls. 5-6.
1931b. A new mayfly genus from North Carolina. *Canad. Ent.*, 63: 103-9, pl. 7.
1931c. Seven new southern species of the mayfly genus *Hexagenia*, with notes on the genus. *Ann. Ent. Soc. Amer.*, 24: 591-620, pl. 1.
1932a. *Neocloeon*, a new mayfly genus (Ephemerida). *J. N. Y. Ent. Soc.*, 40: 365-72, pl. 14.
1932-1933. Mayflies of North Carolina. *J. Elisha Mitchell Sci. Soc.*, 47: 85-161, 162-236, pls. 1-8; 48: 141-206, 1 pl.
1933. Heptagenine mayflies of North America. *J. N. Y. Ent. Soc.*, 41: 105-25.
1934. New North American species of mayflies. *J. Elisha Mitchell Sci. Soc.*, 50: 189-254, 1 pl.
1937. Notes on mayflies of the southeastern states (Ephemeroptera). *J. Elisha Mitchell Sci. Soc.*, 53: 27-86, 1 pl.
1938. Mayflies of Puerto Rico. *J. Agri. Univ. of Puerto Rico*, 22: 5-42, pls. 1-3.

ULMER, G.
1919. Neue Ephemeropteren. *Ach. Naturg. Berlin.* 85 *Abt. A.*, 11: 1-80, 56 figs.
1920. Trichopteren and Ephemeropteren aus Höhlen. *Deutsche ent. Zeitschr. Berlin*, pp. 303-9.
1932-1933. Aquatic insects of China. Art. 6, revised key to the genera of Ephemeroptera. *Peking Nat. Hist. Bull.*, 7: 195-218, pls. 1-2.

UPHOLT, W. M.
1937. Two new mayflies from the Pacific coast. *Pan-Pacific Ent.*, 13: 85-88.

WALSH, B. D.
1862. List of the Pseudoneuroptera of Illinois contained in the cabinet of the writer, with descriptions of over 40 new species, and notes on their structural affinities. *Proc. Acad. Nat. Sci. Phila., 2nd Ser.*, pp. 361-402.
1863. Observations on certain N. A. Neuroptera, by H. Hagen, M. D., of Koenigsberg, Prussia; translated from the original French MS and published by permission of the author with notes and descriptions of about twenty new N. A. species of Pseudoneuroptera. *Proc. Ent. Soc. Phila.*, 2: 169-79, 188-210.
1864. On the pupa of the ephemerous genus *Baetisca*. *Proc. Ent. Soc. Phila.*, 3: 200-206, 1 fig.

WODSEDALEK, J. E.
1912a. Palmen's organ and its function in nymphs of the Ephemeridae, *Heptagenia interpunctata* (Say) and *Ecdyurus maculipennis* (Walsh). *Biol. Bull.*, 22: 253-72, pls. 1-3.
1912b. Natural history and general behavior of the Ephemeridae nymphs *Heptagenia interpunctata* (Say). *Ann. Ent. Soc. Amer.*, 5-6: 31-40.

INDEX

(The principal reference for each species is listed first.)